Springer Proceedings in Energy

The series Springer Proceedings in Energy covers a broad range of multidisciplinary subjects in those research fields closely related to present and future forms of energy as a resource for human societies. Typically based on material presented at conferences, workshops and similar scientific meetings, volumes published in this series will constitute comprehensive state-of-the-art references on energy-related science and technology studies. The subjects of these conferences will fall typically within these broad categories:

- Energy Efficiency
- Fossil Fuels
- Nuclear Energy
- Policy, Economics, Management & Transport
- Renewable and Green Energy
- Systems, Storage and Harvesting
- Materials for Energy

eBook Volumes in the Springer Proceedings in Energy will be available online in the world's most extensive eBook collection, as part of the Springer Energy eBook Collection. To submit a proposal or for further inquiries, please contact the Springer Editor in your region:

Kamiya Khatter (India)
Email: kamiya.khatter@springer.com
Loyola D'Silva (All other countries)
Email: loyola.dsilva@springer.com

CH_4 concentration for a confined geographical area by using a sensor installed at an appropriate height, upstream wind. External tracer can be used to monitor change in concentration of tracer and CH_4. By taking measurements of the concentration of background methane combined with downwind measurements of methane alone, methane emissions can also be calculated using inverse dispersion modeling. Recent advancement in technology includes aircraft-based measurements using sensors, and the use of infrared thermography as an indicator for heat and methane production in dairy cattle. Due to its higher accuracy and predictability in CH_4 measurements, the respiration chamber technique is still regarded as the gold standard method [6]. The International Panel on Climate Change has established a method for estimating GHGs emitted by landfills that have been widely used by researchers. In this study of Madhya Pradesh, the first biogas plant's biogas is collected from the anaerobic decomposition of wet waste which is generated on a daily basis from adjoining vegetable markets or mandi, and this is used to generate electricity to power street-lights in the vicinity as shown in a process flow diagram. The IPCC pointers describe two main strategies:

(A) The default IPCC methodology that's supported the theoretical gas yield (a mass balance equation).
(B) Theoretical first-order kinetic methodologies, through the IPCC pointers, introduce the "First-Order Decay Model" (FOD).

The main distinction between the above-mentioned methods is that method A describes the exponential decrease of a substance over time whereas decay involves tracking the inflow and outflow rates of a substance to determine its decay in method B [7]. Provided that the yearly amounts associated with the nursing composition of waste disposed of likewise as disposal practices are nearly constant for long periods, the tactic A can turn out fairly sensible estimates of the yearly emissions. Increasing amounts of waste disposed of will cause an overestimation, and decreasing amounts correspondingly an underestimation, of yearly emissions. Methodology B provides a more correct estimate of the yearly emissions. Several countries may, however, have issues obtaining the mandatory knowledge and data (historical data on point disposal, rate constant for the decay) to determine the right basis for emission inventories with acceptable accuracy [8].

1.2 Methods

1.2.1 Deciding Project Boundary

The project boundary defines the region inside which emission reductions arise. Emission reductions need to arise at the project or end result from the assignment. In this study, the project boundary includes biogas plants including a digestor and a biogas-based electricity generator. The project boundary shall embody all GHG emissions below the management of the project contributors which can be vast and fair as a result of the CDM project activity. Deciding the project boundary is the initial step to studying emissions in a particular region.

1.2.2 Baseline Emission Estimation

The baseline is the situation that represents the GHG emissions that may occur within the absence of the projected CDM project activity. The project participants will either use the approved methodologies or propose a replacement methodology for determinative baseline scenario. In this study, baseline emission is considered as per IPCC methodology. The default method is based on the equation [3, 8] (see Fig. 1.1):

Methane emissions in biomethanation process (T/year)

$$= (MSW_T * MSW_F * MCF * DOC * DOC_F * F * 16/12 - R) * (1 - OX)$$

where

MSW_T: total MSW generated (T/year)

MSW_F: fraction of MSW disposed to solid waste disposal sites

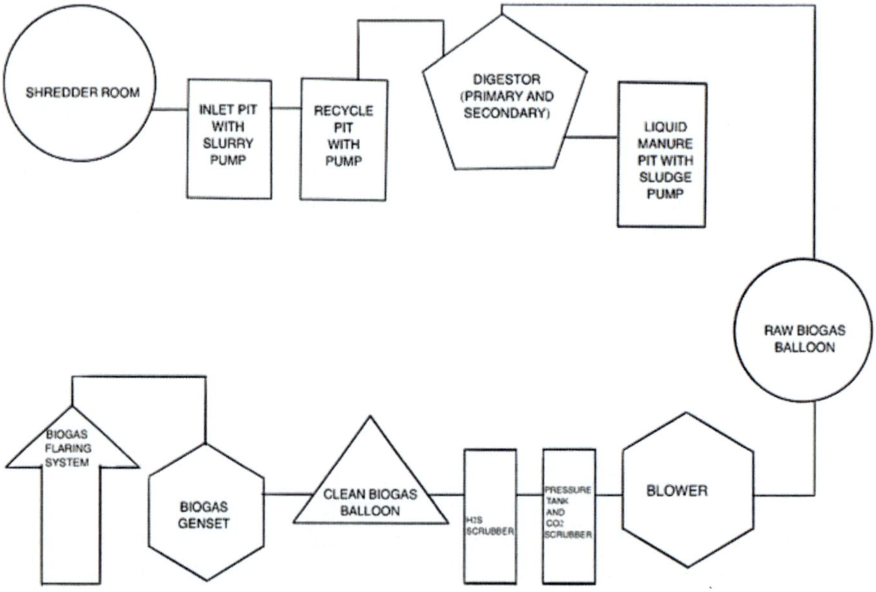

Fig. 1.1 Process flow diagram of plant

MCF: methane correction factor (fraction)

DOC: degradable organic carbon (fraction) (kg C/kg SW)

DOC_F: fraction DOC dissimilated

F: fraction of CH_4 in landfill gas (IPCC default is 0.5)

16/12: conversion of C to CH_4

R: recovered CH_4 (T/year)

OX: oxidation factor (IPCC default is 0).

1.2.3 Project Emission

The project emissions are the one which consist of emission caused by project activities such as diesel or other fossil fuel-based energy consumption in operations of machines and plant; they are further based on electricity consumption, flaring, and methane for digestor; most of the values are taken as standard assumptions of IPCC guidelines based on the scenario [9] (see Fig. 1.2).

Fig. 1.2 Shredder and Collection by operator. Photograph taken by Authors on 19-10-2021 at biogas plant Bhopal. Released to ICRABR 2022

1.2.4 Leakage Emission

Leakage refers to any GHG emissions that occur outside the project boundary, as a result of the project.

$LE_{AD,y}$ = Leakage emissions associated with the anaerobic digester in year y(tCO_2e).

$LE_{storage,y}$ = Leakage emissions associated with storage of digestate in year y(tCO_2e).

$LE_{COMP,y}$ = Leakage emissions associated with composting digestate in year y(tCO_2e).

In the project's case, $LE_{Storage,y}$ is considered zero as the storage is not un-aerated.

1.3 Results

Project activity supports mainly three out of 17 Sustainable Development Goals. SDG 8 Decent work and Economic growth is represented by employing 2 waste collectors and 1 plant manager creating organized sector job, SDGs 7 affordable and clean energy is represented by quantity of biogas captured and used to generate energy and SDG 13 Climate action which is analyzed in this study. Analysis of total GHG emission are quantified based on GHGs emission reduction methodologies of IPCC and total baseline emission is found to be 408.3 tCO_2e/year, excluding project emission due to lack of data availability of project emission during the survey. While the net GHG emission reduction due to project activity and leakage emission is taken to be zero. Any emission from wastewater and flaring gas is not considered in this study and is taken to be zero as per methodologies. The daily waste input capacity of the plant is 5 tons per day, however, the annual average amount of waste input is 2 tons per day. As per the design of the plant, biogas must be capable of producing 100–250 units of electricity per day excluding electricity for operational use. One ton of waste can generate 65–80 m^3 biogas and a one-meter cube of biogas can produce 1.2–1.3 units of electricity using a single 48 kW or 62.5 KVA natural gas generating model generator. The plant has a digestor of capacity 360 cubic meters with 2 raw biogas collection balloons of 80 cubic meters each, but the plant is operating at very low input due to the effect of the recent pandemic (see Fig. 1.3).

1.4 Conclusion

In this study, it can be concluded that similar plants using anaerobic decomposition of biodegradable byproducts can be developed to mitigate CH_4 emissions. However significant challenges in quantifying CH_4 emission reduction are the monitoring parameters such as feedstock, its quality, and contamination. Based on field visits,

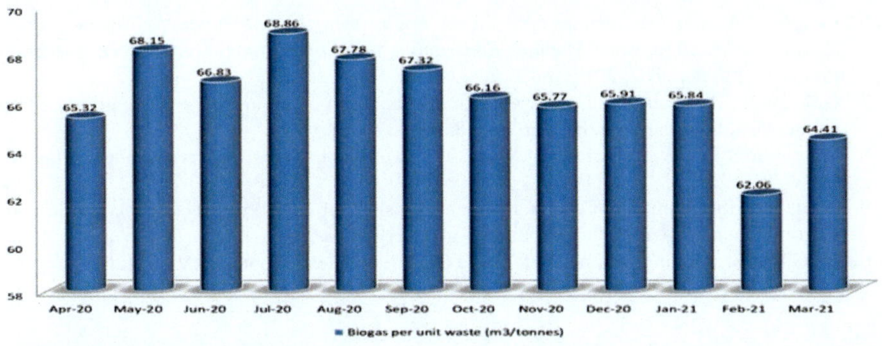

Fig. 1.3 Biogas generation per unit waste

it was found the capacity of biogas plants gets reduced due to the accumulation of sand and other inert material on the biodigester bed, therefore increasing the cost of maintenance and operations. This biogas plant was operational at a 60% lower functional capacity (5TPD). Since the waste source is only from an adjacent vegetable market, it affected the consumption behavior of the consumers during COVID restrictions.

A significant challenge is the availability of good quality feedstock for which waste transfer stations should work in coordination to provide the required waste within the city. A sustainable circular economy can be created through biomass utilization by recycling organic residues including nutrients in order to bring them back to society as energy and fuel. The uncertainties in the estimates of CH_4 emissions from waste are large, no matter the method used. The information on the composition and quantity of waste disposed of at landfills remains usually supported by rough estimates. Statistics on municipal and industrial waste management are presently growing in several countries, and future emissions are based on more reliable data. Due to difference in waste characteristics, SWDS practices involve collecting and selecting emission factors based on the composition of waste, taking into account national circumstances, to ensure clear and comparable coverage of emissions.

References

1. Hannah Ritchie and Max, India: CO_2 Country Profile (2020). https://ourworldindata.org/co2/country/india
2. Ministry of Environment, Forests and Climate Change, Lok Sabha (2021). https://loksabhaph.nic.in/Questions/QResult15.aspx?qref=30895&lsno=17
3. CDM Methodology Booklet Clean Development Mechanism United Nations Framework Convention on Climate Change (2020).
4. S. Bakkaloglu et al., Quantification of methane emissions from UK biogas plants. Waste Manag. **124**, 82–93 (2021)
5. J.P. Goopy, C. Chang, N. Tomkins, A Comparison of Methodologies for Measuring Methane Emissions from Ruminants (2016)

6. *Chapter 3 National Academies of Sciences, Engineering, and Medicine, Improving Characterization of Anthropogenic Methane Emissions in the United States* (The National Academies Press, Washington, DC, 2018). https://doi.org/10.17226/24987
7. C.K. Singh, A. Kumar, S.S. Roy, Quantitative analysis of the methane gas emissions from municipal solid waste in India. Sci. Rep. **8**, 2913 (2018)
8. Infinite Solution. Waste to Energy VCS 2093 (2021). https://registry.verra.org/app/projectDetail/VCS/2093
9. Guidelines for Estimating Greenhouse Gas Emissions of ADB Projects (2017). https://www.adb.org/documents/guidelines-estimating-ghgenergy-projects
10. GHG Platform India Building Sustainable GHG Estimates: Reporting (V3.0) (2017)

Chapter 2
Review of Anaerobic Digestion of Landfill Leachate and its Co-digestion Potential

Devnita Polley and Sudhir Jain

Abstract A landfill is a cost-effective solution to dispose of municipal solid waste; leachate is its secondary pollutant. Around 60–70% of the municipal solid waste generated in India each day is organic, resulting in a large volume of leachate, which includes a significant quantity of organic matter; anaerobic digestion is a comparatively good treatment option. Landfill leachate characteristics are influenced by the type of trash produced, climatic circumstances, temperature, and other factors and because of the heterogeneous mixture adding co-substrates is a cost-effective option for uniform biogas generation from anaerobic digestion. This paper deals with the various landfill leachate characteristics with anaerobic digestion treatment of mono leachate and its co-digestion potential. Monodigestion of leachate yields 24 L/kgVS of CH_4 consumed with 37% of removal efficiency of VS, whereas co-digestion gave a higher CH_4 output of 317 L/kg VS consumed with an 80% VS elimination efficiency.

Keywords Municipal solid waste · Anaerobic digestion · Landfill leachate · Co-digestion

2.1 Introduction

95% of municipal solid waste collected around the world is disposed of in landfills [1]. According to the Central Pollution Control Board (CPCB), India produced 152,076.7 tonnes of municipal solid waste (MSW) per day (TPD) in 2018–2019, with an average waste of 0.11 kg per capita per day, in which only 149,748.6 TPD (80%) was collected, 55,759 TPD (22%) was handled or processed, and roughly 50,161.33 TDP (60–70% organic component) was landfilled [2]. As the organic content is around 60–70% in MSW reported in India, it produces a high amount of leachate consisting of rich organic properties which has the potential to create a considerable quantity of biogas through anaerobic digestion. The biogas generated can be utilised

D. Polley (✉) · S. Jain
College of Technology and Engineering, Maharana Pratap University of Agriculture and Technology, Udaipur, India
e-mail: devnita.polley@gmail.com

N. Gakkhar et al. (eds.), *Recent Advances in Bio-Energy Research*, Springer Proceedings in Energy, https://doi.org/10.1007/978-981-99-5758-3_2

for power generation. Biogas generates 2.14 kWh of power per m^3 of feed [3]. Not only energy generation, anaerobic digestion has the potential to remediate pollutants such as COD, BOD, TKN, pH, and others [4]. Studies found anaerobic digestion can remove 90% of organic matter and fulfil the conditions set out in the "Municipal Solid Wastes (Management and Handling) Rules, 2016" for the disposal of treated landfill leachates. Monodigestion of leachate yields 350–480 ml/g VS under conventional experimental conditions [5]. By combining landfill leachate with agricultural waste or any other easily biodegradable material, the biomethanation process will be accelerated, resulting in increased methane generation. Co-digestion of landfill leachate with various materials like pineapple peel [5], sugarcane bagasse fly ash [6], and crude residual glycerin [3] has shown good results for biogas production, whereas agricultural waste is abundantly available in India.

Leachate pollutes groundwater as it seeps through the soil. For example, the leachate produced at the Ariyamangalam dumpsite in Tiruchirappalli, Tamil Nadu, India, has significantly affected the groundwater nearby [7]. TDS found in the groundwater is quite higher, and the Cl content is in the higher range. The authors also reported toxic heavy metals in leachate-polluted groundwater, including Pb, Zn, Cu, Mn, and Cd. Therefore, leachate treatment is a significant area of research [7].

Leachate is divided into 3 groups according to its age: young, medium, and old or stabilised leachate [4]. Young leachate is defined as below one year old; medium leachate is defined as one to five years old; and stabilised leachate is defined as over five years old [8]. Stabilised leachate seems to be the hardest to treat out of the three since it breaks slowly, whereas young and medium leachates respond well to biological treatment [9, 10]. In contrast, it is discovered that stabilised leachate can be effectively treated using physicochemical wastewater treatment technologies [7]. However, disposing of the sludge or concentrate left over after leachate treatment is a significant additional challenge. In this paper, the various properties of landfill leachate are presented along with anaerobic digestion potential and co-digestion ability in summarised form.

2.2 Landfill Systems

An engineering facility for disposing of MSW that is built and operated with public health and environmental issues in mind is referred to as a sanitary landfill [11]. Monofills are landfills that only accept one type of trash, such as ash, asbestos, and other similar pollutants [12]. Landfills that are utilised to dispose of hazardous materials known as secure landfills. Uncontrolled land disposal sites, sometimes known as waste dumps, are areas where trash is dumped in an unstructured manner on or into the ground [4]. Due to financial restrictions, open dumps are the most common form of MSW disposal in developing nations. Waste is thrown into wetlands for land development in Kolkata, Mumbai, Chennai, and Colombo. Ocean dumping is illegal by legislation in African nations [5]. However, the practice continues to be outlawed in several African coastal cities [4]. Other sophisticated waste disposal methods such as anaerobic digestion, composting, and incineration can be selected based on the kind of waste in Northern European nations such as Germany, where the waste sorting system is quite efficient [4]. Since the previous few decades, there has been an enormous increase in trash production that is correlated with trends in urbanisation and population growth. Management of municipal solid waste (MSW) has become more difficult in India's developing cities. Due to the COVID-19 pandemic in India, there has been a dramatic increase in the production of biomedical waste in recent years, along with exponential population expansion, high urban population density, diversified culture, changing eating habits, and lifestyle changes '[13, 14]. The dynamics of waste creation have evolved over time in "transformed cities," which have experienced fast industrial expansion and population growth over the last. Research has been done in India and across the world to evaluate the severity of soil and groundwater pollution inside and around MSW dumpsites in metropolitan cities to comprehend their detrimental effects on human health [14]. The results of these study activities indisputably show that harmful trace elements, such as As, Cr, Pb, Cu, Ni, Zn, and Hg, are present in both the soil and the groundwater. Heavy metal contamination has already been a significant issue in metropolitan areas [14].

Because of its massive population and growing use of information and communication technologies across all industries, India has an alarmingly high quantity of e-waste. Not only that, but due to urbanisation in rural regions, the number of other electronic devices (such as refrigerators, microwave ovens, air conditioners, colour televisions, DVD players, and MP3 players) are also rising quickly [15]. According to [16], Mumbai leads all other cities in the generation of e-waste. Delhi, Bangalore, Chennai, Kolkata, Ahmedabad, Hyderabad, Pune, Surat, and Nagpur are next in line. About 17 lakh tonnes of e-waste were produced in 2014, according to estimates from the Manufacturer's Association for Information Technology Industry (MAIT) performance annual review [15]. The hazardous waste rule of 1989 states that e-waste should not be considered hazardous until it is established that it has a higher concentration of specific dangerous compounds. Although electronic waste involving PCBs and CRTs invariably exceeds these limits, there are a few grey areas that require attention. India also lacks a suitable, standardised method for disposing of e-waste. Due to this circumstance, the majority of e-waste is dumped in landfills, which causes one of India's fastest-growing environmental problems [16]. The issues are exacerbated by foreign as well as domestic e-waste. Even though the Ministry of Environment and Forest has to give particular approval before importing e-waste, it is getting imported illegally.

Integrated solid waste management (ISWM), 2016, provided a framework to guide the selection of the most appropriate treatment technologies for MSW management. ISWM factories frequently use pre-processing facilities to separate organics from recyclables and other high-calorie waste. A study conducted by CPCB NEERI is presented in Fig. 2.1; from the chart, the fractions of solid waste are seen, and the organic fraction is found to be 47.4%, which is favourable for the MSW to produce a high amount of leachate. Organic waste is degraded aerobically to produce manure or anaerobically to generate power. Ministry of New Renewable Energy, 2021, reported the energy potential of Urban solid waste to be 1247 MW and urban liquid waste

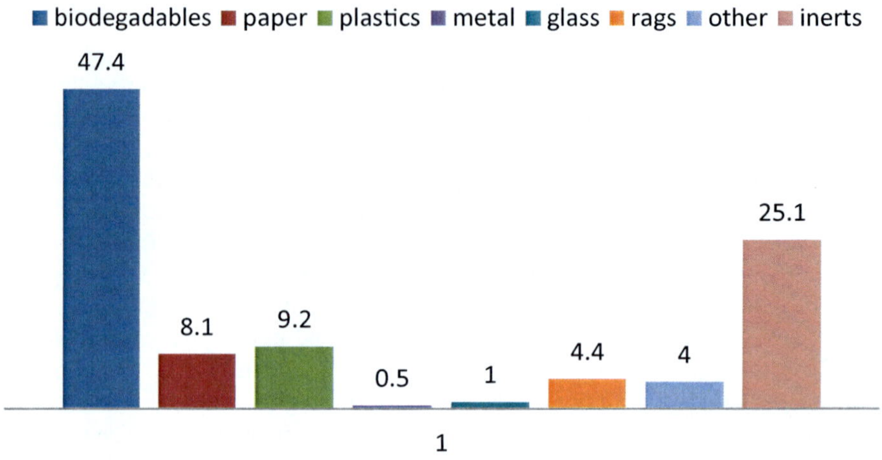

Fig. 2.1 Typical MSW generation fractions in India. *Source* [18]

to be 375 MW [17]. Separated recyclables are sent to wholesalers for shipment to recycling facilities. High-calorie wastes are baled or processed so that they may be used as fuel or co-processed in cement plants [2].

2.3 Landfill Leachate Characteristics

2.3.1 Leachate Formation

The composition of landfill leachate varies greatly depending on the stage of waste development, i.e., aerobic, anaerobic acid, methanogenic, and stabilisation phases [4, 19, 20]. The synthesis of CO_2 results from the rapid depletion of O_2 during the initial aerobic stage, which occurs in newly dumped waste. The waste's temperature tends to rise in this stage, according to a lot of research [19, 20]. Owing to moisture loss during the compaction and precipitation, a huge volume of leachate is produced during this stage [21]. Once oxygen supplies are reduced enough for fermentation to occur, waste enters the anaerobic phase, also known as the second stage [19, 20]. This stage is dominated by hydrolytic, fermentative, and acetogenic bacteria, which causes a buildup of carboxylic acid and a decrease in pH. This phase is also called acid phase and BOD/COD in this stage is reported to be 0.4–0.7 [19, 20]. The third phase is the methanogenic phase. At the start of this stage, pH of the waste reaches a neutralised level which promotes the development of methanogenic bacteria.

During this stage, bacteria that produce methane convert acids created during the acidic stage into CH_4 and CO_2, increasing the rate at which methane is produced. During the stable methanogenic phase or stabilisation phase, the rate of methane

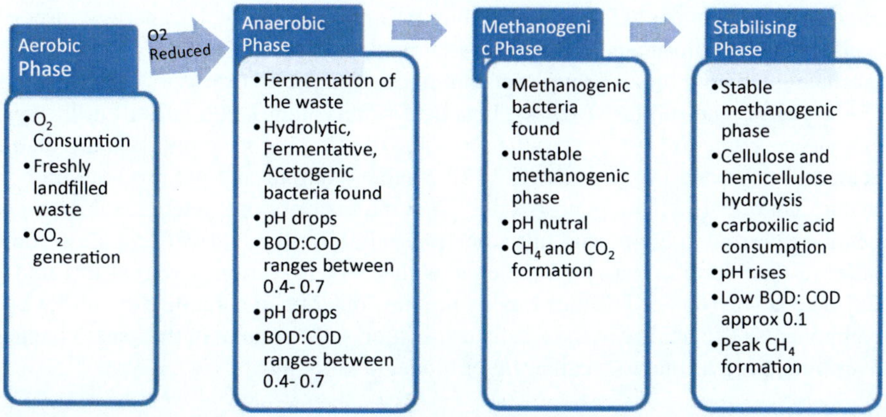

Fig. 2.2 Leachate formation process in Landfills. *Source* [20]

generation peaks when carboxylic acids decrease, and it starts to slow down. The hydrolysis of cellulose and hemicellulose during the acid phase is proportional to the rate of CH_4 production. As pH levels rise, they eventually stabilise at a few milligrammes per litre in a steady state. The BOD: COD ratio frequently falls to as low as 0.1 because carboxylic acid is digested as rapidly as it is produced [19–21]. The leachate formation steps are summarised in Fig. 2.2.

The anaerobic biological conditions are again divided into four phases. The first phase in anaerobic degradation is acid fermentation, which is followed by hydrolysis, acidogenesis, and acetogenesis [12]. Researchers denoted waste age, waste type and content, site hydrogeology, seasonal weather variation, dilution by rainfall, precipitation, and the degree of decomposition within the landfill are all variables that impact the parameters of municipal landfill leachate [4].

2.3.2 Leachate Properties

Around 200 toxic chemicals have previously been detected in landfill leachate in previous investigations [22]. Landfill leachate is one of the most complicated wastewaters, containing inorganic salts, heavy metals, a large number of biodegradable organics, and refractory components such as humic compounds, among other things [23].

Leachate contaminants are divided into four categories by Luo et al. [4]: (1) Organic compounds, e.g., BOD, COD; (2) macro inorganic components such as ammonia (NH_4^+–N), sodium (Na^+), potassium (K^+), calcium (Ca^{2+}), magnesium (Mg^{2+}), manganese (Mn^{2+}), iron (Fe^{2+}), chloride (Cl^-), sulphate (SO_4^{2-}), and hydrogen carbonate (HCO_3^-); (3) heavy metals such as chromium (Cr^{3+}), nickel (Ni^{2+}), copper (Cu) (4), Xenobiotic compounds, e.g., aromatic hydrocarbons, phenols, chlorinated aliphatics, pesticides, and plasticisers [12, 19]. Landfill leachate is divided into 3 groups depending on age: young (<5 years), intermediate (5–10 years), and old (>10 years). Leachate from young landfills (still in the acid stage), according to [24], is classified by a lower pH, a higher proportion of volatile acid, and degraded organic material [8]. Methane output and pH are both high in mature landfills (methanogenesis stage), and the majority of biodegradable components are present in humic and fulvic sections, although there is a difference in some other research due to waste characteristics that rely on various external factors [8, 25]. The key drivers of leachate quality are the content and solubility of solid waste components. The quality of the leachate will alter as the content of the waste changes over time, such as due to weathering or biodegradation [26].

2.4 Anaerobic Digestion of Landfill Leachate

Due to its reliability, simplicity, and high cost-effectiveness, biological treatment is commonly used for the removal of the bulk of leachate containing high concentrations of BOD. Anaerobic digestion reduces the amount of organic waste discharged in landfills in industrialised nations, is ecologically friendly, and produces valuable byproducts [27]. Ghani and Idris [28] built laboratory-scale digesters to investigate the impact of leachate chemical oxygen demand strength on biogas (methane) production. The results reveal biochemical oxygen demand in the effluent was reduced by up to 80%; the performance of other parameters such as chemical oxygen demand, total suspended solids, and volatile suspended solids, which contribute 33–46%, 21–37%, and 20–35%, respectively, was slightly reduced [28].

Biogas and sludge can be utilised as by-products to replace fossil fuels and chemical fertilisers [4]. Anaerobic digestion not only resulted in renewable biogas production but also improved effluent quality. Mohamed et al. [29], investigated batch anaerobic digestion of municipal solid waste at room temperature for 60 days using two distinct inoculums (cow dung and sewage sludge) at varied inoculum concentrations of 10%, 20%, and 30%. The biogas yields from the reactors using cow dung as inoculums were 337.365 ml/gVS, 481.95 ml/gVS, and 567 ml/gVS, respectively; similarly, sewage sludge as inoculums yielded 214.775 ml/gVS, 321.198 ml/gVS, and 383.52 ml/gVS. 65.45%, 80.06%, 86.655%, 60.25%, 65.89%, and 72.78% [29].

As landfill leachate is a heterogeneous mixture and its properties depend on various external factors, pretreatment of landfill leachate is useful for consistent biogas production as well as the treatment of the leachate. Yarimtepe and Oz [30] investigated the use of an ultrasonic method for leachate as a pretreatment before anaerobic batch reactors in order to boost the hydrolysis rate and biogas output. When compared to raw leachate at 400 W/L, the ratio of sCOD/tCOD was raised by 32% after 45 min of sonication. These findings revealed that ultrasonic pretreatment has a considerable impact on the conversion of particles to soluble organic matter. Biogas generation and methane concentration rise by 33% and 15% in pretreated and raw leachate, respectively, during anaerobic batch experiments [30]. Anaerobic biological processes provide a number of benefits over aerobic biological processes, including (1) much-reduced sludge creation, (2) lower organics stabilisation, and (3) energy production owing to methane recovery [31]. (1) Anaerobic digestion (AD); (2) anaerobic filter (AF); (3) up-flow anaerobic sludge blanket (UASB); and (4) anaerobic ammonium oxidation are examples of anaerobic treatment of landfill leachate [4].

2.4.1 Co-digestion

For constant biogas production, co-digestion materials are added in the anaerobic digestion process along with leachate. The effects of pH and initial organic loading rate on volatile fatty acid (VFA) and biogas generation must be evaluated during landfill leachate treatment by single and two-stage anaerobic digestion processes [12]. The yields of methane and VFA are 0.21–0.34 L CH_4/(g COD removed) and 0.26–0.36 g VFA/(g COD removed), respectively (g COD removed). There has been a 21% increase in COD elimination overall [19]. de Castro et al. [3], used the modified Gompertz model to assess the performance of anaerobic co-digestion of different concentrations of industrial landfill leachate associated with crude residual glycerin in terms of methanogenic potential, COD removal, accumulated methane production, and the effects of the factors (food/microorganism ratio and percentage of glycerin added to the leachate) and their interactions on kinetic parameters of methane. The results showed a significant influence on the response variables: methanogenic potential, COD elimination, cumulative CH_4 production, and maximum predicted CH_4 production, with a 95% confidence interval [3].

Sudibyo et al. [6] investigated the use of natural zeolite and sugarcane bagasse fly ash (BFA) as immobilisation media to improve the digestion rate of landfill leachate in an anaerobic packed bed reactor. On the basis of the experiment data, the Contois and Haldane growth kinetic models were evaluated. For both acidogenic and methanogenic phases, Contois provided the greatest fit. According to a statistical examination of Contois kinetic parameters using the Pearson correlation coefficient, zeolite, as an immobilisation media, had a stronger beneficial effect on the performance of anaerobic digestion of a leachate than BFA [6].

The landfill leachate is already in anaerobic condition so it needs less time for methane formation, compared to cow dung. An experiment was done using hydraulic retention durations of 23 and 12 h to conduct an interventional investigation with a wide range of organic loading rates of 0.93–25 g L^{-1} day^{-1} [32]. It was reported Chemical Oxygen Demand (COD) was 1.85–25 g L^{-1} at the start. Depending on the loading rates used, the organic matter removal efficiency ranged from 76 to 81%. At a loading rate of 19.65 g L^{-1} day^{-1}, the highest volumetric methane production rate of 5.7 l CH_4 L^{-1} day^{-1} was reached. During biodegradation, almost 85% of the organic matter was transformed into methane.

For the treatment of landfill leachate and acid mine drainage, the authors [8, 33] built a laboratory-scale up-flow anaerobic sludge blanket (UASB) reactor (AMD). Chemical oxygen demand (COD), sulphate, and metal ions were all investigated. During the UASB's start-up phase, the highest COD and sulphate removal efficiency was 75% and 69%, respectively. At a 20-h HRT, the greatest removal efficiency for COD and sulphate was 83% and 78%, respectively. In the UASB, the methane generation process was interrupted against the sulphate reduction process.

2.5 Biogas Yield

The correct assessment of the volume of producible biogas and its methane content is one of the most important components of an anaerobic digester design. According to Karthikeyan and Visvanathan [34], by creating the right external conditions, depending on the organic substrate properties and pretreatment stages, the yield can be maximised. Under conventional experimental circumstances, leachate monodigestion yields up to 350–480 ml/g VS of methane [5]. Another study indicates that the UASB reactor's intake chemical oxygen demand load is 1.08 kg m^3 day^{-1} at 10-day HRT, with a 30% chemical oxygen demand elimination rate and a 40% sulphate elimination rate. The amount of biogas produced overall was 875 mL/day, while the amount of CH_4 produced was approximately 578 mL/day [33].

The quantity of COD in the leachate has a significant influence on the amount of biogas produced and the amount of methane produced. Other studies revealed that for every tonne of COD added to leachate during AD treatment, more than 387 m^3 biogas was produced, with a methane concentration of around 65% and COD removal of 70% at an OLR of 17 kg m^{-3} day^{-1} [19]. In addition, AD is combined with other technologies to improve the efficacy of landfill leachate treatment in terms of pollutant removal and biogas production [35].

Biogas production rates in the mesophilic temperature range of 25–35 °C are connected to the volatile solid content of the feed because methane is created by the breakdown of VSS [28]. The authors [36] reported that the anaerobic digestion of grass silage produced biogas outputs of 178 mL, 157 mL, 208.6 mL, and 134 mL at 35 °C, 40 °C, 45 °C, and finally 50 °C, respectively. The peak temperature for methane generation within the thermophilic range was observed at 47.5–57.5 °C, with decreased methane found up to 67.5 °C [37]. The authors also discovered that, despite the fact that microbial populations did not adapt above 67.5 °C, methanogenic activity restarted when the trash pile cooled [37]. Anaerobic digestion of leachate with a temperature range of 35–45 °C for 1.0 days results in a biogas output of 0.057–2.372 mol CH_4/g COD, according to Kahar et al. [38].

2.6 Future Gap and Challenges

The anaerobic biomethanation process has evolved quickly in recent years and is notably well adapted to wastewater with high organic content. However, it has a long retention time, limited pollutant removal, and is more susceptible to temperature changes. Leachate treatment using an anaerobic biological process is superior, but it is more expensive to build, operate, and manage, and since water treatment facilities will eventually be abandoned when landfills close, it is important to make a wise choice. Leachate biotechnology, however, may advance in maturity and have several opportunities in terms of growth and applicability as technology and society [39]. Biological treatment alone may be not feasible because of the heterogeneous and unstable properties of landfill leachate. Along with the biological treatment, exploration of technically and economically workable procedures, process combinations, and patterns in the different procedures and project complexity with research.

2.7 Conclusion

Because of its high organic content, anaerobic digestion is excellent for both landfill leachate treatment and biogas generation. The manufacturing and processing sectors around the globe have been growing over time, contributing to the enormous solid waste output. Waste output is expected to sharply increase over the next 20 years, which will lead to a rise in leachate infiltration. Concerns have been raised about the employment of biological treatment, membrane technology, or other integrated technologies as a tool to mitigate environmental pollution in light of the growing disparity and limited success of remediation in field applications. Although a number of issues and difficulties have been found and explained, widespread advancement in this field is anticipated in the future.

References

1. J. Gao et al., The present status of landfill leachate treatment and its development trend from a technological point of view. Rev. Environ. Sci. Biotechnol. **14**(1), 93–122 (2015). https://doi.org/10.1007/s11157-014-9349-z
2. Swachh Bharat Mission Municipal Solid Waste Management Manual Part I: An Overview. www.swachhbharaturban.gov.in
3. T.M. de Castro et al., Anaerobic co-digestion of industrial landfill leachate and glycerin: methanogenic potential, organic matter removal and process optimization. Environ. Technol. (United Kingdom) **41**(20), 2583–2593 (2020). https://doi.org/10.1080/09593330.2019.1575915
4. H. Luo, Y. Zeng, Y. Cheng, D. He, X. Pan, Recent advances in municipal landfill leachate: a review focusing on its characteristics, treatment, and toxicity assessment. Sci. Total Environ. **703** (10 Feb 2020). https://doi.org/10.1016/j.scitotenv.2019.135468

5. S. Jaroenpoj, B. Eng, Biogas production from co-digestion of landfill leachate and pineapple peel (2014)
6. H. Sudibyo et al., Anaerobic digestion of landfill leachate with natural zeolite and sugarcane bagasse fly ash as the microbial immobilization media in packed bed reactor. Acta Polytech. **58**(1), 57–68 (2018). https://doi.org/10.14311/AP.2018.58.0057
7. T. Sruthi, R. Gandhimathi, S.T. Ramesh, P.V. Nidheesh, Stabilized landfill leachate treatment using heterogeneous Fenton and electro-Fenton processes. Chemosphere **210**, 38–43 (2018). https://doi.org/10.1016/j.chemosphere.2018.06.172
8. A. Mojiri et al., Treatment of landfill leachate with different techniques: an overview. J. Water Reuse Desalin. **11**(1), 66–96 (2021). https://doi.org/10.2166/wrd.2020.079
9. M. Kumari, P. Ghosh, I.S. Thakur, Landfill leachate treatment using bacto-algal co-culture: an integrated approach using chemical analyses and toxicological assessment. Ecotoxicol. Environ. Saf. **128**, 44–51 (2016). https://doi.org/10.1016/j.ecoenv.2016.02.009
10. P. Lakshmikanthan, G.L. Sivakumar Babu, Performance evaluation of the bioreactor landfill in treatment and stabilisation of mechanically biologically treated municipal solid waste. Waste Manag. Res. J. Int. Solid Wastes Public Clean. Assoc. ISWA **35**(3), 285–293 (Mar 2017). https://doi.org/10.1177/0734242X16681461
11. G. Singh, S.K. Arya, A review on management of rice straw by use of cleaner technologies: abundant opportunities and expectations for Indian farming. J. Clean. Prod. **291**, 125278 (Apr 2020). https://doi.org/10.1016/j.jclepro.2020.125278
12. M. Sanjay, D. Amit, S.N. Mukherjee, Review Paper (SS-1) applications of adsorption process for treatment of landfill leachate
13. S. Chakraborty, K. Majumdar, M. Pal, P. Kumar Roy, Assessment of bio-gas from municipal solid waste for generation of electricity-a case study of Agartala city (2019). http://www.ripublication.com
14. M. Choudhury et al., Investigation of groundwater and soil quality near to a municipal waste disposal site in Silchar, Assam, India. Int. J. Energy Water Resour. **6**(1), 37–47 (2022). https://doi.org/10.1007/s42108-021-00117-5
15. Department of Electronics and Information Technology, Electronics and information technology annual report 2012–13, pp. 1–140, 2012. http://deity.gov.in/sites/upload_files/dit/files/AnnualReport2012-13.pdf
16. A. Kumar, 2021 Kumar, vol. 12, no. 1 (2021)
17. MNRE, *Current status | Ministry of New and Renewable Energy* (Government of India, 2021)
18. CPCB, Central Pollution Control Board (CPCB): Consolidated... -Google Scholar (2017)
19. P. Mandal, B.K. Dubey, A.K. Gupta, Review on landfill leachate treatment by electrochemical oxidation: drawbacks, challenges and future scope. Waste Manag. **69**, 250–273 (2017). https://doi.org/10.1016/j.wasman.2017.08.034
20. D. Polley, S. Jain, A comprehensive review on landfill leachate characteristics and biogas generation potential from anaerobic digestion of landfill leachate, **11**(8), 445–461 (2022). https://doi.org/10.21275/SR22727111054
21. P. Kjeldsen, M.A. Barlaz, A.P. Rooker, A. Baun, A. Ledin, T.H. Christensen, Present and long-term composition of MSW landfill leachate: a review. Crit. Rev. Environ. Sci. Technol. **32**(4), 297–336 (2002). https://doi.org/10.1080/10643380290813462
22. B.P. Naveen, D.M. Mahapatra, T.G. Sitharam, P.V. Sivapullaiah, T.V. Ramachandra, Physico-chemical and biological characterization of urban municipal landfill leachate. Environ. Pollut. **220**, 1–12 (2017). https://doi.org/10.1016/j.envpol.2016.09.002
23. A.R. Salehiyoun, M. Sharifi, F. Di Maria, H. Zilouei, M. Aghbashlo, Effect of substituting organic fraction of municipal solid waste with fruit and vegetable wastes on anaerobic digestion. J. Mater. Cycles Waste Manag. **21**(6), 1321–1331 (2019). https://doi.org/10.1007/s10163-019-00887-5
24. M. Vaccari, T. Tudor, G. Vinti, Characteristics of leachate from landfills and dumpsites in Asia, Africa and Latin America: an overview. Waste Manag. **95**, 416–431 (15 Jul 2019). https://doi.org/10.1016/j.wasman.2019.06.032

25. B. Wang, Z. Mi, I. Nistor, X.C. Yuan, How does hydrogen-based renewable energy change with economic development? Empirical evidence from 32 countries. Int. J. Hydrogen Energy **43**(25), 11629–11638 (2018). https://doi.org/10.1016/J.IJHYDENE.2017.03.059
26. N. Preston et al., How can the integration of renewable energy and power-to-gas benefit industrial facilities? From techno-economic, policy, and environmental assessment. Int. J. Hydrogen Energy **45**(51), 26559–26573 (2020). https://doi.org/10.1016/J.IJHYDENE.2020.07.040
27. M.F.M.A. Zamri, et al., A comprehensive review on anaerobic digestion of organic fraction of municipal solid waste. Renew. Sustain. Energy Rev. **137**, 110637 (Mar 2021). https://doi.org/10.1016/j.rser.2020.110637
28. W.A.W.A.K. Ghani, A. Idris, Preliminary study on biogas production of biogas from municipal solid waste (MSW) leachate. J. Eng. Sci. Technol. **4**(4), 374–380 (2009)
29. J. Mohamed, A. Ali, S. Mohan, T. Velayutham, S. Sankaran, Comparative study of biogas production from municipal solid waste using different innoculum concentration on batch anaerobic digestion (2016). www.ajouronline.com
30. C.C. Yarimtepe, N.A. Oz, Enhanced biogas production from landfill leachate by low frequency ultrasound, 225–234 (2015)
31. C.-Y. Lin, F.-Y. Chang, C.-H. Chang, Co-digestion of leachate with septage using a UASB reactor. Bioresour. Technol. **73**(2), 175–178 (2000). https://doi.org/10.1016/S0960-8524(99)00166-2
32. A.A.M. Hassanein, L. Qiu, P. Junting, G. Yihong, F. Witarsa, A.A. Hassanain, Simulation and validation of a model for heating underground biogas digesters by solar energy. Ecol. Eng. **82**, 336–344 (2015). https://doi.org/10.1016/j.ecoleng.2015.05.010
33. S. Zhou, J. Wang, S. Peng, T. Chen, Z. Yue, Anaerobic co-digestion of landfill leachate and acid mine drainage using up-flow anaerobic sludge blanket reactor. Environ. Sci. Pollut. Res. **28**(7), 8498–8506 (2021). https://doi.org/10.1007/s11356-020-11207-y
34. O.P. Karthikeyan, C. Visvanathan, Bio-energy recovery from high-solid organic substrates by dry anaerobic bio-conversion processes: a review. Rev. Environ. Sci. Biotechnol. **12**(3), 257–284 (2013). https://doi.org/10.1007/s11157-012-9304-9
35. M. Somani, M. Datta, S.K. Gupta, T.R. Sreekrishnan, G.V. Ramana, Comprehensive assessment of the leachate quality and its pollution potential from six municipal waste dumpsites of India. Bioresour. Technol. Reports **6**(March), 198–206 (2019). https://doi.org/10.1016/j.biteb.2019.03.003
36. N. T. Sibiya, E. Muzenda, and H. B. Tesfagiorgis, Effect of temperature and pH on the anaerobic digestion of grass silage, no. September 2015 (2014)
37. S. Schupp, Q. Cheng, D.F. Call, M.A. Barlaz, Evaluation of the temperature range for biological activity in landfills experiencing elevated temperatures (2020). https://doi.org/10.1021/acsestengg.0c00064
38. A. Kahar, I., J. Hermana, The effects of temperature-pH on biochemical degradation at leacha00te treatment in anaerobic bioreactor, **10**(4), 172–181 (2017)
39. Y. Peng, Perspectives on technology for landfill leachate treatment. Arab. J. Chem. **10**, S2567–S2574 (2017). https://doi.org/10.1016/j.arabjc.2013.09.031

Chapter 3
Nanobiotechnological Routes in Lignocellulosic Waste Pre-treatment for Bio-renewables Production

Madan L. Verma, Prateek Kumar, and Heena Chandel

Abstract Reservoirs of fossil fuel are depleting rapidly to meet increasing energy demands which necessitate the intense need for alternative sources of energy. The utilization of non-renewable energy immensely affects the ecological balance by releasing the hazardous pollutants responsible for global warming. Thus, there is a huge demand for renewable sources of energy. The lignocellulose biomass is a promising base of renewable sources of energy. Various conventional pre-treatment methods were routinely utilized for the disintegration of lignocellulosic biomass used for biofuel production. But certain disadvantages have been associated with these methods such as expensive mechanical equipment, toxic chemicals, and the release of a high number of inhibitory compounds causing environmental pollution. To overcome all these drawbacks, the interventions of nanotechnology pace the production of biorefinery. Reutilization of nanomaterials offers a cost-effective and economically feasible methodology. Nanomaterial immobilized various enzymes such as saccharolytic and lignolytic enzymes for the pre-treatment of lignocellulose biomass via hydrolyzation which enhances the production of biofuel and various by-products. The significant techniques of nanobiotechnology provide a cost-efficient hyperproduction of biofuel by utilizing lignocellulose biomass. The present paper discusses the various conventional and modern nanotechnological-based pre-treatment methods employed for the production of renewable energy.

Keywords Lignocellulosic biomass · Biomass pre-treatment · Nanomaterial · Biofuel

---

M. L. Verma (✉) · P. Kumar · H. Chandel
Department of Biotechnology, School of Basic Sciences, Indian Institute of Information Technology Una, Bhadsali 177209, Himachal Pradesh, India
e-mail: madanverma@iiitu.ac.in; madanverma@gmail.com

N. Gakkhar et al. (eds.), *Recent Advances in Bio-Energy Research*, Springer Proceedings in Energy, https://doi.org/10.1007/978-981-99-5758-3_3

3.1 Introduction

Fossil fuel involves coal, peat, petrol, diesel, gas, and coke synthesized by normal cycles, such as anaerobic degradation of buried dead creatures, containing energy emerging in primitive photosynthesis, which requires a long period of time for the formation [1]. About 70% of worldwide carbon emission gases are discharged as carbon dioxide gas because of the excessive combustion of fossil fuels. The current status of fossil fuels is at risk all over the world. It is evident from the survey of current conditions and excessive composition of fossil fuels that the reserves of fossil fuels may deplete soon. Hence, there is an urgent need to foster a sustainable source of energy. If this trend of depletion of fossil fuels continues, then soon it may lead to a state where there are no fossil fuels in the future. Biofuel production is one of the viable choices for the production of sustainable bioenergy [2, 3]. Presently, only a minimal percentage of biofuels are utilized in the transport industries. But it is predicted that more than 30% of the fuel demand will be fulfilled by biofuels in 2050 [4]. Researchers examined that the residuals derived primarily from farming (66%) followed by timberland (34%) constitute a sum of 12,693 petajoule LB. All these studies revealed that agricultural waste and forest waste are the best sources of biofuel production [5]. Among the various forms of biomass, forest waste is an effective biomass feedstock for biofuel and biochemical production. Forest waste is classified as 'modern biomass' that substitutes conventional fuel sources [6]. So, their waste resources require innovative bioprocessing for the production of biofuels.

The lignocellulose biomass (LB) has significant advantages as they are utilized as a fossil fuel alternative, enhancing the production of biofuels [7]. Lignocellulose biomass consists of three main components—cellulose, hemicellulose, and lignin (Fig. 3.1). Lignocellulose biomass is the most important raw material for the production of renewable energy. Annually, around 200 billion tons of lignocellulose biomassare utilized for the production of biofuel. It shows significant advantages such as being environmentally benign, cost-effective, easily accessible, and containing high levels of sugars which convert into biofuel [8]. The recalcitrant LB is a major challenge for the production of bioenergy. Various conventional pre-treatment methods were utilized such as physicochemical, biological, and combined pre-treatment methods. But all these methods have certain disadvantages like complexity in the pre-treatment process, costly chemical substances, and the release of hazardous compounds causing pollution in the environment [9].

Presently, nanobiotechnology and microbial technology significantly improve the production of biofuel. In the microalgal technique, various microbes were used such as cyanobacteria, microalgae, kelp, and other microbial groups for biofuel production. Biofuel production by utilizing these raw materials was considered as a third generation of biofuel [3]. The intervention of nanotechnology immensely improves the production of biofuel. Various nanocatalysts based on metal oxides such as calcium oxide, magnesium oxide, strontium, and titanium oxide have significantly improved the catalytic efficiency utilized during the production of biodiesel [10, 11]. Different nanoparticles were used for immobilizing the enzyme for hydrolyzing the

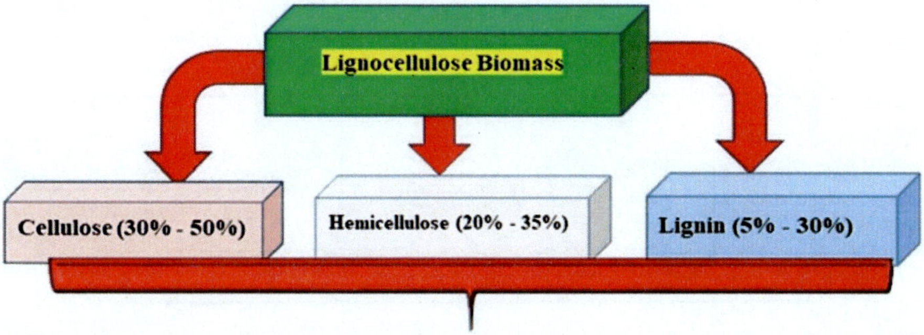

Fig. 3.1 Chemical composition of lignocellulose biomass

Table 3.1 Chemical composition of lignocellulose biomass

Lignocellulose biomass	Cellulose (%)	Hemicellulose (%)	Lignin (%)	References
Corn stover	30.9	17.4	15.2	[14]
Sorghum straw	37.7	28.1	21.5	[15]
Water hyacinth	33	24	9.9	[16]
Sugarcane straw	36.9	19.7	13.5	[17]
Rice straw	37	24	14	[18]

LB. Magnetic nanoparticles play a major role in immobilization of enzymes. It also improves the chemistry of LB at a molecular level. The high potential of recyclability and reusability of magnetic nanoparticles by applying a magnetic field enhance their utilization in the pre-treatment of biomass. Nanotechnology-based pre-treatment methods have significant advantages such as high reusability and costefficiency and limit the use of harmful chemicals causing the environmental pollution [12, 13].

The main purpose of this paper is to provide an updated overview of alternative sources of renewable energy. Nanobiotechnology is a technique which can boost the sustainable production of biofuels/biochemicals. Here, we discussed various lignocellulosic biomass materials and their potential applications to improve biofuel production through nanobiotechnological intervention (see Table 3.1).

3.2 Conventional Pre-treatment Methods for Lignocellulose Biomass

LB is regarded as a potential substrate for bioenergy production because of its natural abundance. LB mainly consists of cellulose, hemicellulose, and lignin. Lignin connects with cellulose and hemicellulose by linking with ether and ester bonds. The structure of LB material is highly complex. So, to expose the cellulosic material from

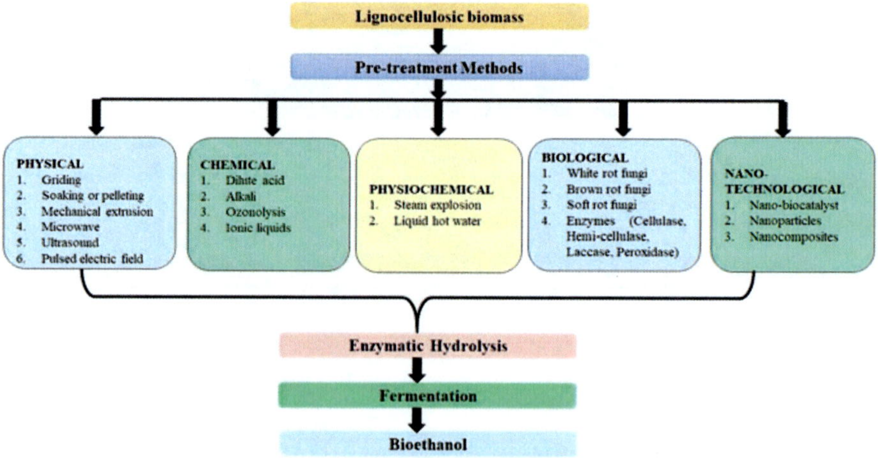

Fig. 3.2 Different pre-treatment methods applied to the lignocellulosic biomass processing

the lignocellulosic matrix, a highly efficient pre-treatment method is required. The main motive of pre-treatment methods is to enhance the digestibility of recalcitrant biomass. Pre-treatments also offer other advantageous features in the bioprocess such as minimal loss of polymers, and limited use/release of harsh chemicals/inhibitory compounds [18, 19]. A wide range of most commonly employed pre-treatment methods such as physicochemical, biological, and combined methods were used for the disintegration of LB for the production of biofuel (Fig. 3.2). All these pre-treatment methods with their pros and cons are summarized in Table 3.2.

All these pre-treatment methods have certain drawbacks such as incomplete digestibility of LB, utilization of toxic chemical substances, production of inhibitory compounds, time-consuming process, the high load of organic catalyst and costly equipment. All these limitations limit biofuel production. Thus, the intervention of nanobiotechnology is required to boost the sustainable production of renewable energy by their significant properties.

3.3 Role of Nanotechnology in the Pre-treatment of LB for the Production of Renewable Energy

Presently, nanotechnology-based pre-treatment methods are not explored at a large scale. A wide range of nanomaterials such as nanoparticles, magnetic nanocatalysts, carbon nanotubes, polymers, and nano-shear hybrid alkaline procedures were utilized for the pre-treatment of LB. Nanomaterials have high penetrating power which helps to penetrate the LB material and expose the matrix content for further enzymatic

Table 3.2 Various pre-treatment methods for LB with their pros and cons

Pre-treatment methods	Advantage	Disadvantage	References
Physical pre-treatment			
Grinding	Decreases the size of LB and also reduces the cellulose crystallinity	Expensive method	[20]
Mechanical Extrusion	Converts the LB into simpler form and also does not release the inhibitory substances which resist the treatment process	Needs high amount of energy	[21]
Microwave	Disintegrates the LB in a very short time span and also limits the production of inhibitory compounds	Expensive method	[22]
Ultrasound	Maximal frequency of ultrasound waves disintegrates the LB which reduces the cellulose crystallinity and enhances the pore size volume	Costly method	[23]
Pulsed electric field	Needs limited amount of energy to disintegrate the LB and this exposure increases the permeability	More research is required for this method	[24]
Freezing	Uses limited amount of harmful chemicals which decreases the pollution and enhances the productiveness	Costly method	[25]
Pyrolysis	A wide range of feedstock is used and releases char on decomposition of cellulose	Needs high temperature	[26]
Physiochemical pre-treatment			
Steam explosion	Recycling is cost-effective, needs limited amount of energy which reduces the environment pollution	Production of inhibitory compounds	[27]
Liquid hot water	No need of any chemicals and the process of neutralization. Highly cost-efficient	Release of inhibitory compounds directly proportional to the temperature	[28]
Ammonia pre-treatment	Limits the production of inhibitory compounds	Costly method	[29]

(continued)

Table 3.2 (continued)

Pre-treatment methods	Advantage	Disadvantage	References
Carbon dioxide explosion	Requires a low temperature, does not release toxins, and low cost of carbon dioxide	Expensive method due to the high cost of reactors	[26]
Chemical pre-treatment			
Dilute acid	Releases xylose from hemicellulose disintegration	Costly method and releases toxic compounds	[30]
Alkaline	Improves the surface area and removes the lignin content	Downstream processing is expensive and also requires high amount of water	[31]
Ozonolysis	Limits the release of toxic residues	Expensive process due to utilization of ozone	[32]
Ionic Liquid	Needs minimal amount of energy and efficiently dissolves the maximal amount of cellulose under gentle conditions	Expensive process due to high cost of ionic liquids, produces inhibitors	[33]
Organsolv	The bonds are breaks present between the hemicellulose and cellulose	Elimination of solvent is necessary	[34]
Deep eutectic solvents	Cost-efficient, biocompatible, eco-friendly, and biodegradable process	Recovery and reutilization are necessary	[35, 36]
Biological pre-treatment	Needs minimal level of energy, decomposition of cellulose and lignin occurs	Minimal level of biodegradation rate and also consumes more time	[37]
Combined pre-treatment	Highly efficient process than others	Costly process as it is requires different chemicals	[25]

hydrolysis of LB for the production of renewable energy such as bioethanol, biohydrogen, biogas, and biodiesel [38, 39]. The different nanoparticles/nanocomposites used in the bioprocessing of LB were described in Table 3.3.

3.3.1 Nanoparticles in the Pre-treatment of LB

The use of acid-functionalized magnetic nanoparticles is one of the most promising techniques for the pre-treatment of LB. It is a strong acid catalyst which uses an acid

Table 3.3 Role of nanomaterial in the pre-treatment of Lignocellulose biomass

Nanomaterial	Substrate	Properties	References
Magnetic nanoparticles	Corn stover	Around 41% of delignification rate	[40]
Magnetite nanoparticles	Sugarcane bagasse	Around 68 g/L of reducing sugar was observed	[41]
Calcium alginate beads	Sugarcane bagasse	About 30% of cellulose digestibility was observed	[42]
Magnetic nanoparticles	Hemp hurd	High rate of disintegration, i.e., around 80%	[38]
Carbon nanoparticles	Pomelo peel	Biohydrogen (402 ml) was achieved after 84 h of anaerobic fermentation	[43]

in the decomposition of LB material. They alter the structure of LB at a molecular level. Their magnetic nature plays a significant role in recovery and reusability which makes this process cost-effective. Various examples of magnetic nanoparticles are silica-protected cobalt-bearing spinel ferrite nanoparticles, alkyl-sulfonic acid functionalized nanoparticles, and silica-propyl sulphonic functionalized nanoparticles that fully hydrolyzed the LB and converts into the simple sugar form which further converts into biofuel [44, 45].

Nanoparticles are used in the immobilization of enzymes which function in enzymatic hydrolysis [46]. The hydrolyzing process of cellulase immobilized on magnetic nanoparticles is described in Fig. 3.3.

Various nanomaterials are used in immobilizing the lignolytic and saccharolytic enzymes for the disintegration of LB. Their reusability, release of minimal levels of

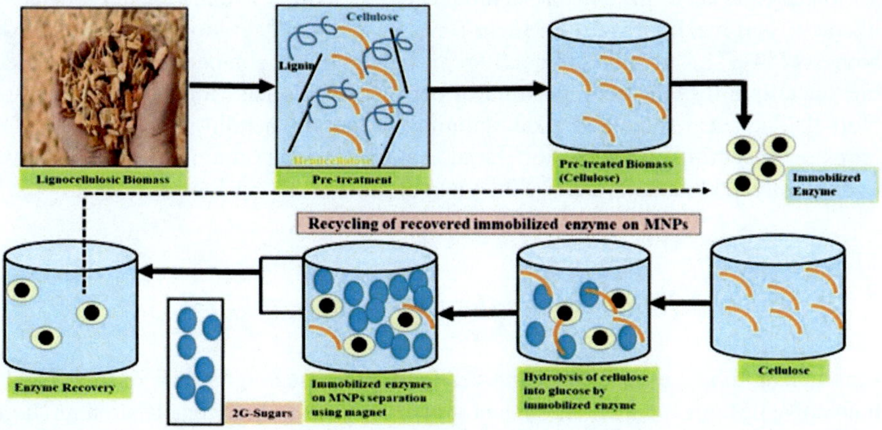

Fig. 3.3 Hydrolysis lignocellulosic biomass using cellulase immobilized on magnetic nanoparticles

Table 3.4 Immobilization of enzymes on nanoparticles used in pre-treatment of LB

Nano-support	Enzyme	Method for immobilization	Activity	References
Magnetic iron oxide nanoparticle	Cellulase	Co-precipitation	65.55% of binding efficiency and 50.34% of reusability after 4 cycles	[49]
Magnetic gold mesoporous silica nanoparticles	Cellulase	Co-precipitation	50% of binding efficiency and 28% of reusability after 5 cycles	[50]
Copper ferrite magnetic nanoparticles	Laccase	Co-precipitation	43.28% of delignification rate and around 70% of residual activity up to 6 cycles	[51]
Chitosan-coated magnetic nanoparticles	Cellulase	Covalent binding	Maximal yield of sugar was about 20 g/L	[52]
Chitosan-coated magnetic nanoparticles	Laccase	Cross-linking	Around 84% of delignification rate	[53]

inhibitory compounds, and use of eco-friendly compounds enhance the production of biofuel [47, 48]. The various nanoparticles used in the immobilization of enzymes in the pre-treatment of LB are described in Table 3.4.

3.3.2 Nanoscale Shear Hybrid Alkaline Method

Various factors are utilized in this technique like a volatile solvent, a chemical catalyst like acid, and maximal speed of shear force for the disintegration of lignocellulose biomass [54]. This method efficiently removes the lignin content. Researchers used this technique for the decomposition of corn stover biomass for the production of biofuel. The shear force and alkali solution in equimolar ratio were employed for biomass processing for a very short duration at 25 °C. A maximum degree of cellulose was hydrolyzed that will improve the yield of biofuel production considerably [55].

3.3.3 Carbon Nanotubes in the Pre-treatment of LB

Carbon nanotubes are recently employed in biofuel research because of their highly innovative design, and their significant properties such as high tensile strength, high durability, and thermal and electrical stability. Multi-walled carbon nanotubes are significantly used in immobilizing enzymes. Carbon nanotubes bound enzymes show dynamic and consistent nature at high temperature, and their value of recovery and reusability is also very high [48]. Various researchers immobilized the cellulase on multi-walled carbon nanotubes for the pre-treatment of LB. About 85% of enzymatic activity was maintained after multiple usage for hydrolyzing the cellulose.

3.4 Conclusion

The population in the world is expected to multiply in the years to come eventually increasing the demand for the utilization of fossil fuels in various societal applications. Recently, the utilization of magnetic nanoparticles has increased the effectiveness of pre-treatment methods, which help in the development of cost-effective renewable sugars production. These sugars are the building block of biofuels and biochemical production through biotechnological and chemical transformation. To sum up, the use of nanomaterials in the pre-treatment of lignocellulosic biomass for economic 2G sugar production followed by their biotransformation into renewable fuels and chemicals eventually strengthened the renewable economy.

Acknowledgements The authors are grateful to the Director, Indian Institute of Information Technology Una, Himachal Pradesh, India, for providing the necessary facility to pursue the present work. The present work has been supported by a funding agency, and we gratefully acknowledge the financial support from Himachal Pradesh Council for Science, Technology & Environment (HIMCOSTE Sanction No. STC/F(8)-6/2019 (R&D 2019-20)-377).

References

1. N. Muradov, Low to near-zero CO_2 production of hydrogen from fossil fuels: Status and perspectives. Int. J. Hydrogen Energy **42**(20), 14058–14088 (2017)
2. A. Datta, A. Hossain, S. Roy, An overview on biofuels and their advantages and disadvantages **31**, 1851–1858 (2019)
3. Z. Yin, L. Zhu, S. Li, T. Hu, R. Chu, F. Mo, D. Hu, C. Liu, B. Li, A comprehensive review on cultivation and harvesting of microalgae for biodiesel production: Environmental pollution control and future directions. Biores. Technol. **301**, 122804 (2020)
4. P. Murali, K. Hari, D. Puthira Prathap, An economic analysis of biofuel production and food security in India. Sugar Tech **18**(5), 447–456 (2016)
5. J. Gao, A. Zhang, S.K. Lam, X. Zhang, A.M. Thomson, E. Lin, K. Jiang, L.E. Clarke, J.A. Edmonds, P.G. Kyle, S. Yu, An integrated assessment of the potential of agricultural and forestry residues for energy production in China. Gcb Bioenergy **8**(5), 880–893 (2016)
6. E. Toklu, Biomass energy potential and utilization in Turkey. Renew. Energy **107**, 235–244 (2017)
7. N.S. Hassan, A.A. Jalil, C.N. Hitam, D.V.N. Vo, W. Nabgan, Biofuels and renewable chemicals production by catalytic pyrolysis of cellulose: a review. Environ. Chem. Lett. **18**(5), 1625–1648 (2020)
8. E. Ahmad, T.S. Khan, M.I. Alam, K.K. Pant, M.A. Haider, Understanding reaction kinetics, deprotonation and solvation of brønsted acidic protons in heteropolyacid catalyzed synthesis of biorenewable alkyl levulinates. Chem. Eng. J. **400**, 125916 (2020)

9. S. Kassaye, K.K. Pant, S. Jain, Hydrolysis of cellulosic bamboo biomass into reducing sugars via a combined alkaline solution and ionic liquid pretreament steps. Renew. Energy **104**, 177–184 (2017)
10. M.L. Verma, M. Puri, C.J. Barrow, Recent trends in nanomaterials immobilised enzymes for biofuel production. Crit. Rev. Biotechnol. **36**(1), 108–119 (2016)
11. M.L. Verma (Ed.), *Nanobiotechnology for Sustainable Bioenergy and Biofuel Production* (CRC Press, 2020)
12. J. Jia, W. Zhang, Z. Yang, X. Yang, N. Wang, X. Yu, Novel magnetic cross-linked cellulase aggregates with a potential application in lignocellulosic biomass bioconversion. Molecules **22**(2), 269 (2017)
13. M.L. Verma, N.M. Rao, T. Tsuzuki, C.J. Barrow, M. Puri, Suitability of recombinant lipase immobilised on functionalised magnetic nanoparticles for fish oil hydrolysis. Catalysts **9**(5), 420 (2019)
14. M. Hijosa-Valsero, J. Garita-Cambronero, A.I. Paniagua-García, R. Díez-Antolínez, A global approach to obtain biobutanol from corn stover. Renew. Energy **148**, 223–233 (2020)
15. M. Dong, S. Wang, F. Xu, J. Wanf, Q. Li, J. Chen, Pretreatment of sweet sorghum straw and its enzymatic digestion: insight into the structural changes and visualization of hydrolysis process. Biotechnol. Biofuels **12**, 276 (2019)
16. E.A. Omondi, P.K. Ndlba, P.G. Njuru, Characterization of water hyacinth (*E. crassipes*) from Lake Victoria and ruminal slaughterhouse waste as co-substrates in biogas production. SN Appl. Sci. **1**, 848 (2019)
17. L.M.S. Menandro, H. Cantarella, H.C.J. Franco, O.T. Kölln, M.T.B. Pimenta, G.M. Sanches, Comprehensive assessment of sugarcane straw: implications for biomass and bioenergy production. Biofuels Bioprod. Biorefinery **11**, 488–504 (2017)
18. B.S. Dhanya, A. Mishra, A.K. Chandel, M.L. Verma, Development of sustainable approaches for converting the organic waste to bioenergy. Sci. Total Environ. **723**, 138109 (2020)
19. D. Paul, A. Arora, M.L. Verma, Advances in microbial biofuel production. Front. Microbiol. 2768 (2021)
20. A. Arora, P. Nandal, J. Singh, M.L. Verma, Nanobiotechnological advancements in lignocellulosic biomass pretreatment. Mater. Sci. Energy Technol. **3**, 308–318 (2020)
21. J.Y. Zhu, X.J. Pan, Woody biomass pretreatment for cellulosic ethanol production: technology and energy consumption evaluation. Biores. Technol. **101**(13), 4992–5002 (2010)
22. X. Lu, B. Xi, Y. Zhang, I. Angelidaki, Microwave pretreatment of rape straw for bioethanol production: focus on energy efficiency. Biores. Technol. **102**(17), 7937–7940 (2011)
23. S.K. Khanal, D. Takara, S. Nitayavardhana, B.P. Lamsal, P. Lamsal, Ultrasound applications in biorenewables for enhanced bioenergy and biofuel production, in *Green Chemistry for Environmental Sustainability* (CRC Press, Rajasthan, 2010), pp. 303–313
24. V.K. Garlapati, A.K. Chandel, S.J. Kumar, S. Sharma, S. Sevda, A.P. Ingle, D. Pant, Circular economy aspects of lignin: towards a lignocellulose biorefinery. Renew. Sustain. Energy Rev. **130**, 109977 (2020)
25. D. Kumari, R. Singh, Pretreatment of lignocellulosic wastes for biofuel production: a critical review. Renew. Sustain. Energy Rev. **90**, 877–891 (2018)
26. B. Kumar, N. Bhardwaj, K. Agrawal, V. Chaturvedi, P. Verma, Current perspective on pretreatment technologies using lignocellulosic biomass: an emerging biorefinery concept. Fuel Process. Technol. **199**, 106244 (2018)
27. V. Rooni, M. Raud, T. Kikas, Technical solutions used in different pretreatments of lignocellulosic biomass: a review. Agron. Res. **15**(3), 848–858 (2017)
28. Y. Zheng, J. Zhao, F. Xu, Y. Li, Pretreatment of lignocellulosic biomass for enhanced biogas production. Prog. Energy Combust. Sci. **42**, 35–53 (2014)
29. C. Zhao, Q. Shao, S.P. Chundawat, Recent advances on ammonia-based pretreatments of lignocellulosic biomass. Biores. Technol. **298**, 122446 (2020)
30. P. Kumar, D.M. Barrett, M.J. Delwiche, P. Stroeve, Methods for pretreatment of lignocellulosic biomass for efficient hydrolysis and biofuel production. Ind. Eng. Chem. Res. **48**(8), 3713–3729 (2009)

31. N. Mosier, C. Wyman, B. Dale, R. Elander, Y.Y. Lee, M. Holtzapple, M. Ladisch, Features of promising technologies for pretreatment of lignocellulosic biomass. Biores. Technol. **96**(6), 673–686 (2005)
32. K.S. Baig, J. Wu, G. Turcotte, Future prospects of delignification pretreatments for the lignocellulosic materials to produce second generation bioethanol. Int. J. Energy Res. **43**(4), 1411–1427 (2019)
33. Z. Usmani, M. Sharma, P. Gupta, Y. Karpichev, N. Gathergood, R. Bhat, V.K. Gupta, Ionic liquid based pretreatment of lignocellulosic biomass for enhanced bioconversion. Biores. Technol. **304**, 123003 (2020)
34. R. Millati, R. Wikandari, T. Ariyanto, R.U. Putri, M.J. Taherzadeh, Pretreatment technologies for anaerobic digestion of lignocelluloses and toxic feedstocks. Biores. Technol. **304**, 122998 (2020)
35. D. Tian, Y. Guo, J. Hu, G. Yang, J. Zhang, L. Luo, Y. Xiao, S. Deng, O. Deng, W. Zhou, F. Shen, Acidic deep eutectic solvents pretreatment for selective lignocellulosic biomass fractionation with enhanced cellulose reactivity. Int. J. Biol. Macromol. **142**, 288–297 (2020)
36. H. Xu, J. Peng, Y. Kong, Y. Liu, Z. Su, B. Li, X. Song, S. Liu, W. Tian, Key process parameters for deep eutectic solvents pretreatment of lignocellulosic biomass materials: a review. Biores. Technol. **310**, 123416 (2020)
37. A.K. Kumar, S. Sharma, Recent updates on different methods of pretreatment of lignocellulosic feedstocks: a review. Bioresour. Bioprocess. **4**(1), 1–19 (2017)
38. R.E. Abraham, M.L. Verma, C.J. Barrow, M. Puri, Suitability of magnetic nanoparticle immobilised cellulases in enhancing enzymatic saccharification of pretreated hemp biomass. Biotechnol. Biofuels **7**(1), 1–12 (2014)
39. M.L. Verma, R. Rajkhowa, X. Wang, C.J. Barrow, M. Puri, Exploring novel ultrafine Eri silk bioscaffold for enzyme stabilisation in cellobiose hydrolysis. Biores. Technol. **145**, 302–306 (2013)
40. Z. Gou, N.L. Ma, W. Zhang, Z. Lei, Y. Su, S. Chunyu, G. Wang, H. Chen, S. Zhang, G. Chen, Y. Sun, Innovative hydrolysis of corn stover biowaste by modified magnetite laccase immobilized nanoparticles. Environ. Res. **188**, 109829 (2020)
41. N. Srivastava, A. Alhazmi, A. Mohammad, S. Haque, M. Srivastava, D.B. Pal, R. Singh, P.K. Mishra, N.V. Dai Viet, T. Yoon, V.K. Gupta, Biohydrogen production via integrated sequential fermentation using magnetite nanoparticles treated crude enzyme to hydrolyze sugarcane bagasse. Int. J. Hydrog. Energy (2021). In Press
42. S.G.C. de Almeida, G.F. de Mello, M.G. do Santos, D.D.V. da Silva, E.C. Giese, M. Hassanpour, Z. Zhang, K.J. Dussán, Saccharification of acid–alkali pretreated sugarcane bagasse using immobilized enzymes from *Phomopsis stipata*. 3. Biotechnology **12**(1), 1–13 (2022)
43. P. Samaddar, Y.S. Ok, K.H. Kim, E.E. Kwon, D.C. Tsang, Synthesis of nanomaterials from various wastes and their new age applications. J. Clean. Prod. **197**, 1190–1209 (2018)
44. N. Singh, S. Singh, M. Mathesh, Recent trends in nanobiocatalysis for biofuel production, in *Nanobiotechnology for Sustainable Bioenergy and Biofuel Production* (CRC Press, 2020), pp. 150–176
45. M. Puri, C.J. Barrow, M.L. Verma, Enzyme immobilization on nanomaterials for biofuel production. Trends Biotechnol. **31**(4), 215–216 (2013)
46. N. Singh, B.S. Dhanya, M.L. Verma, Nano-immobilized biocatalysts and their potential biotechnological applications in bioenergy production. Mater. Sci. Energy Technol. **3**, 808–824 (2020)
47. M.L. Verma, M. Naebe, C.J. Barrow, M. Puri, Enzyme immobilisation on amino-functionalised multi-walled carbon nanotubes: structural and biocatalytic characterisation. PLoS ONE **8**(9), 73642 (2013)
48. M.L. Verma, R. Chaudhary, T. Tsuzuki, C.J. Barrow, M. Puri, Immobilization of β-glucosidase on a magnetic nanoparticle improves thermostability: application in cellobiose hydrolysis. Biores. Technol. **135**, 2–6 (2013)
49. P. Kaur, M.S. Taggar, A. Kalia, Characterization of magnetic nanoparticle–immobilized cellulases for enzymatic saccharification of rice straw. Biomass Convers. Biorefinery **11**(3), 955–969 (2021)

50. E. Poorakbar, A. Shafiee, A.A. Saboury, B.L. Rad, K. Khoshnevisan, L. Ma'mani, H. Derakhshankhah, M.R. Ganjali, M. Hosseini, Synthesis of magnetic gold mesoporous silica nanoparticles core shell for cellulase enzyme immobilization: improvement of enzymatic activity and thermal stability. Process Biochem. **71**, 92–100 (2018)
51. K.S. Muthuvelu, R. Rajarathinam, R.N. Selvaraj, V.B. Rajendren, A novel method for improving laccase activity by immobilization onto copper ferrite nanoparticles for lignin degradation. Int. J. Biol. Macromol. **152**, 1098–1107 (2020)
52. E. Zanuso, H.A. Ruiz, L. Domingues, Magnetic nanoparticles as support for cellulase immobilization strategy for enzymatic hydrolysis using hydrothermally pretreated corn cob. Biomass Bioenergy Reserv. (2020)
53. S. Shanmugam, S. Krishnaswamy, R. Chandrababu, U. Veerabagu, A. Pugazhendhi, T. Mathimani, Optimal immobilization of *Trichoderma asperellum* laccase on polymer coated $Fe_3O_4@SiO_2$ nanoparticles for enhanced biohydrogen production from delignified lignocellulosic biomass. Fuel **273**, 117777 (2020)
54. S. Sahay, Impact of pretreatment technologies for biomass to biofuel production, in *Substrate Analysis for Effective Biofuels Production. Clean Energy Production Technologies*, N, Srivastava, M. Srivastava, P. Mishra, V. Gupta (Springer, Singapore, 2020), pp. 173–216
55. W. Wang, S. Ji, I. Lee, Fast and efficient nanoshear hybrid alkaline pretreatment of corn stover for biofuel and materials production. Biomass Bioenergy **51**, 35–42 (2013)

Chapter 4
Comparative Study of Enriched Biogas Bottling Cylinder in the Presence of Distinct Filler at Low Pressure

Sameer Ahmad Khan, Komalkant Adlak, Ram Chandra, and Virendra Kumar Vijay

Abstract Biogas production is a very retro technology that requires some technological advancement to compete with the recent fuel demand. Only biogas is not sufficient; it has to be enriched before applying to recent applications, i.e., mobile and stationary. In stationary applications, mainly household or domestic fuel requirement is focused on. Low-pressure high-volume storage of enriched biogas is such an advancement in the biogas sector. Enriched biogas can be compressed under high pressure (200 bar) to increase the storage capacity or increase energy density. To make it fit for domestic usage, the enriched biogas must be bottled at low pressure (20 bar). Our work shows a possibility in the same direction, i.e., storing the enriched biogas at low pressures. The appropriate experiments were performed on the storage cylinder in two ways, one is simple compression, and, in another method, the cylinder was filled with adsorbing material (activated carbons produced from biomass). Three different materials, i.e., activated biochar derived from coconut shell procured from NORIT Americas Inc., pigeon pea stalk biochar, and bamboo biochar developed within the lab at 500 °C temperature in an inert environment, were used as filler for the bottling cylinder. The desired characterization of raw material and biochar was also performed. Permissible results are found during this study showing that activated biochar is best suited as filler for bottling cylinders to store the enriched biogas.

Keywords Enriched biogas · Biochar · Pyrolysis · Thermogravimetric analysis · And Adsorbed biogas cylinder (ABG)

S. A. Khan (✉) · K. Adlak · R. Chandra · V. K. Vijay
Centre for Rural Development and Technology, Indian Institute of Technology Delhi, Hauz Khas, New Delhi 110 016, India
e-mail: sksam786@gmail.com

N. Gakkhar et al. (eds.), *Recent Advances in Bio-Energy Research*, Springer Proceedings in Energy, https://doi.org/10.1007/978-981-99-5758-3_4

4.1 Introduction

Biochar is a product produced from the pyrolysis of organic matter with the presence of a high amount of carbon that can be used as an adsorbent [1]. Biochar based on agricultural/forest residue can be produced to work as an adsorbent. The pigeon pea stalk is one such agricultural residue. The pigeon pea (*Cajanus cajan L.*) is also known as *Arhar* or *Toor* in the local language of India. Pigeon pea stalk could be one such agricultural residue. India accounts for about 90% of the world's supply of pigeon pea. India harvests pigeon pea in an area of 3.6 MHA (million hectares) that produces 3.3 million tons of pigeon pea in a year which places India in the second position throughout Asia and has a total harvest area of 4.3 MHA [2]. One hectare of land under the production of pigeon pea produces 50–60 quintals (q) of fuel sticks or agricultural residue [3]. This results in the production of a large quantity of feedstock for biochar. Another feedstock for biochar is from forest residue, i.e., Bamboo. Bamboo forests spread over 36 MHA globally, which is about 0.92% of the world's forest area. India is the second major bamboo-producing country with 16 MHA (22.46%) of a total forest area out of a total of 71.2 MHA. Bamboo is a cheap, renewable abundant resource that outgrows most plants [4]. Each biomass is used to produce biochar. Biochar produced can be used for various applications; one such application is the storage of enriched biogas or biomethane through adsorption.

Physical adsorption is where a molecule known as adsorbate adheres to the surface of another atom also known as adsorbent, with weak van der Waals forces [5]. In the gas adsorption process, the gas molecule is adsorbed on the surface of the microporous material. When raw biogas is processed through any enrichment process, the biogas obtained has more methane percentage than the biogas used as feed known as enriched biogas [6]. According to Indian standards when methane percentage in enriched biogas reaches up to a minimum of 90%v/v, it is known as biomethane [7]. The energy density of enriched biogas is low and in this case, it is found to be 37.5 kJ/m^3 which is nearly nine times and six times less than petrol and LPG, respectively (considering the calorific value of petrol and LPG as 48 MJ/kg and 45 MJ/kg, respectively); to increase the energy density of biomethane, it can be either transformed to compressed biogas/compressed biomethane (CBG) or liquefied biogas/liquified biomethane (LBG) [8, 9]. The liquefaction process is cost-effective and high compression is not safe. Another way to increase the energy density of biomethane at low pressure is by adsorption process, and this type of biomethane is also called Adsorbed Biogas/Adsorbed Biomethane (ABG). In an ABG system, biomethane is adsorbed on activated charcoal, which is packed in the storage vessel, to achieve the desired energy density. ABG has the potential to replace compressed natural gas both in mobile and stationary applications. In this, adsorption of enriched biogas was studied on biochar produced through bamboo and pigeon pea stalks. Obtained results were compared to the ABG system filled with commercially available activated coconut biochar [10].

4.2 Material and Methods

4.2.1 Sample Preparation

Pigeon pea was bought from a rural region of Uttar Pradesh, whereas bamboo was received from Gramodaya Parisar part of IIT-Delhi. Sample preparation for proximate analysis on the as-received basis of each feedstock was done by removing the dirt as much as possible and cut down to small pieces and then it was powdered, whereas, for a dry basis, the powdered sample was dried for 2–4 h at 105 °C in a hot air oven. This powdered sample was also used for carbon hydrogen and @nitrogen (CHN) analysis, SEM, thermal gravimetric analysis (TGA), and Fourier-transform infrared spectroscopy (FTIR). Small pieces of the sample of pigeon pea stalk and bamboo char were prepared by increasing their surface area as much as possible. The sample was powdered using a mixer of 750 W. The powdered sample was washed one time with distilled water and dried for 12 h at 105 °C [11]. FTIR, iodine number, and BET test of biochar sample were also performed. Granular coconut biochar was brought from Norit Americas Inc., which was steam-activated. The size of granules was between 0.42 and 2 mm. The coconut biochar sample was powered and dried for its characterization.

4.2.2 Biochar Preparation

A sample for pyrolysis was prepared, first by removing the dirt. Then pieces of pigeon pea stalk and bamboo were cut down to approximately 10 cm in length. After cutting, these pieces were washed with distilled water and dried for 12 h at 105 °C in a hot air oven [11]. Biochar was prepared from the dried sample by slow pyrolysis as this process is best for producing good-quality biochar [10]. Nitrogen gas was introduced in the furnace 30 min before starting the pyrolysis with the help of a nitrogen cylinder connected to the furnace as shown in Fig. 4.1. The sample was pyrolyzed in the presence of nitrogen gas at an average heating rate of 10 °C/min with a peak temperature of 500 °C, and residence time was taken as 2 h [11, 12]. Biochar obtained was crushed manually to reduce the size. The maximum percentage of biochar has a size of less than 10 mm.

4.2.3 Sample Characterization

Proximate Analysis of each feedstock was performed on an as-received basis and on a dry basis; for ash content, ASTM D1102-84(2013) was preferred, moisture content was found with the help of ASTM E871-82(2013), volatile matter follows ASTM E872-82(2006) @and proximate analysis of biochar was conducted according to

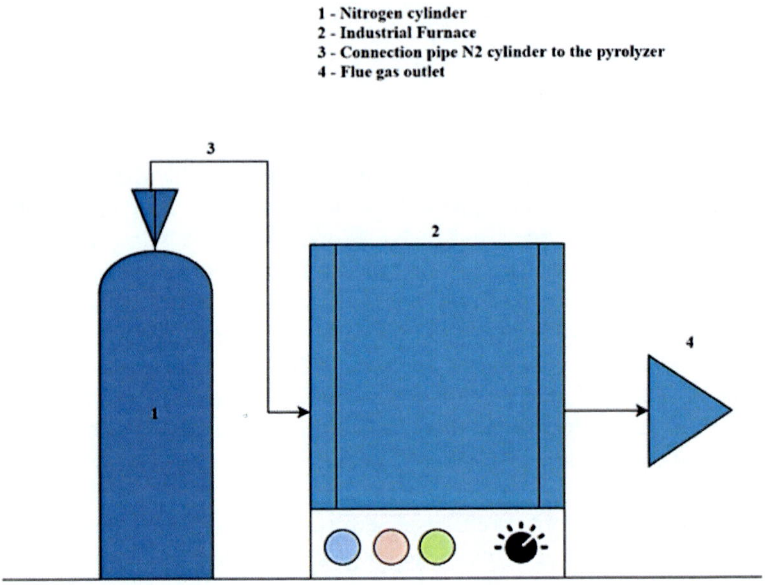

Fig. 4.1 Biochar production setup

ASTM D1762-84(2013). Ultimate analysis of the sample was performed with an elemental analyzer currently available at Center for Rural Development and Technology (CRDT)-Indian Institute of Technology Delhi (IIT-D), which determines the carbon, hydrogen, nitrogen, and oxygen percentage. Oxygen content was obtained by subtracting each element percentage from 100%. Thermal gravimetric analysis (TGA) of the sample was performed at CSIR-NIEST Jorhat, Assam. Thermal analysis is used to measure the mass change of a sample being heated at a constant rate of 10 °C/min in an inert environment. It helps to understand the nature of the feedstock and how samples behave at various temperatures. This will help to understand how feedstock will behave under pyrolysis and at what temperature breakdown of various components of biomass will occur [13].

FTIR analysis of the raw sample and its char was carried out with Thermo-scientific IS 50NIR at the Central research facility of IIT-Delhi. FTIR produces markedly superior spectra and can provide more precise information concerning the oxidation of carbons and the formation of carbon–oxygen surface groups. This technique can also allow measurements of lower concentrations of surface functional groups [14]. SEM analysis was performed at the Central Research Facility at IIT-Delhi. Iodine number and BET, both values define the adsorption capacity of a porous material. The iodine number of biochar and activated biochar can be found with the help of ASTM D4607-14. This iodine solution is titrated against standardized sodium thiosulfate with starch as an indicator before adsorption for standardization of iodine. Adsorption of iodine is performed in an acidic environment to remove any sulfur within the sample. After the adsorption of iodine, the filtrate is titrated against the standardized sodium thiosulfate. The amount of iodine present before and after the adsorption in the solution is used to evaluate the iodine number for the sample. Brunauer–Emmett–Teller equation or BET analysis of biochar was carried out in the Lab of Chemical Engineering at IIT-Delhi on BELSORP-maxII. The experiment was carried out in the presence of nitrogen at 77.36 K.

4.2.4 Adsorption Experiment

The adsorption experiment was performed at room temperature 25 ± 3 °C with enriched biogas produced within the Biogas Laboratory CRDT at IIT-Delhi. Biogas was produced using kitchen waste within the IIT-Delhi campus [15]. Biogas was enriched with the help of a water-scrubbing column [16]. Enriched biogas was analyzed with the help of a biogas analyzer from GEOTECH "BIOGAS 5000". Enriched biogas was stored in a rubber balloon of 6 m^3 which was further processed for compression with the help of a Coltri compressor (Made in Italy) into a CNG cylinder. This cylinder was removed from the compressor and connected to the ABG cylinder that was stationed on an electronic balance. ABG cylinder has a capacity of 2.831 L which was fitted with a thermocouple and a pressure gauge as shown in Fig. 4.2. ABG cylinder was filled with the desired sample as fillers and before the start of the experiment, the cylinder was vacuumed for 30 min with the help of a vacuum pump [17]. Enriched biogas was introduced to the ABG cylinder in a controlled way using the manually operated valve. The fueling time for the cylinder was taken around three minutes. The experiment was performed for the 5th cycle for each filler. One cycle consists of charging and discharging the cylinder with enriched biogas. The residual amount of enriched biogas after the 5th cycle was noted down.

4.3 Result and Discussion

4.3.1 Proximate and Ultimate Analysis

Under rudimentary tests, proximate analysis plays an important role as it is comprised of moisture content, ash content, volatile matter, and fixed carbon. For biochar, it is very crucial to know the element percentage. The ultimate analysis is a very quick way to determine the elements within any organic material, mainly carbon plays a huge role in an adsorption process. In proximate analysis, ash content increases for biochar in comparison to the raw biomass samples as shown in Table 4.1. Also, the overall carbon content increases in the biochar of both samples, i.e., pigeon pea and bamboo. An increase in carbon content improves the adsorption capacity.

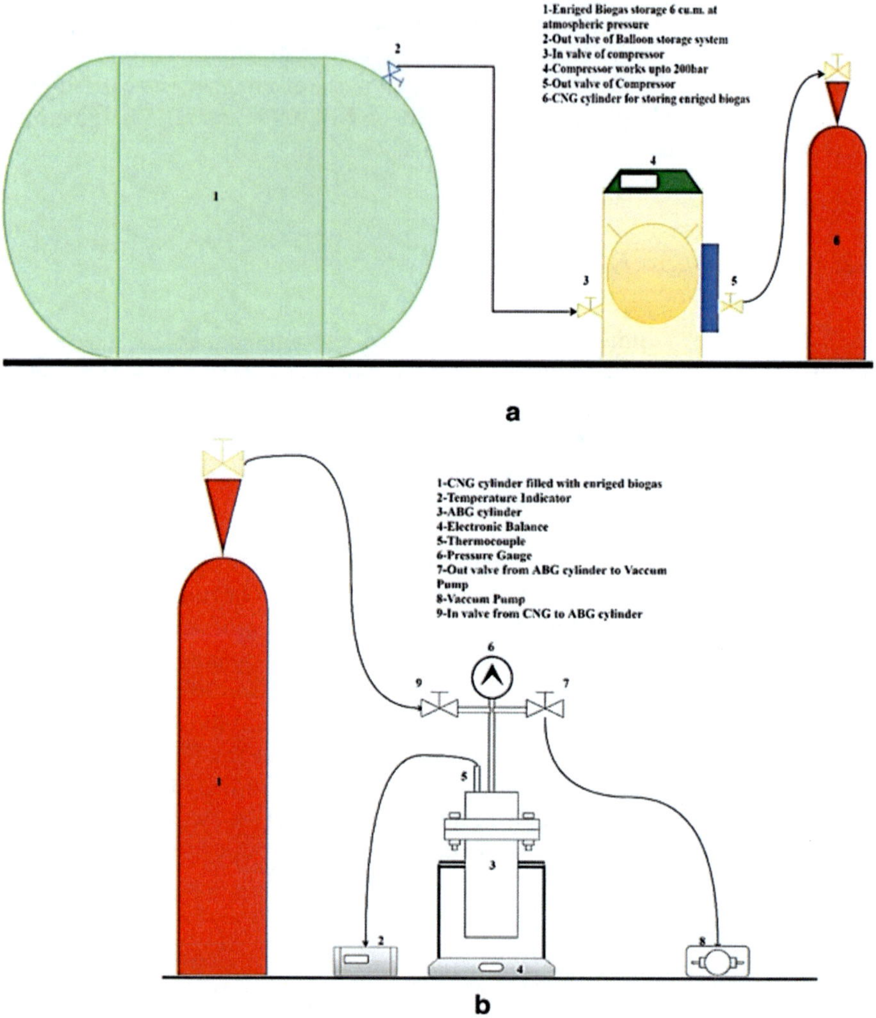

Fig. 4.2 **a** Showing setup for compressing enriched biogas into CNG cylinder, **b** Experimental setup for adsorption of enriched biogas

4.3.2 Thermal Gravimetric Analysis (TGA)

TGA has been applied to extract information about physical phenomena such as desorption and chemical phenomena [18]. Differential thermogravimetric (DTG) shows the mass loss rate curve at 10 °C/min in the presence of nitrogen for pigeon pea stalk and bamboo as shown in Fig. 4.3. Three distinct regions can be observed from the DTG curve: region 1 shows the dehydration or removal of moisture which is below 200 °C, mass loss occurs between 200 and 350 °C due to the degradation

Table 4.1 Proximate and ultimate analysis of pigeon pea stalk, bamboo, and their biochar with ultimate analysis of coconut-activated biochar

Sample	Proximate analysis				Ultimate analysis			
	Ash content % by wt.	Volatile matter % by wt	Fixed carbon % by wt. *	Moisture content % by wt.	Carbon content % by wt.	Hydrogen content % by wt.	Nitrogen content % by wt.	Oxygen content % by wt.*
Pigeon pea stalk (a.r.b)	1.08 ± 0.02	74.63 ± 0.08	12.71 ± 0.08	11.57 ± 0.02	–	–	–	–
Pigeon pea stalk (d.b)	1.12 ± 0.01	84.26 ± 0.1	14.61 ± 0.11	–	38.18	1.022	0.279	60.51
Pigeon pea stalk biochar (d.b)	4.2 ± 0.16	30.74 ± 0.18	65.06 ± 0.33	–	70.93	2.801	1.007	25.26
Bamboo (a.r.b)	1.84 ± 0.04	69.62 ± 0.09	14.63 ± 0.03	14.12 ± 0.11	–	–	–	–
Bamboo (d.b)	2.66 ± 0.01	80.52 ± 0.13	16.81 ± 0.12	–	50.86	2.168	1.32	45.65
Bamboo biochar (d.b)	10.42 ± 0.36	19.73 ± 0.19	69.84 ± 0.18	–	74.13	5.527	0.148	20.19
Coconut-activated biochar (d.b)	–	–	–	–	85.59	2.195	0.6	11.61

a.r.b = as received basis, d.b = dry basis, * = subtraction method of hemicellulose and cellulose, and region 3 corresponds to the decomposition of lignin [19]. The region above 200 °C is categorized as active and passive pyrolysis. Devolatilization of hemicellulose and cellulose is responsible for active pyrolysis, whereas lignin degradation corresponds to passive pyrolysis [20]. On complete degradation, i.e., beyond region 3, ash of the sample remains [21]. Bamboo has a higher content of lignin than pigeon pea stalk [11], but the maximum degradation of pigeon pea stalk sample takes place around 800 °C whereas for bamboo it is around 650 °C. It can be said that a more complex form of lignin is present within the pigeon pea stalk than in bamboo.

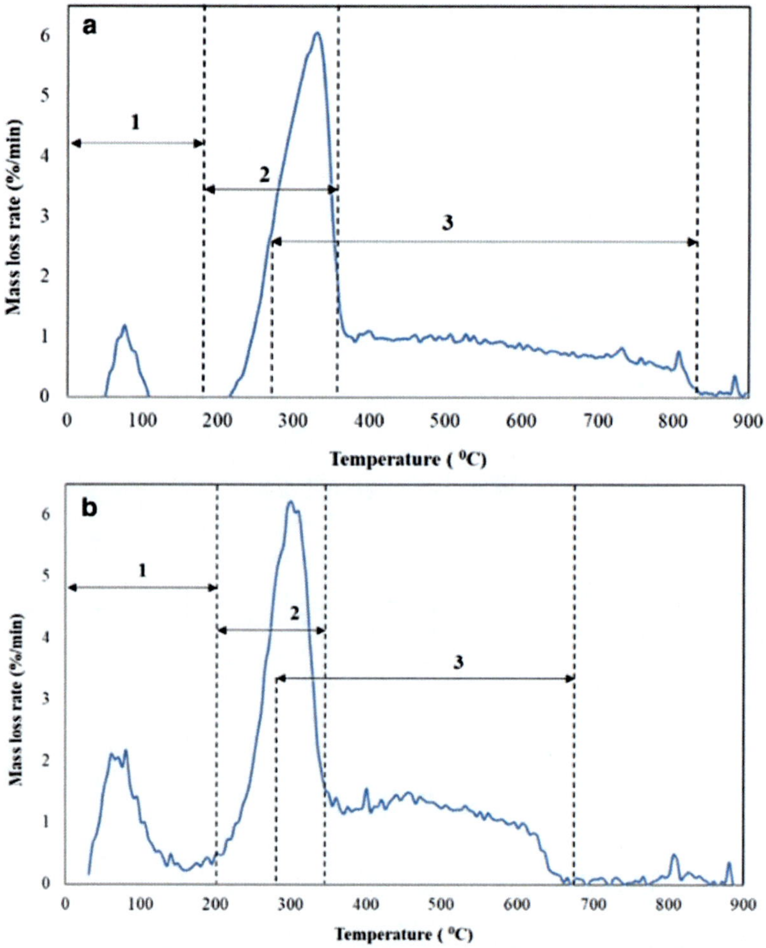

Fig. 4.3 Differential thermogravimetric analysis of pigeon pea stalk (**a**) and bamboo (**b**)

4.3.3 *Fourier-Transform Infrared Spectroscopy (FTIR)*

FTIR of the sample is performed to find the major functional groups that are present in the raw sample and biochar. FTIR of the sample was generated for wavenumber in between 400 and 4000 cm^{-1}. A significant change in the functional group was identified for pyrolyzed sample. The graph produced between transmittance and wavenumber is shown in Fig. 4.4. A peak at a particular wavenumber corresponds to a distinct functionality as mentioned in Table 4.2 [22].

When comparing the raw material to its biochar, the reduction of the hydroxyl group is around 3500 cm^{-1}, indicating that the sample has been dehydrated, as illustrated in Fig. 4.4a and b. The existence of aliphatic hydrocarbons is indicated by a peak near 2900 cm^{-1} in the raw sample, but the absence of this peak in the biochar indicates that the component was degraded during pyrolysis. In the raw biomass and biochar sample, an extensive stretch of aromatic hydrocarbon was discovered between 1500 cm^{-1} and 1700 cm^{-1}. Coconut biochar does not exhibit a peak in this range, which suggests that aromatic hydrocarbons are not present. The presence of a phenolic ring can be seen at a peak of 600 cm^{-1} in raw bamboo, bamboo char, and coconut char.

4.3.4 *Scanning Electron Microscope (SEM)*

SEM is used to read the surface morphology of various samples. SEM of the biochar is used to see the porous structure created on the surface. There is a drastic change in surface morphology on comparing the raw sample to its biochar. The macroporous structure had formed in both biochars. In the case of coconut biochar, not much of a macroporous structure can be seen as shown in Fig. 4.5.

4.3.5 *Iodine Number and Brunauer–Emmett–Teller (BET) Test*

Iodine number is the amount of iodine adsorbed by an adsorbent per gram which gives an idea of the adsorption capacity of a porous material. The value of the iodine number for biochar is low compared to activated biochar as mentioned in Table 4.3. The generated graph with the BET method is very helpful for explaining macroporous or non-porous surfaces as shown in Fig. 4.6. It not only gives an idea of surface area but also tells us about the adsorption characteristics [14]. The adsorption-based gas storage system is mainly based on physisorption since the desorption process will take place by reducing the pressure or increasing the temperature. Six types of adsorption isotherm are available according to IUPAC classification and in the case of biochar, it was best described by TYPE IV adsorption isotherm [24]. Before the experiment,

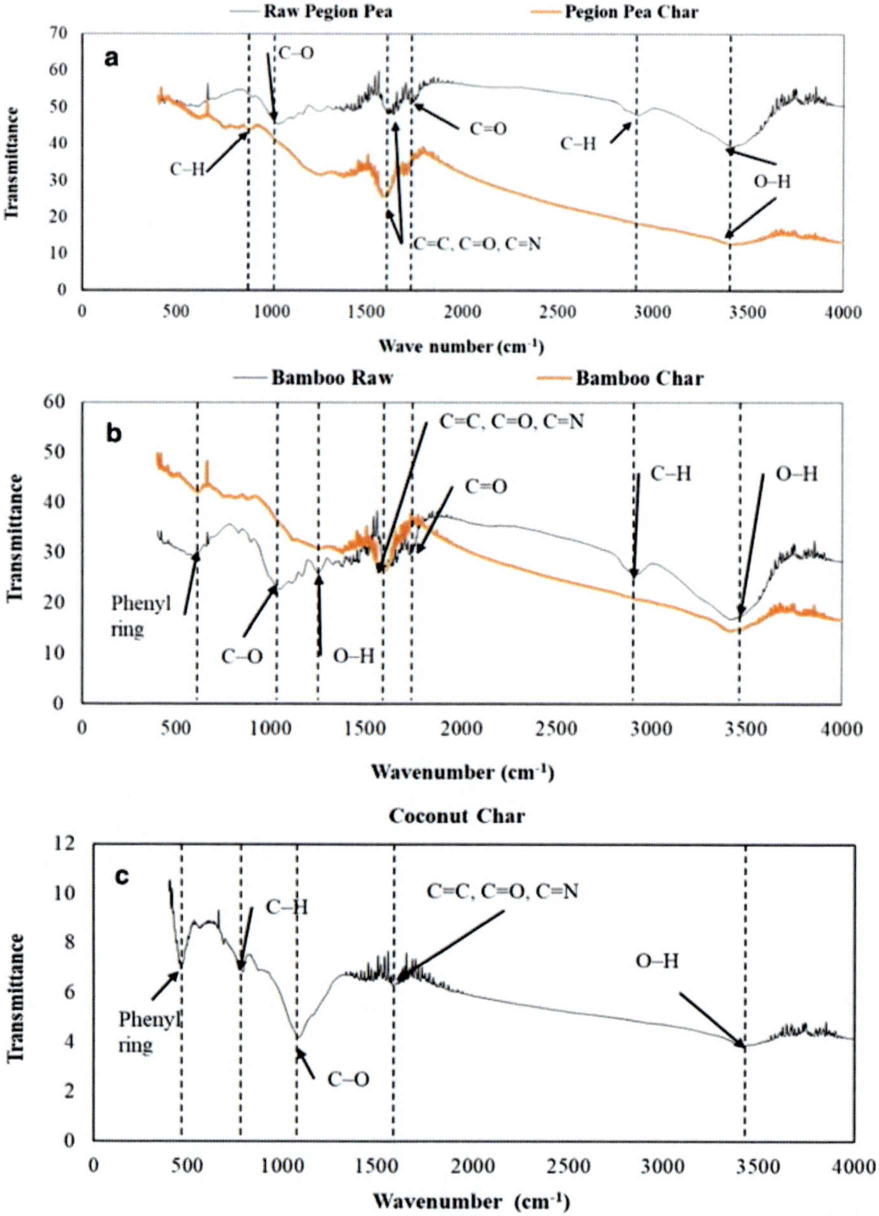

Fig. 4.4 FTIR analysis of **a** pigeon pea stalk and its biochar, **b** bamboo and its biochar, and **c** coconut-activated biochar

Table 4.2 Different functional groups corresponding to different wavenumbers [22]

Wavenumber (cm^{-1})	Characteristic vibrations (functionality)
3500–3200	O–H stretching (water, hydrogen-bonded hydroxyl)
2935; 2885	C–H stretching (aliphatic CHx; 2935-asymmetric, 2885-symmetric)
1740–1700	C=O stretching (carboxyl, aldehyde, ketone, ester)
1653–1645	N–C=O (amide I: carbonyl stretching vibration in peptide bond)
1600–1595	CC=C, C=O, C=N (aromatic components conjugated ketones and quinones, amide, amine)
1514	C=C, N–H (secondary aromatic amines, pyridine rings)
1440	C=C stretching; –C–H2 bending (lignin carbohydrate)
1375	O–H bending (phenolic; ligneous syringyl)
1110	Symmetric C–O stretching (C–O–C in lignocelluloses)
1100–950	P–O (asymmetric and symmetric stretching of PO2 and P(OH)2 in phosphate)
1030	Symmetric C–O stretching (cellulose; hemicellulose; methoxy groups of lignin)
885	C–H bending (aromatic CH out-of-plane deformation)
781	Pyridine (pyridine ring vibration and C–H deformation)
635	Phenyl ring vibration [23]

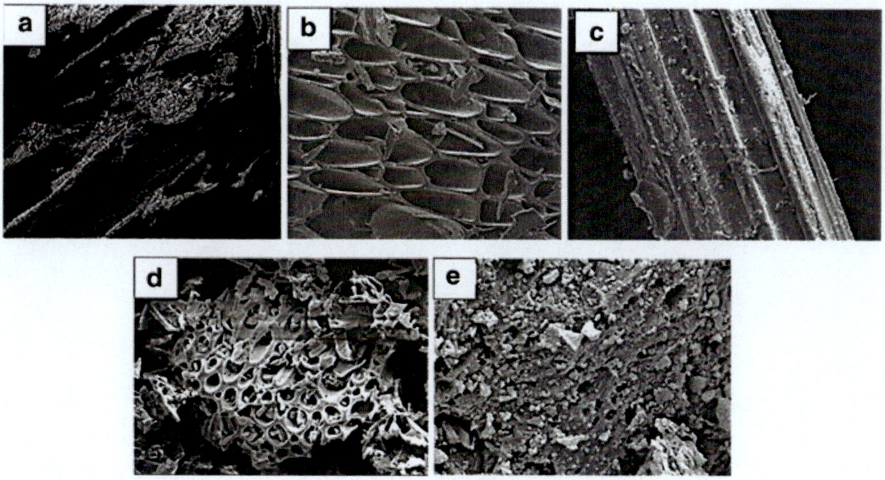

Fig. 4.5 **a** SEM of raw pigeon pea stalk, **b** SEM of pigeon pea biochar, **c** SEM of raw bamboo, **d** SEM of bamboo biochar, and **e** Coconut-activated biochar

a vacuum pump was utilized to remove the air from the ABG cylinder. The removal of air was performed for half an hour. Experiments on ABG cylinder were performed in both ways with and without biochar. The maximum amount of gas that was stored in the ABG cylinder was 75 g at 20 bar pressure using granular coconut-activated biochar as mentioned in Table 4.4. The residual amount of gas that was present at the end of the 5th cycle was 0 g for granular coconut-activated biochar as shown in Table 4.5. The result clearly shows that the adsorption capacity of activated biochar with high surface area shows the maximum storing amount at 20 bar pressure.

Table 4.3 Iodine number and surface area of different biochar

Sample	Pigeon pea biochar	Bamboo biochar	Coconut-activated biochar
Iodine number (mg/g)	106.4	87.14	920.04
BET surface area (m^2/g)	85.67	30.01	957
Total pore volume (cm^3/g)	0.0697	0.0272	0.5427
Mean pore diameter (nm)	3.2586	3.634	2.2662

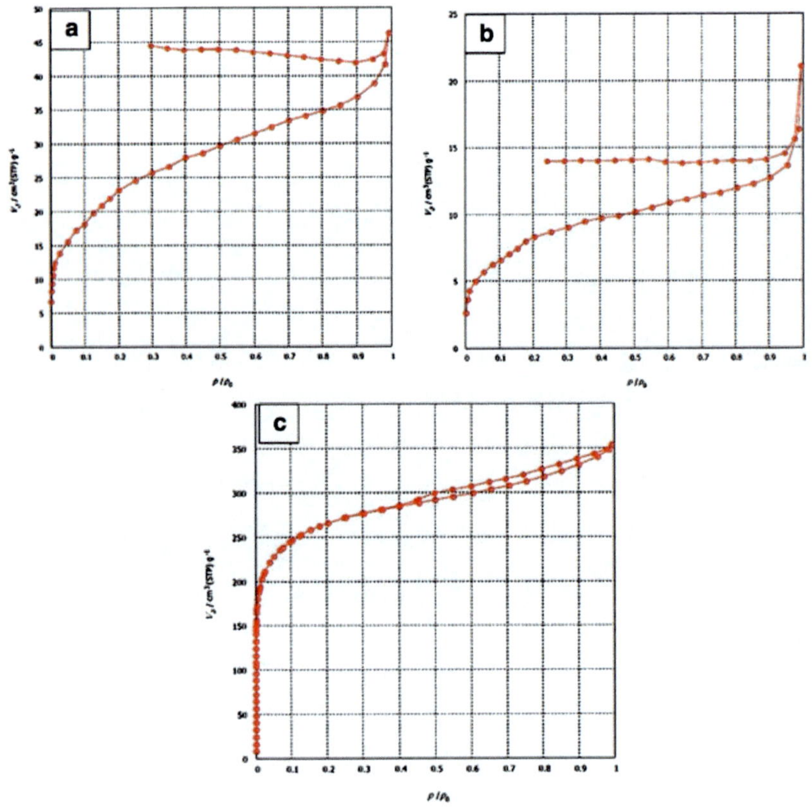

Fig. 4.6 The graph represents adsorption isotherm for **a** pigeon pea biochar, **b** bamboo biochar, and **c** coconut-activated biochar from NORIT

Table 4.4 Amount of enriched biogas gas stored when ABG cylinder was filled with distinct filler

Distinct filler	Maximum storage at 20 bar pressure	Gravimetric capacity (g/kg)	Volumetric capacity (g/L)
Without any adsorbent	44	2.05	15.54
Filled with pigeon pea stalk char	54	2.51	19.07
Filled with bamboo char	48	2.23	16.96
Filled with coconut-activated biochar	75	3.49	26.49

Table 4.5 Maximum amount enriched biogas resides within cylinder after 5th cycle with distinct filler

Distinct filler	Maximum amount of enriched biogas resides within cylinder after 5th cycle (g)
Without any adsorbent	0
Filled with pigeon pea stalk char	6
Filled with bamboo char	4
Filled with coconut-activated biochar	0

4.4 Conclusion

This study indicates the possibility of storing more enriched biogas at low pressure with the help of a porous material. Considering the case of biochar and activated biochar, activated biochar shows enhanced storing capacity. This is due to its high surface area. The cost of the empty cylinder is less than that of a cylinder filled with porous material until and unless ABG can store the amount of enriched biogas or biomethane at 20 bar pressure same as the amount that is stored at 200 bar pressure. On achieving this goal, this technology can be further applied to mobile and stationary applications. The rural aspect of this application will be storing biomethane that is produced within a rural area.

This research signifies that the amount of methane that can be stored with the help of an adsorbent will be higher than an empty cylinder under similar conditions. Based on this study, the author has figured out some research gaps and future research direction. The quality of biochar can be optimized to improve the adsorption by varying the pyrolysis temperature and activation. More closely packed porous material such as monolith can be used for the study.

References

1. J.S. Cha, S.H. Park, S.C. Jung et al., Production and utilization of biochar: a review. J. Ind. Eng. Chem. **40**, 1–15 (2016). https://doi.org/10.1016/j.jiec.2016.06.002
2. ICRISAT Crops: PigeonPea (2022). http://www.icrisat.org/what-we-do/crops/PigeonPea/PigeonPea.htm
3. Pigeon pea—Vikaspedia (2022). https://vikaspedia.in/agriculture/crop-production/package-of-practices/pulses/pigeon-pea
4. A.D. Sawarkar, D.D. Shrimankar, A. Kumar et al., Commercial clustering of sustainable bamboo species in India. Ind. Crops Prod. **154**, 112693 (2020). https://doi.org/10.1016/j.indcrop.2020.112693
5. M. Suzuki, *Adsorption Engineering* (Kodansha Ltd., Tokyo, Japan, 1990)
6. R. Kapoor, P. Ghosh, M. Kumar, V.K. Vijay, Evaluation of biogas upgrading technologies and future perspectives: a review. Environ. Sci. Pollut. Res. (2019)
7. IS 16087 (2013): Biogas (Biomethane)-Specification [PCD 3: Petroleum L and their RP (2013) Burea of Indian standards
8. P. Singh, H. Gundimeda, Life cycle energy analysis (LCEA) of cooking fuel sources used in India households. Energy Environ. Eng. **2**, 20–30 (2014). https://doi.org/10.13189/eee.2014.020103
9. C.O. Osueke, T.M.A. Olayanju, C.A. Ezugwu et al., Comparative calorific evaluation of biomass fuel and fossil fuel. Int. J. Civ. Eng. Technol. **9**, 1576–1590 (2018)
10. S.A. Khan, T.C. D'Silva, S. Kumar, et al., Mutually trading off biochar and biogas sectors for broadening biomethane applications: a comprehensive review. J. Clean. Prod. **318**, 128593 (2021). https://doi.org/10.1016/j.jclepro.2021.128593
11. S.S. Sahoo, V.K. Vijay, R. Chandra, H. Kumar, Production and characterization of biochar produced from slow pyrolysis of pigeon pea stalk and bamboo. Clean. Eng. Technol. **3**, 100101 (2021). https://doi.org/10.1016/j.clet.2021.100101
12. S.K. Katyal, P.V.R. Iyer (2010) Thermochemical characterization of pigeon pea stalk for its efficient utilization as an energy source, 8312
13. T.J. Hopkins, Carbonization of wood for advanced materials **35**, 259–266 (1997)
14. R.C. Bansal, M. Goyal, *Activated Adsorption Carbon* (Taylor & Francis Group, LLC, Boca Raton, Florida, USA, 2005)
15. A. Isha, S. Kumar, B. Jha et al., Development of stabilization methods using a pilot scale anaerobic digester for seasonal variations in kitchen wastes for improved methane production with zero breakdowns. Clean. Eng. Technol. **1**, 100015 (2020)
16. R. Kapoor, P.M.V. Subbarao, V.K. Vijay et al., Factors affecting methane loss from a water scrubbing based biogas upgrading system. Appl. Energy **208**, 1379–1388 (2017). https://doi.org/10.1016/j.apenergy.2017.09.017
17. Z. Zakaria, T. George, The performance of commercial activated carbon absorbent for adsorbed natural gas storage. Int. J. Res. Rev. Appl. Sci. **9**, 225–230 (2011)
18. A.W. Coats, J.P. Redfern, Thermogravimetric analysis. a review. Analyst **88**, 906–924 (1963). https://doi.org/10.1039/AN9638800906
19. S.A. Khan, D. Kumar, S. Kumar, et al., Recent advances in fast pyrolysis and oil upgradation, in *Thermochemical and Catalytic Conversion Technologies for Future Biorefineries*, ed by P. Verma, vol. 1 (Springer Nature Singapore, Singapore, 2022), pp. 297–344
20. K. Slopiecka, P. Bartocci, F. Fantozzi, Thermogravimetric analysis and kinetic study of poplar wood pyrolysis. Appl. Energy **97**, 491–497 (2012)
21. X. Huang, J. Cao, X. Zhao et al., Pyrolysis kinetics of soybean straw using thermogravimetric analysis. Fuel **169**, 93–98 (2016)
22. K.B. Cantrell, P.G. Hunt, M. Uchimiya et al., Bioresource technology impact of pyrolysis temperature and manure source on physicochemical characteristics of biochar. Bioresour. Technol. **107**, 419–428 (2012)

23. R. Kumar, Kamakshi, M. Kumar, K. Awasthi, UV-irradiation assisted functionalization and binding of Pd nanoparticles in polycarbonate membranes for hydrogen separation. Environ. Sci. Pollut. Res. (2020). https://doi.org/10.1007/s11356-020-11106-2
24. Z. Nie, Y. Lin, X. Jin, Research on the theory and application of adsorbed natural gas used in new energy vehicles: a review. Front. Mech. Eng. **11**, 258–274 (2016)

Chapter 5
Design and Techno-economic Analysis of a Biogas Plant as an Alternative Heat Source in the Food Processing Industry

Raman Kumawat, Lata Gidwani, and Kunj Bihari Rana

Abstract Biogas has been a vital source of energy for ages. Biogas is the fourth largest contributor to renewable energy and contributes around 11% of India's and 15% of global renewable energy power generation. Recent advancements in technology make it viable to use biogas as an alternative heat source for small and medium-scale industries. This study focuses on designing a biogas plant to meet up the energy requirements of Bikaji Foods International Ltd., a food processing industry located in Bikaner, Rajasthan. The highest livestock population increase rate (31.85%) in the state makes Bikaner a perfect site for the plant setup. The region has a hot desert climate with an average annual temperature of 29 °C. The plant design is based on volumetric analysis which incorporates economical as well as technical aspects. The currently operating coal-powered (Indonesian coal) setup in the industry needs 41,840 MJ of heat energy on a daily basis. This energy requirement can be met up by producing 3,032 m^3 of biogas daily. Replacing the Indonesian coal-powered setup with the biogas one will significantly reduce the plant's operating cost and will also improve the ecosystem and respiration rate in the region. The study resulted in a capital recovery factor (CRF) of 0.11 and a levelized cost of energy (LCOE) of ₹14.72 m^{-3}, at the 10% discount rate. The maximum net present value at a 6% discount rate is ₹433.68 million, whereas the minimum net present value is ₹219.92 million and ₹314.59 million at discount rates of 14% and 10%, respectively. The digestate obtained after the completion of anaerobic digestion can be used as an organic fertilizer.

R. Kumawat (✉)
Department of Renewable Energy, Rajasthan Technical University, Kota, Rajasthan 324010, India
e-mail: raman.uce.ret@gmail.com

L. Gidwani
Department of Electrical Engineering, Rajasthan Technical University, Kota, Rajasthan 324010, India

K. B. Rana
Department of Mechanical Engineering, Rajasthan Technical University, Kota, Rajasthan 324010, India

N. Gakkhar et al. (eds.), *Recent Advances in Bio-Energy Research*, Springer Proceedings in Energy, https://doi.org/10.1007/978-981-99-5758-3_5

Keywords Biogas · Biogas plant design · Levelized cost of energy · Net present value · Techno-economic analysis · Renewable energy

5.1 Introduction

The burning of fossil fuels for power generation adversely affects the environment and damages the ozone layer and imbalances the composition of atmospheric gases. Fossil fuels are widely used as a primary heat source in food processing industries. It is harmful to the atmosphere and one of the prime contributors to global warming issues. This can be avoided by meeting the power demand by using renewable energy sources [1]. Food processing industries have fixed energy demand. Thus, biogas-powered setups can easily replace the existing coal-powered setups. It is feasible in a developing nation like India since it already has a large livestock population. It can also increase the decarbonization of the food industry by cutting down carbon emissions significantly. Biogas can be a useful source of energy for such industries. It has enormous benefits such as eco-friendly fuel and generating many types of fertilizers with much transcendence as by-products [2].

Bikaji Foods International Limited, Bikaner, Rajasthan, was established in the year 1986. At present, the industry's thermal facilities have four coal-fired boilers, with an energy output of 12,552 MJ, 12,552 MJ, 8,368 MJ and 8,368 MJ, respectively. Currently, the burner and boiler are sufficient for operating the biogas without major changes in the system. Ghazi University's 126 MW energy demand is approximately equivalent to 23,923 cubic meters of biogas. The total capacity of 16 boilers is 26265 KW which has been used for heating purposes [3]. Techno-economic analysis of 3,101 m^3 per day biogas production, and municipal solid waste-based biogas plant considering geographical conditions, temperature and climate change [4]. Techno-economic analysis of 320.76 KWh per day heat generation from 2713 m^3 volume anaerobic digestion biogas power plant, and payback time of 5.3 years and high NPV indicate good economic results [5].

Biogas is commonly used for household activities such as cooking and lighting in developing countries. On the other hand, developed countries are focusing more on the commercial use of biogas such as for industrial purposes, power generation and transport [6]. Biogas production by anaerobic digestion is a sophisticated process to convert biodegradable matter into biogas in the absence of oxygen. This is a four-step microbiological process that consists of hydrolysis, acidogenesis, acetogenesis and methanogenesis [7]. The general composition of biogas is methane (40–60%), CO_2 (35–55%), moisture (1–5%), H_2S (0.1–3%), N_2 (0–5%), O_2 (<2%), H_2 and N_2(0–500 ppm) [6].

Biogas potential has been dependent on the availability of biomass and livestock in a nation. India has a total installed capacity of 10,577.45 MW bio-power. Rajasthan has an installed capacity of 125.080 MW bio-power. The total estimated potential of a family/small-type biogas plant is 1,025 MW in Rajasthan [8]. According to the 20th livestock census, in the state of Rajasthan, the total number of cattle (indigenous

Table 5.1 Number of equipment in the industry

Fryers	Nos
Bhujia making	50
Rasgulla making	45
Total	95

and exotic) is 13.93 million [9]. The estimated annual available cow dung is 50.844 Million tonnes in Rajasthan.

The study aims to analyze the technical and economic performance of the biogas plant. Technical performance is based on design features, and economic performance is based on the net present value at various discount rates and the payback period of the proposed biogas plant [5].

5.2 Energy Demand of the Industry

An interrogation-based detailed survey of the production hall, frying lines and roasting lines was carried out to estimate the net daily energy demand of Bikaji Foods International Limited, Bikaner. The total number of equipment that uses heat energy in this industry is summarized in Table 5.1. The total heat energy needed for all the operations is estimated to be 41,840,000 kJ per day (data as received from the industry).

5.3 Energy Equivalence (Replacing Coal with Biogas)

This industry is currently based on a coal-powered setup. In this study, a coal-powered setup is being replaced with a biogas-powered one. It will improve the sustainability of the environment and will be economically beneficial to the industry. The heating value of biogas is 23,000 kJ m^{-3} and burner efficiency has been taken as 60% (assumed):

$$\begin{aligned}\text{Equivalation biogas energy (E)} &= \frac{\text{Total energy required}}{\text{Heating value} \times \text{Burner efficiency}}. \qquad (5.1)\\ &= 3031.884\text{–}3032\,\text{m}^3\ \text{per day}\end{aligned}$$

5.4 Designing of Biogas Power Plant

Three techniques of biogas production are widely accepted in India. These are balloon type, fixed dome type and floating drum type. The floating drum technology has been selected for designing the large-scale biogas plant. It provides gas at constant pressure and is easy to construct and maintain [10, 11].

The floating drum-type biogas plant consists of a cylindrical well-type digester tank and a moveable gas holder. The moveable gas holder has an external and internal guide frame that provides stability. The digester is made of quarry stone and brick masonry while the gasholder drum is made of steel.

The slurry (cow dung and water) is prepared in the mixing tank and is fed into the digester through the inlet pipe. This slurry remains unused for about 60 days. During this period, anaerobic fermentation converts organic matter into biogas in the absence of oxygen. Generated biogas will be collected in the gas holder, and the drum will start rising. At a certain limit, pressure applied to the slurry increases. This results in the shifting of the digested slurry from the digester to the outlet slurry tank through the outlet pipe (Fig. 5.1).

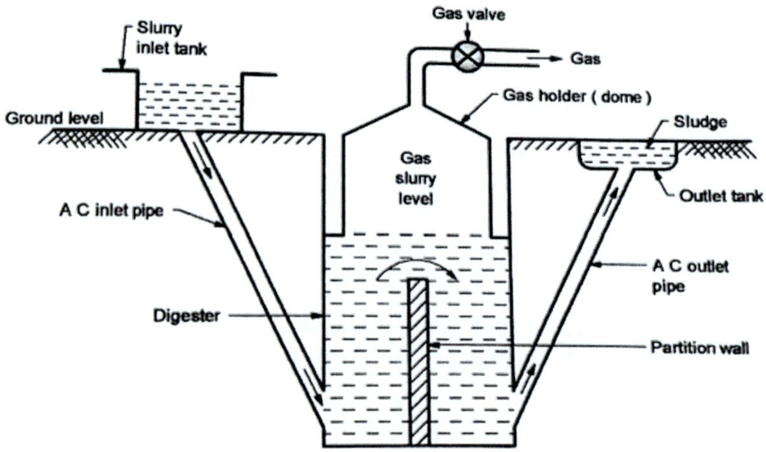

Fig. 5.1 Floating drum-type biogas plant

5.4.1 Mathematical Analysis

5.4.1.1 Designing of Gasholder Drum

The gas holder drum is made up of steel. The thickness of the steel sheet varies between 2–4 mm for the side and 2 mm for the top [12]. Some assumptions are taken into consideration to estimate the volume of the gasholder:

cow dung produced (10 kg per day per cow); collectable cow dung (70%); the weight of dry solid mass in cow dung (18%); production of biogas (0.34 m^3 kg^{-1}); the fraction of gas to be stored in the gas holder (0.6) [13].

At an equilibrium state, the production and consumption of biogas must match. The number of cows required (N) to fulfill the demand for dung for biogas production:

$$\begin{aligned} &0.34 \times 0.18 \times 0.70 \times 10 \times N = 3032 \\ &N \sim 7078 \end{aligned} \tag{5.2}$$

Total available cow dung (CW_t) and total collectable cow dung (CW_c) kg per day are

$$CW_t = 10 \times N \tag{5.3}$$

$$CW_C = 10 \times 0.70 \times N \tag{5.4}$$

$$\text{Biogas production(BP)} = 10 \times 0.70 \times 0.18 \times 0.34 \times N \tag{5.5}$$

$$\text{Volume of biogas holder}\left(V_{gh}\right) = 0.6 \times \text{BP} \tag{5.6}$$

where V_{gh} is the volume of biogas holder in m^3.

5.4.1.2 Designing of Digester

The anaerobic digestion process is completely done in the digester where the microorganisms decompose cow dung into biogas in the absence of oxygen. It should be taken care that there is no gas leakage and no damage due to water. The following assumptions are used for estimating the volume of the digester: the density of slurry (1090 kg m^{-3}); fixed retention time (FRT) (50 days); temperature (35 °C); pH value range (6.6–8.0); collectable cow dung (70%) [13].

For the highest biogas yield, cow dung and water can be mixed in the same ratio of 1:1 [14, 15]:

$$\text{Daily feeding of slurry}\left(\text{m}^3\,\text{day}^{-1}\right) = \frac{\text{Dung} + \text{Water}}{\text{Density of slurry}} \quad (5.7)$$

$$\text{Volume of slurry in the digester}\,(\text{V}_{\text{sd}}) = \text{Fixed retention time} \times \text{daily feeding of slurry} \quad (5.8)$$

where V_{sd} is the volume of slurry in the digester in m^3.

Approximately, 90% volume is occupied by the slurry, then the estimated volume of the digester (V_{d}):

$$\text{V}_{\text{d}}\left(\text{m}^3\right) = \frac{\text{V}_{\text{sd}}}{0.9} \quad (5.9)$$

$$\text{Total plant volume}\left(\text{V}_{\text{pt}}\right) = \text{V}_{\text{gh}} + \text{V}_{\text{d}} \quad (5.10)$$

The site selected for the biogas plant is in the Karni extension, RICCO industrial area Bikaner, Rajasthan. The average annual temperature of the site is 29 °C as shown in Table 5.2. The geographical location of the site lies between 28.01762° North latitude and 73.31495° East longitude. The total number of cows (exotic and indigenous) is 11,94,729 in Bikaner, Rajasthan [9]. The estimated available cow dung is 11,947.30 tonnes per day. There are many gaushalas (cow shelters) located near this site with enough cows [16] (Fig. 5.2 and Table 5.3).

The number of exotic (crossbreed) and indigenous cows are exponentially increasing in the state of Rajasthan. These numbers are highest in Bikaner. The livestock population increase rate in Bikaner from 2012 to 2019 is 31.85%. All the gaushalas (cow shelters) are located close to the site selected for the biogas plant.

Table 5.2 Annual weather temperature report for Bikaner (source: world weather online)

Month	Temperature (°C)
January	17
February	21
March	27
April	32
May	36
June	37
July	31
August	34
September	33
October	31
November	25
December	20
Annual average temperature	29

Fig. 5.2 Satellite view of Bikaji Foods International Limited

Table 5.3 List of gaushalas (cow shelters) located near the site

Gaushalas	Allocated code, no. of cows	Collectable cow Dung per day (kg)
1. Shri Agrasen Jeev Jantu Kalyan Evem Gou Sewa Samiti, Kanasar, Bikaner	GP-08-18, 1666	11,662
2. Shri Karni Gaushala, Deshnok, Bikaner	GP-08-52, 1642	11,494
3. Shri Ganga Jubli Panjarapol Gaushala, Sheetla Gate, Bikaner	GP-08-27, 2286	16,002
4. Shri Murli Manohar Goshala, Bhinasar, Bikaner	GP-08-68, 3584	25,088
Total	9178	64,246

A total of 64,246 kg of cow dung can be obtained from the selected cow shelters per day, but 49,546 kg of cow dung is required per day to meet the energy demand. Adequate cow dung is available to meet the energy demand of the industry under consideration. It is also a good initiative for the use of cow dung waste and a healthy environment. Gaushala owners can use the revenue generated from selling cow dung to look after their cows. Two temperature ranges are highly economical and feasible compared to intermediary ranges for large-scale biogas production.

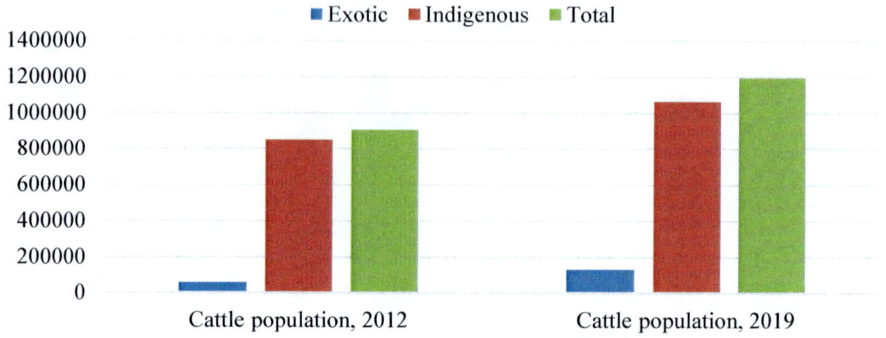

Fig. 5.3 Exotic and indigenous cattle population of Bikaner, Rajasthan

These are optimal mesophilic (30–37 °C) and optimal thermophilic (50–55 °C) for industrial-level biogas production [17, 18] (Fig. 5.3).

Based on the demand price of energy, the estimated required number of cows is 7,078, and 70,780 kg of cow dung is produced ona dailyy basis. The estimated total collectable dung is 49,546 kg per day, which is 70% of the total dung produced. According to this above estimate, the biogas production is 3,032 m^3 per day. The daily feeding of slurry to the digester through the inlet pipe is 91 m^3. The total volume of the biogas plant is 6,876 m^3. The volume of the biogas holder is 1,820 m^3 and that of the digester is 5,056 m^3 as shown in Table 5.4 [22].

The total capital cost of the biogas plant is ₹60 million. In this research, the design of a floating drum-type biogas plant has been prepared. This is manufactured by Dreamtech Machineries. In this system, the construction material of the floating drum is stainless steel or powder-coated mild steel. This is an automatically operated type of plant. The cost of cow dung with transportation is ₹1.5 kg^{-1} so the total cost of cow dung is ₹27.13 million. Operation and maintenance cost is an initial 10% of total capital cost, then the total O&M cost is ₹6 million. If the digester solution is prepared in the ratio of 1:1 of cow dung and water, the annual water cost for this is 0.11 million. Both skilled and unskilled workers are required to operate a biogas

Table 5.4 Biogas plant design (volumetric analysis)

Parameters	Value	Unit
Total available cow dung (CW_t)	70,780	kg
Total collectable cow dung (CW_c)	49,546	kg per day
Biogas production (BP)	3032	m^3 per day
Volume of biogas holder (V_{gh})	1820	m^3
Daily feeding of slurry	91	m^3 per day
Volume of slurry in the digester (V_{sd})	4550	m^3
Volume of the digester (V_d)	5056	m^3
Total plant volume (V_{pt})	6876	m^3

Table 5.5 Cost analysis of biogas plant (annual)

Items	Cost (₹)
Capital cost of biogas plant[a]	60 million
Cow dung cost (with transportation)[b]	27.13 million
Operation and management cost [c]	6 million
Labor cost (unskilled) (Rs. 300 per day)[d]	0.10 million
Labor cost (skilled) (Rs. 500 per day)[d]	0.18 million
Water cost[e]	0.11 million
Total input cost	93.52 million

[a] Ref. [8]
[b] Ref. [19]
[c] Ref. [21]
[d] Based on market survey
[e] Ref. [20]

plant, with a total annual cost of ₹0.28 million. The total cost of the biogas plant based on all the analyses and estimates is ₹93.52 million as shown in Table 5.5. The annual production of biogas is 1,106,680 m^3, which is equivalent to 475,872.40 kg of liquefied petroleum gas (1 cubic meter of biogas is equivalent to 0.43 kg of LPG) [23]. This is equal to about 10,018 commercial 47.50 kg cylinders of LPG. The price of a 47.50 kg commercial LPG cylinder in Bikaner, Rajasthan, is ₹5091.50 in February 2022. Thus, the annual amount of LPG is ₹52,010,315.94.

The biogas-digested slurry is the most effective organic fertilizer for the healthy growth of crops and increases the fertility of the soil [24]. The average composition of biogas digesting solution is 1.5% nitrogen, 1.1% phosphorus and 1% potassium, and is also rich in nutrients. An average of 30% slurry is considered digested slurry as on average 1 kg of cow dung produces 0.30 kg of slurry [25]. It will generate a 14,864 kg per day amount of digested slurry and the annual income from the digested slurry is ₹27.13 million, if sold at a rate of ₹5 kg^{-1}.

5.5 Economic Analysis

5.5.1 Capacity Recovery Factor

The capacity recovery factor is defined in terms of the discount rate and the duration of the plant's life cycle:

$$\text{Capacity recovery factor (CRF)} = \frac{i(1+i)^n}{(1+i)^n - 1} \tag{5.11}$$

i = Discount rate; n = Useful life of biogas plant.

Table 5.6 Capacity recovery factor and levelized cost of energy for the different discount rates and different plant life[a]

Discount rate (%)	6	8	10	12	14
Plant life (years)	CRF				
25	0.08	0.09	0.11	0.12	0.14
20	0.09	0.10	0.11	0.13	0.15
15	0.10	0.12	0.13	0.14	0.16
	LCOE (₹/m^3)				
25	12.18	13.02	14.71	15.56	17.25
20	13.03	14.72	13.87	16.41	18.09
15	13.87	15.56	16.41	17.25	18.94

[a] Ref. [26, 27]

The capacity recovery factor for a 20 year life of the biogas power plant at a 10% discount rate is 0.11 as shown in Table 5.6.

5.5.2 *Levelized Cost of Energy*

$$\text{Levelized cost of energy (LCOE)} = \frac{(CC_p \times CRF) + O\&M}{EG_p} \tag{5.12}$$

CC_p = overall cost of biogas plant, CRF = capacity recovery factor,

O&M = operation & management cost, and EG_p = annual energy generated by biogas power plant.

The levelized cost of energy is equal to the average price that consumers must pay in order to pay properly with a rate of return equal to the investor discount rate for capital, operation and management and fuel costs [28]. The levelized cost of energy for this biogas power plant is ₹14.72 m^{-3} at a discount rate of 10%. LCOE (₹/m^3) has been evaluated at different capacity recovery factors and different discount rates for better economic analysis of the biogas plant as shown in Table 5.7.

5.5.3 *Net Present Value (NPV)*

Net present value is the sum of the requisite annual cash flow in the current currency [29]:

$$NPV = -C_0 + \sum_{k=0}^{n} \frac{CF_k}{(1+i)^k} \tag{5.13}$$

Table 5.7 Economics parameters for biogas power plant

Particular	Value
Discount rate (i)	6%, 8%, 10%, 12%, 14%
Inflation rate (j)	6%
Effective discount rate (i_e)[a]	0%, 0.29%, 0.57%, 0.85%, 1.14%
Plant's life (years)	20
Biogas price per m^3 (₹)[b]	24
Total Input cost (₹)	93.52 million

[a] Effective discount rate$(i_e) = \frac{i-j}{1+j}$

[b] Ref. [30].

$$CF_k = P_k - V_k - LC \tag{5.14}$$

C_0 = Investment cost, K = Plant's life, CF_k = Annual cash flow,

P_k = Income from biogas power plant, and V_k = Operation & management cost.

LC = Labor cost.

5.6 Results and Discussion

A biogas power plant designed for the Bikaji food industry has been analyzed. This type of energy source is not only used to reduce the use of fossil fuels, but it is also a big step toward the sustainability of the environment. In this study, an attempt to identify the energy needs of the Bikaji foods industry has been done by using a biogas-powered plant. All the technical and financial aspects have been analyzed. The total estimated energy required for Bikaji Foods International Ltd. for all operations is 41,840 MJ per day. The estimated equivalation energy (coal replaced with biogas) is 3,032 m^3 per day. Cow dung required for a 6,876 m^3 volume biogas power plant to meet the energy demand and to make digester slurry was 49,546 kg per day. The total number of cows that can meet the demand for cow dung is 7,078.

The capital cost of the biogas power plant is ₹60 million. The total annual fixed and variable input cost was ₹93.52 million. The total annual income from the biogas power plant is ₹53.69 million which includes income generated from both slurry and biogas. The capacity recovery factor is 0.11 and the levelized cost of energy is ₹14.72 m^{-3} at the discount rate of 10% considering the plant's life of 20 years. The country's inflation rate is 6%, which is an important factor in the project's financial vitality. The results of the financial analysis of the biogas power plant are shown in Table 5.8. The net present value of the biogas plant is ₹433.68 million at a 6% discount rate, the net present value at a 10% discount rate is ₹314.59 million, and the net present value at a 14% discount rate is ₹219.92 million considering the plant

Table 5.8 Financial analysis of biogas power plant

Discount rate (i)	Net present value	Payback period
6%	433.68 million	2 years
8%	388.81 million	
10%	314.59 million	
12%	250.30 million	
14%	219.92 million	
Effective discount rate (i_e)		
0.29%	825.37 million	2 years
0.57%	750.44 million	
0.85%	719.44 million	
1.14%	million	

life of 20 years. The payback period of the plant is 2 years. The net present value at the 0.29% effective discount rate is ₹825.37 million, the net present value at the 0.57% effective discount rate is ₹750.44 million, and the net present value at the 1.14% effective discount rate is ₹698.28 million. The payback period is the same as the discount rate.

5.7 Conclusion

The food processing industry is one of the emerging sectors in India. The food processing sector of India is one of the largest in the world. The share of agriculture and allied sectors to the total economy is 20.2%. It provides 11.6% of total employment. It contributes to 10.4% of India's exports. This sector is expected to grow as big as $535 billion by 2025–26. In anaerobic digestion systems, carbon dioxide removed from raw biogas reduces carbon emissions. It contributes to greenhouse gas mitigation and is environmentally friendly.

This plant setup currently uses coal as a fuel for heating purposes. For climate and environmental sustainability, we can switch the energy source from conventional to non-conventional. This research paper sums up the technical and financial vitality of a 6,876 m^3 volume biogas power plant which meets the energy demand of Bikaji Foods International Limited at 41,840 MJ per day. To meet the energy demand, a biogas plant with a total volume of 6,876 m^3 has been designed. The volume of the floating drum is 1,820 m^3 and the digester volume is 5,056 m^3.

A financial analysis of biogas plants with different discount rates and different effective discount rates has been done. The most efficient discount rate for the proposed plant is 10% and the effective discount rate is 0.57%. The net present value at a 10% discount rate is ₹314.59 million. The country's inflation rate is currently 6%. The payback period of the biogas power plant at a discount rate of 10% is 2 years.

References

1. N.A. Pambudi, R. Laurensia, D.S. Wijayanto, V.L. Perdana, M. Fasola, M. Imran, L.H. Saw, R. Handogo, Exergy analysis of boiler process powered by biogas fuel in ethanol production plant: a preliminary analysis. Energy Procedia **142**, 216–223 (2017)
2. R. Kapoor, P. Ghosh, B. Tyagi, V.K. Vijay, V. Vijay, I.S. Thakur, H. Kamyab, D.D. Naguyen, A. Kumar, Advances in biogas valorization and utilization systems: a comprehensive review. J. Clean. Prod. **273**, 123052 (2020)
3. G. Coşar, M. Pooyanfar, E. Amirabedin, H. Topal, Design and economic analysis of a heating/absorption cooling system operating with municipal solid waste digester: a case study of Gazi university. Environ. Clim. Technol. **11**, 12–18 (2013)
4. G. Taleghani, A.S. Kia, Technical-economical analysis of the Saveh biogas power plant. Renew. Energy **30**, 441–446 (2015)
5. A. Akbulut, Techno-economic analysis of electricity and heat generation from farm-scale biogas plant: Çiçekdağı case study. Energy **44**, 381–390 (2012)
6. R. Kapoor, P. Ghosh, M. Kumar, S. Sengupta, A. Gupta, S.S. Kumar, V. Vijay, V. Kumar, V.K. Vijay, D. Pant, Valorization of agricultural waste for biogas based circular economy in India: a research outlook. Biores. Technol. **304**, 123036 (2020)
7. P. Ghosh, M. Kumar, R. Kapoor, S.S. Kumar, L. Singh, V. Vijay, V.K. Vijay, V. Kumar, I.S. Thakur, Enhanced biogas production from municipal solid waste via co-digestion with sewage sludge and metabolic pathway analysis. Biores. Technol. **296**, 122275 (2019)
8. Ministry of new and renewable energy, India. https://mnre.gov.in/bio-energy/current-status, (2022)
9. Department of animal husbandry and dairying, India. https://dahd.nic.in/animal-husbandry-statistics (2019)
10. K. Mushtaq, A.A. Zaidi, S.J. Askari, Design and performance analysis of floating dome type portable biogas plant for domestic use in Pakistan. Sustainable Energy Technol. Assess. **14**, 21–25 (2016)
11. A.K. Kalia, S.P. Singh, Case study of 85 m3 floating drum biogas plant under hilly conditions. Energy Convers. Manage. **40**(7), 693–702 (1999)
12. A. Yasar, S. Nazir, R. Rashid, A.B. Tabinda, M. Nazr, Economic review of different designs of biogas plants at household level in Pakistan. Renew. Sustain. Energy Rev. **74**, 221–229 (2017)
13. B.H. Khan, *Non-Conventional Energy Resources*, 3rd edn. (McGraw Hill Education Pvt. Ltd., Chennai, India, 2016)
14. S.C. Iweka, K.C. Owuama, J.L. Chukwuneke, O.A. Falowo. Optimization of biogas yield from anaerobic co-digestion of corn-chaff and cow dung digestate: RSM and python approach. Heliyon cell Press **7**(11), e08255 (2021)
15. A.R. Pati, S. Saroha, A.P. Behera, S.S. Mohapatra, S.S. Mahanand. The anaerobic digestion of waste food materials by using cow dung: a new methodology to produce biogas. J. Inst. Eng. (India): Ser. E **100**, 111–120 (2019)
16. Directorate of Gopalan, Rajasthan. https://gopalan.rajasthan.gov.in/districtwise-gau-shala.htm, (2022)
17. D. Wu, X. Peng, L. Li, P. Yang, Y. Peng, H. Liu, X. Wang, Commercial biogas plants: review on operational parameters and guide for performance optimization. Fuel **303**, 121282 (2021)
18. M.I. Alfa, H.I. Owamah, A.O. Ogochukwu, S. Gopikumar, S.O. Oyebisi, S.S. Kumar, S. Bajar, O.D. Samuel, S.C. Ilabor, Evaluation of biogas yield and kinetics from the anaerobic co-digestion of cow dung and horse dung: a strategy for sustainable management of livestock manure. Energy, Ecol. Environ. **6**, 425–434 (2020)
19. D. Balussou, A. Kleyböcker, R. McKenna, D. Möst, W. Fichtner. An economic analysis of three operational co-digestion biogas plants in Germany. **3** 23–41 (2012)
20. Public health engineering department, Rajasthan. https://phedwater.rajasthan.gov.in/content/raj/water/public-health-engineering-department/en/home.html (2022)

21. K. Kozłowski, M. Pietrzykowski, W. Czekała, J. Dach, A. Kowalczyk-Juśko, K. Jóźwiakowski, M. Brzoski, Energetic and economic analysis of biogas plant with using the dairy industry waste. Energy **183**, 1023–1031 (2019)
22. A. Shahzad, S. Hanif, Techno-economic feasibility of biogas generation in Attari village, Ferozepur road, Lahore. Environ. Dev. Sustain. **16**, 977–993 (2014)
23. V.K. Singh, R. Kumar, P.L.R. Teja, Performance evolution of various models of biogas technologies in household environment. Int. J. Sci. Technol. Res. **3**(3), 50–54 (2014)
24. A. Yadav, R. Gupta, V. Garg, Organic manure production from cow dung and biogas plant slurry by vermicomposting under field conditions. Int. J. Recycl. Org. Waste Agric. **2**, 21 (2013)
25. S. Kumar, L.C. Malav, M.K. Malav, S.A Khan. Biogas slurry: source of nutrients for eco-friendly agriculture. Int. J. Extensive Res. **2**, 42–46 (2015)
26. O. Capitan, Cost Analysis of energy production from biogas and bio syngas. J. Soc. Sci. Financ. Account. **2**, 32–41 (2020)
27. U.U.R. Zia, R.U. Tanzeel, M. Ali, W.N. Awan, Techno-economic assessment of energy generation through municipal solid waste: a case study for small/medium size districts in Pakistan. Waste Dispos. Sustain. Energy **2**, 337–350 (2020)
28. J.A. Silva-González, A.K. Chandel, S.S.d. Silva, N. Balagurusamy. Biogas in circular bio-economy: sustainable practice for rural farm waste management and techno-economic analyses. Biogas Prod., 389–414 (2021)
29. S. A. Gebrezgabher, M. P. M. Meuwissen, B. A. M. Prins, A. G. J. M. Oude Lansink. Economic analysis of anaerobic digestion-a case of Green power biogas plant in The Netherlands. **57**, 109–115 (2010)
30. International renewable energy agency (IRENA) (2022)

Part II
Chemical Conversion

Chapter 6
Synthesis of Biolubricant by Transesterification of Soybean Oil Using Ni–Al Hydrotalcite as a Catalyst

Sakshi Shrivastava, Pooja Prajapati, Virendra, Priyanka Srivastava, Ajay P. S. Lodhi, Deepak Kumar, Varsha Sharma, S. K. Srivastava, and D. D. Agarwal

Abstract Lubricant is a material used to minimize the friction and wear between interacting surfaces; research in the formulation of biolubricants from vegetable oil has again significant interest due to their good lubricity and eco-friendly properties. This work aims to production of biolubricant base stock trimethylpropane triester (TMPTE) which is derived from soybean oil via a transesterification reaction. A transesterification reaction is completed through two steps. The first step is the production of biodiesel (fatty acid methyl ester). The second step is the reaction of FAME with trimethylolpropane by using Ni–Al hydrotalcite as a catalyst under pressure and at 160 °C temperature. The Ni–Al hydrotalcite was characterized by XRD, FT-IR, TGA, and SEM–EDX. The product Biolubricant was characterized by ^{1}H NMR, FT-IR, and GC. Physiochemical analyses were also done to analyze the product.

Keywords Soybean oil · Transesterification reaction · Hydrotalcites · Biolubricant

6.1 Introduction

Lubricants are the materials which are used to decrease friction, heat, and wear upon moving surfaces, in the form of a liquid film which keeps the two opposite surfaces separated to improve the efficiency and durability of machinery. Africa makes up 6% of all global consumption, which reached nearly 36.1 million tonnes in 2017 [1].

S. Shrivastava (✉) · P. Prajapati · P. Srivastava · V. Sharma · S. K. Srivastava · D. D. Agarwal
School of Studies in Chemistry, Jiwaji University, Gwalior 474001, India
e-mail: shrivastavasakshi1993@gmail.com

Virendra
CIF, Jiwaji University, Gwalior 474001, India

A. P. S. Lodhi · D. Kumar
CART, Indian Institute of Technology, New Delhi, India

N. Gakkhar et al. (eds.), *Recent Advances in Bio-Energy Research*, Springer Proceedings in Energy, https://doi.org/10.1007/978-981-99-5758-3_6

The demand for lubricants is increasing at a speed of 1.6% annually [2]. With an increase in the number of automobiles and mechanical industrialization, the usage of lubricants will keep rising at a speed of 2% annually [3].

Prices for petroleum products are going up because reservoirs are rapidly running out. In addition, petroleum-based lubricants are harmful and non-biodegradable, causing dangers to human health, water quality, and the environment [4]. Biolubricants are regarded as a suitable alternative to petroleum lubricants due to their advantageous and acceptable physical features [4, 5]. Because they are renewable, low volatile, have a wide viscosity range, can endure high temperatures, and have good viscosity–temperature properties [7–9], bio-based lubricants are preferable to mineral lubricants. They are also non-toxic, biodegradable, and do not harm the environment when spilled [10, 11]. They use both animal and vegetable oils as their raw materials [12]. The other four types of biolubricant base oils are poly alkylene glycols, dibasic acid esters, poly alpha olefins, dibasic acid esters, and polyol esters [13].

Triglycerides called ester are made up of three long chains of different fatty acids connected at the hydroxyl group via ester bonds, and they contain glycerol molecules [14]. Vegetable oils contain fatty acids with a variety of unsaturation levels and carbon chain lengths [15]. The most common and unique fatty acid in most vegetable oils is oleic acid (C18:1). Triglycerides from vegetable oils have a stronger attraction due to their being more polar than those found in oils derived from petroleum [16]. As a result of preventing metal-to-metal contact, the polar carboxyl group of fatty acids attaches to the metallic surface and forms a suitable thin lubricating coating to minimize friction and wear [17, 18]. Transesterification is a procedure used to improve thermal performance and oxidative stability [19, 20] by removing hydrogen atoms from the beta-carbon of the oil [21]. The entire reaction took place in just two steps. When beta hydrogens are absent, the lubricant becomes more thermally stable at high temperatures, preventing the formation of free fatty acids through self-polymerization, which is the main benefit of using a polyol in place of glycerol [22].

Many authors used chemical modifications of seed oils with trimethylolpropane via transesterification reaction for the manufacturing of biolubricants by using different catalyst concentrations and at various temperatures. Homogeneous and heterogeneous base/acid catalysts and enzymes are the most common types of transesterification catalysts for synthesized TFATE [23–26]. Using a lesser corrosive heterogeneous catalyst can therefore not only eliminate extra processes such as neutralization and minimize washing through solvents but it can also be recycled by simple filtration [27]. According to previous reports [28–31], a large variety of heterogeneous base catalysts have been successfully discovered for the synthesis of biodiesel.

Their transesterification reaction produces the biolubricant that can be catalyzed by homogeneous or heterogeneous catalysts. It is difficult to separate the catalyst from the products when using a homogeneous catalyst like p-toluene sulfonic acid, H_2SO_4, or phosphoric acid to accelerate the rate of reaction. These catalysts typically cause corrosion in reactors [32]. In contrast, heterogeneous catalysts are non-corrosive, simple to remove from the products, and have less difficulty when removed. They can be produced to have increased activity, selectivity, and lifetimes for the catalysts [33].

Hydrotalcite (HT) is a double-layered and porous di-hydroxyl complex metal oxide that has been used as a catalyst ion-exchange material, composite material, and carrier. However, the use of such a heterogeneous catalyst in biolubricant synthesis, as well as its detailed characterization of catalyst and product, optimal preparation for reaction, and reusability of catalyst, has hardly ever been investigated.

The goal of this study is to synthesize a heterogeneous catalyst to make biolubricants from renewable resources (soybean oil-based biodiesel with TMP as a tri-alcohol) via a transesterification reaction.

6.2 Experimental

6.2.1 Reagents and Chemicals

All chemicals like nickel nitrate hexahydrate [Ni $(NO_3)_2{\cdot}6H_2O$], aluminum nitrate Al $(NO_3)_3$, sodium carbonate (Na_2CO_3), and sodium hydroxide (NaOH) from Rankem with analytical grade were obtained. The HPLC-quality (99% purity) methanol was bought from Merck. Soybean oil was got from the market in Gwalior, Madhya Pradesh, India. Trimethylolpropane was ordered from Sigma Aldrich. Deionized water of Rankem was used throughout the entire experiment.

6.2.2 Hydrotalcite Preparation

The hydrotalcite employed in this study was prepared by using the coprecipitation method, which Climent et al. [41] reported earlier. At room temperature, a solution containing 3 mol Ni $(NO_3)_2{\cdot}6H_2O$ and 1 mol Al $(NO_3)_3{\cdot}9H_2O$ was acquired and mixed with a second solution containing 1 mol NaOH and 1 mol Na_2CO_3, vigorously stirring up to gain the pH 9–10; then the combination was kept at 60 °C, for the next 12 h. The resultant material was filtered and rinsed in distilled water 3 to 4 times up to the pH level was neutralized. The catalyst was dried for 24 h. at 70 °C. After drying up, it was crushed into a powder and calcined at 550 °C for 6 h. inside the muffle furnace. Synthesis of Ni–Al hydrotalcites was characterized by using different methods like XRD. Rikaguminiflex 600 diffracts-meter apply CuKα radiation ($\lambda =$ 1.5406 Å) and a step size of 0.02°, voltage 45 kV, Fourier transform Infrared on Perkin Elmer Equipment, Thermo-gravimetric analysis on TW60 Shimadzu anal. Ltd., and SEM on Zeiss Gemini sem.

6.2.3 Synthesis of the Soybean Oil-Based Biodiesel

Soybean oil-based biodiesel was prepared via transesterification reaction of soybean oil with methanol using 2% w/w calcinated Ni–Al hydrotalcite and placed into a 500 ml round bottom flask attached with a reflux condenser. Firstly, the calculated amounts of catalyst (2%) and the methanol are poured into a flask and then stirred at least 10 min to form a homogeneous mix. Then soybean oil was added into a flask then heated up to 65 °C for 2 h. After the completion of the reaction, Ni–Al hydrotalcite was separated by centrifuging and washed with a mixture of hexane and acetone (hexane:acetone = 1:1). The catalyst was dried at 100 °C for 24 h for further use. The upper FAME layer was separated from the lower glycerol layer using a separating funnel. Gas chromatography-mass spectrometry (Clarus*680 GC, Clarus*SQ8C MS) was used to identify the biodiesel. The structural analysis of products was done by nuclear magnetic resonance (NMR; JEOL ECZ500R/S1 (500 MHz) spectrometer) and Fourier-transformed infrared spectroscopy (Perkin Elmer FT-IR spectrophotometer-105627). Physiochemical analysis was also done.

6.2.4 Transesterification of Soybean Oil-Based Biodiesel with Alcohols

The calculated amount of TMP and calcinated Ni–Al hydrotalcites was discharged into a stainless-steel autoclave reactor, and heated up to 60 °C. Then, biodiesel was poured into the liquid paste and heated at 160 °C. After completion of the reaction, the catalyst was separated using centrifugation at 2900 rpm for 14 min and washed 3 to 4 times with a mixture of hexane and acetone (hexane:acetone = 1:1). Then the catalyst was dried in an oven at 90 °C for 1 day. Recovery of the by-product (methanol) was done in a vacuum using a rotary evaporator at 45 °C. The catalyst was reused three times. After completion of the reaction, gas chromatography-mass spectrometry (Clarus*680 GC, Clarus*SQ8C MS) was used to identify the composition of the biolubricant. The structural analysis of products was done by nuclear magnetic resonance (NMR; JEOL ECZ500R/S1 (500 MHz) spectrometer) and Fourier-transformed infrared spectroscopy (Perkin Elmer FT-IR spectrophotometer-105627). Physiochemical analysis was also done. The friction tests according to ASTM D4172 were performed to analyze the frictional property using a four-ball tester.

6.3 Results and Discussion

6.3.1 Catalyst Characterizations

According to the Ni–Al hydrotalcite, the XRD pattern is represented in Fig. 6.1. Diffraction peaks at 18.2, 23.3, 34.1, 43.4, 47.8, 62, and 66.94 indexed at 003, 006, 012, 015, 018, 110, 113, and 116 planes in Ni–Al CO_3 hydrotalcite with 3:1 metal ratio show purity and crystallinity. The face-centered cubic (FCC) crystalline structure was represented by all of these diffraction peaks (in accordance with the standard spectrum—JCPDS, No. 04-0835). Three times the recycled catalyst was analyzed in this XRD spectra. All the diffraction peaks of samples reduced and move toward to a minor angle, whereas crystal plane diffraction peaks grew longer, representing that crystal particle diameter reduced. This could be due to Ni and Al dispersion throughout the repeating reactions procedure and because the catalyst absorbed FAME, TMPTE, and impurities.

The FT-IR graph of calcined Ni–Al CO_3 hydrotalcite (3:1) was analyzed as shown in Fig. 6.2. This demonstrates that the sample has a pure hydrotalcite structure. Infrared spectra peak at 3363.26 cm^{-1} for hydroxyl group (OH) stretching vibration. H_2O is responsible for the peak recorded at 1738 cm^{-1}. At 1365.91 cm^{-1} represent a strong and sharp band of carbonate anion [34, 35].

Scanning electron microscope results are shown in Fig. 3a, b. The compound has high crystallinity and a lamellar structure. A small needle-shaped particle can be seen. The hexagonal form of the hydrotalcite particles is observed. An EDX test showed that calcined Ni–Al (3:1) catalysts contain nickel and aluminum oxide metals, with Ni (61.37%) and Al (38.63%) as shown in Fig. 6.4 (Table 6.1).

Figure 6.5 depicts a TGA graph of calcined Ni–Al CO_3 hydrotalcite (3:1). It shows a total weight loss of 32%. The weight loss began at 290 °C because of the water loss in an inter-layer region. Later, weight loss was seen around 325 °C because of the de-hydroxylation of OH and the decomposition of the carbonate anion [36, 37].

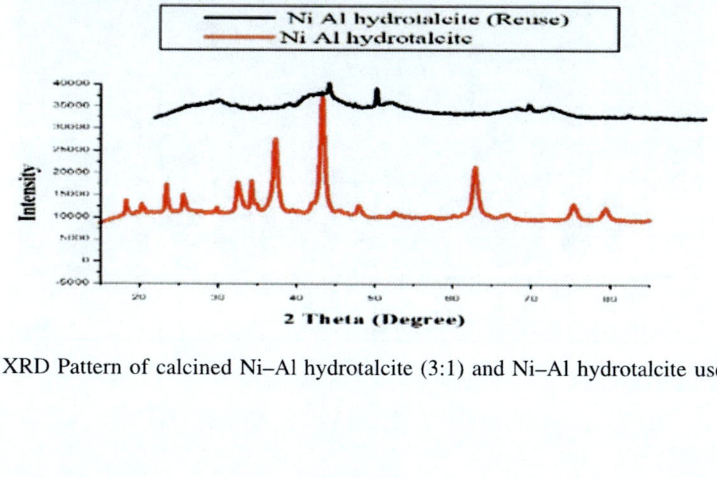

Fig. 6.1 XRD Pattern of calcined Ni–Al hydrotalcite (3:1) and Ni–Al hydrotalcite used up to 3 cycles

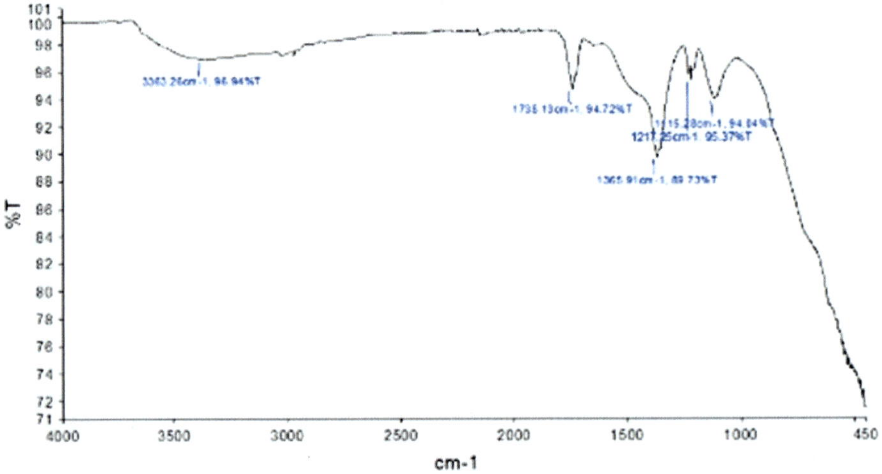

Fig. 6.2 FT-IR spectroscopy of Ni–Al CO_3 hydrotalcite

Fig. 6.3 The SEM image of calcined Ni–Al CO_3 hydrotalcite at low magnification (**a**) and high magnification (**b**)

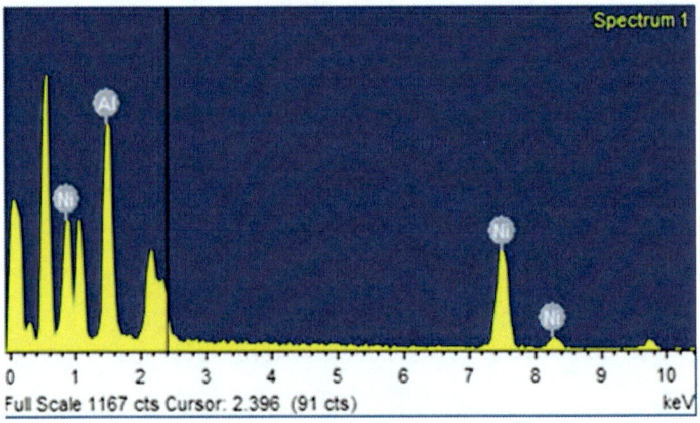

Fig. 6.4 The EDX image of calcined Ni–Al CO_3 hydrotalcite

Table 6.1 Element composition

Elements	Weight %	Atomic %
Ni K	61.37	42.20
Al K	38.63	57.80
Total 100		

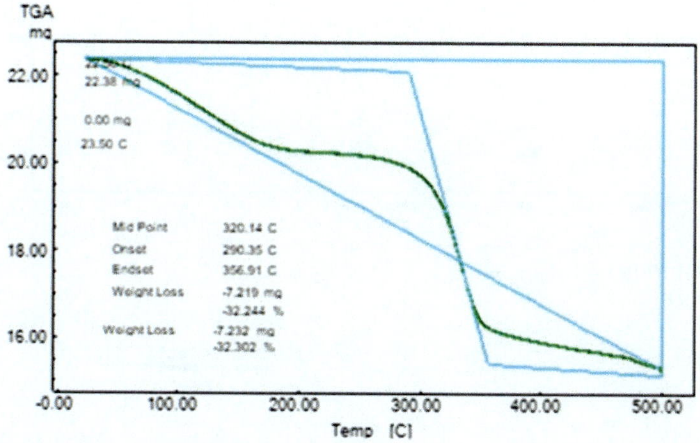

Fig. 6.5 The TGA graph of calcined Ni–Al CO_3 hydrotalcite.

6.3.2 Biodiesel Characterization

The conversion of soybean oil to soybean oil-based biodiesel (FAME) was confirmed by ^{1}H NMR spectroscopy.

^{1}H NMR spectra of soybean oil showed a peak around 4.0–4.3 ppm for triglyceride protons and other hydrocarbon protons in oil (Fig. 6.6), which has totally vanished in soybean oil-based biodiesel, and a new distinguished peak at 3.66 ppm for methoxy protons appeared (Fig. 6.7), which confirms the formation of biodiesel (FAME). While α-CH_2 proton showed a triplet peak at 2.31 ppm [38–40]. The biodiesel yield was 93%.

Commercially available soybean oil has twelve fatty acids. The fatty acid methyl ester obtained from the transesterified soybean oil is depicted in Table 6.2. Gas chromatography spectra of biodiesel are shown in Fig. 6.8.

The formation of soybean oil-based biodiesel (FAME) from soybean oil can also be confirmed by FT-IR spectroscopy as seen in Fig. 6.9. The band present at 1744 cm^{-1} showed characteristics of a peak of C=O stretching vibration because of glyceride linkage in fatty acid of oil, which has totally vanished in soybean oil-based biodiesel, and a new distinguished peak at 1739 cm^{-1} appeared because of vibration of methyl esters in biodiesel as seen in Fig. 6.9. Strong absorption bands at 2923 and 2854 cm^{-1} present in oil and 2924 and 2854 cm^{-1} present in biodiesel are because

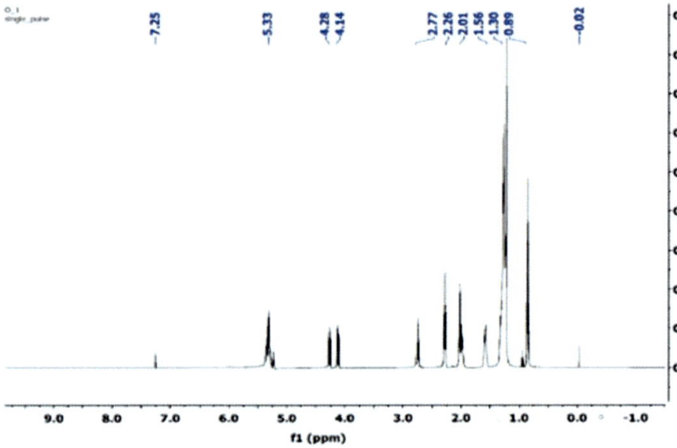

Fig. 6.6. ^{1}H NMR of soybean oil

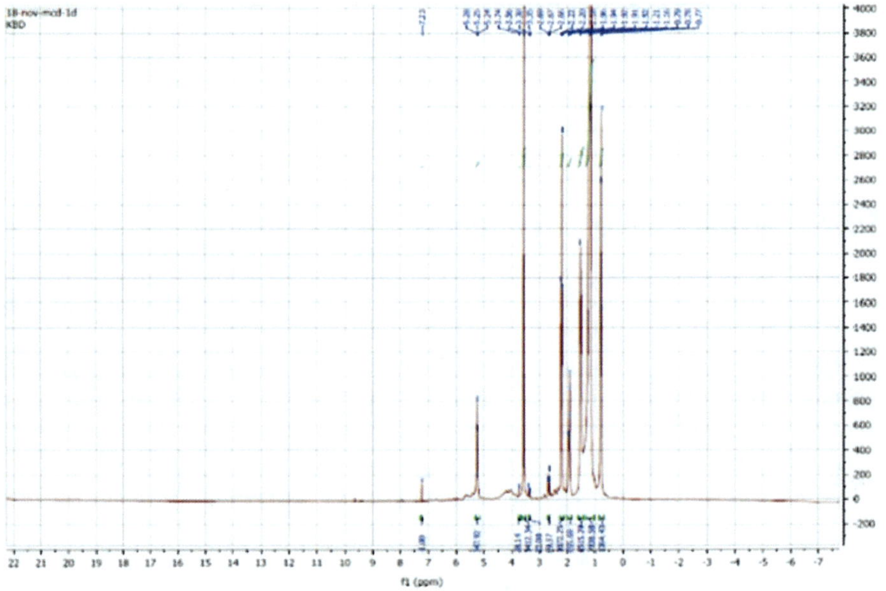

Fig. 6.7. ^{1}H NMR of soybean oil-based biodiesel

of C–H stretching vibrations. The significant bending vibrations of the CH_3 band are 1463 and 1377 cm^{-1} present in the oil, whereas it is present at 1456, 1436, and 1362 cm^{-1} in biodiesel. The stretching vibrations of C–O present in triglyceride are shown at 1160 and 1098 cm^{-1} in oil and at 1196, and 1170 cm^{-1} in the biodiesel, respectively.

Table 6.2 Mass % composition of Biodiesel synthesized from soybean oil

S. no.	Compound	Composition (%)
1	Methyl palmitate C16:0	11.57
2	Methyl stearate C18:0	3.8
3	Methyl oleate C18:1	44.2
4	Methyl linoleate C18:2	30.05
5	Methyl linolenate C18:3	4.7
6	Methyl arachidate C20:0'	0.31
7	Methyl behenate C22:0'	0.58

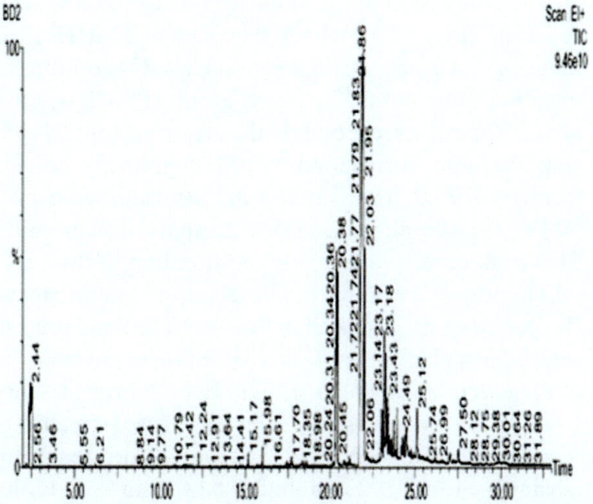

Fig. 6.8 GC spectra of soybean oil-based biodiesel

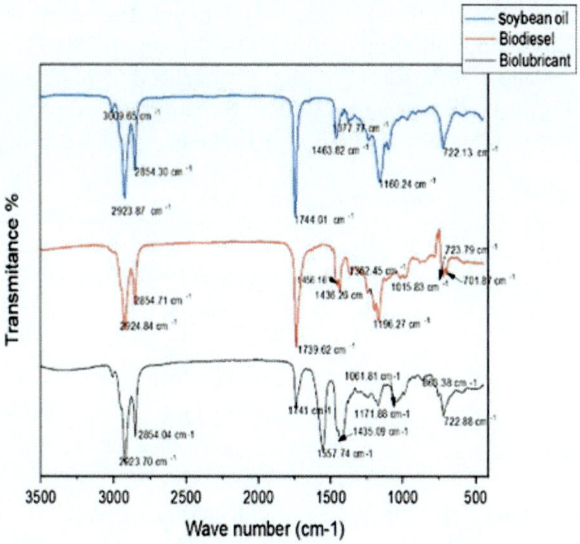

Fig. 6.9 FT-IR spectra of Soybean oil, soybean oil-based biodiesel, and its biolubricant

The physiochemical parameters of the soybean oil-based biodiesel (FAME) are shown in Table 6.3.

6.3.3 Biolubricant Characterization

The best chemical conditions for biolubricant formation were a 5% w/w catalyst concentrate of Ni–Al hydrotalcites; a 4:1 FAME/alcohol molar ratio; temperature of 160 °C; for 2 h.

The conversion of the biodiesel into a biolubricant was confirmed by the ^{1}H NMR spectrum (Fig. 6.10). ^{1}H NMR spectra of TMPTE reveal the presence of peaks *in* the range of 4–4.2 ppm attributed to ($-CH_2-O-$) protons of alcohol [41, 42] which verify the replacement of the hydroxyl group of TMP by the fatty acid of the oil. The signal near 2.37 ppm corresponds to the acidic proton ($CH_2C=O$). The presence of a signal near 3.60 ppm corresponds with H-bonding to unreacted hydroxyls, confirming the formation of monosubstituted and disubstituted trimethylolpropane esters. In the ^{1}H NMR of biodiesel, a peak near 5.25 ppm is also present in the ^{1}H NMR of biolubricant. This peak corresponds to the unsaturation of the fatty acid chain in soybean oil.

The conversion of the biodiesel into a biolubricant was also confirmed by using FT-IR spectroscopy, and seen in Fig. 6.9. The band present at 1739 cm^{-1} showed a characteristic peak of methyl ester vibration in soybean biodiesel, which has completely disappeared in soybean oil-based biolubricant, and a newly prominent peak at 1742.95 cm^{-1} appears due to the presence of stretching vibrations of carboxylic esters (C=O) in the biolubricant [43]. Strong absorption bands at 2924 cm^{-1} and 2854 cm^{-1} in biodiesel and 2923.64 and 2854.02 cm^{-1} in biolubricant are shown because of C–H stretching vibrations. The bending vibrations of CH_3 appear at 1456, 1436, and 1362 cm^{-1} in FAME[44*], whereas they are 1463.56 cm^{-1} in the biolubricant. Stretching vibration of the C–O–C group of TMPTE appears near 1162–1008 cm^{-1}.

The composition of the biolubricant can be confirmed by GC as shown in Fig. 6.11. By observing the chromatogram, we concluded that trimethylolpropane has completely reacted with soybean oil-based biodiesel and been converted into

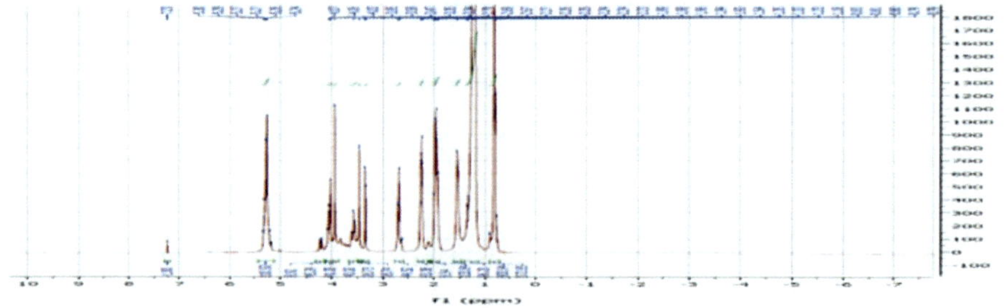

Fig. 6.10. ^{1}H NMR of soybean oil-based biolubricant

trimethylolpropane trimester. The appearance of four peaks at the end of the chromatogram corresponded to TMPTE 36, TMPTE 37, TMPTE 38, and TMPTE 39 [45].

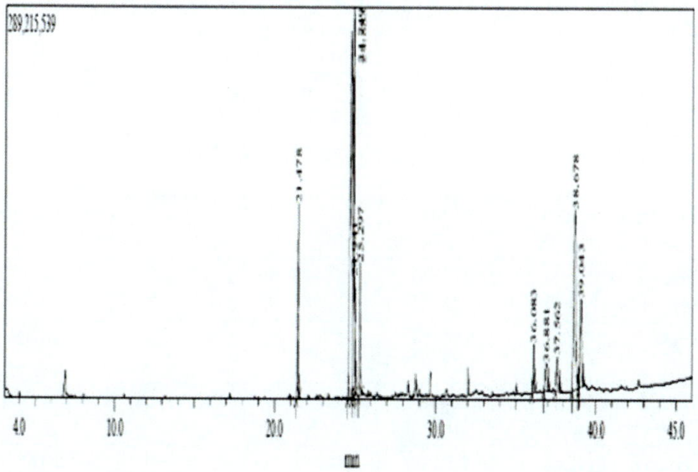

Fig. 6.11 GC spectra of biolubricant

Figure 6.12 shows the frictional coefficient curves of the biolubricant using a four-ball friction tester. The average friction coefficient of soybean oil-based TMPTE is 0.11 within 60 min [46]. The results indicated the wear resistance of biolubricant is similar to base oil lubricant present in market [47].

The physiochemical properties of the soybean fatty acid-based TMPTE are shown in Table 6.3.

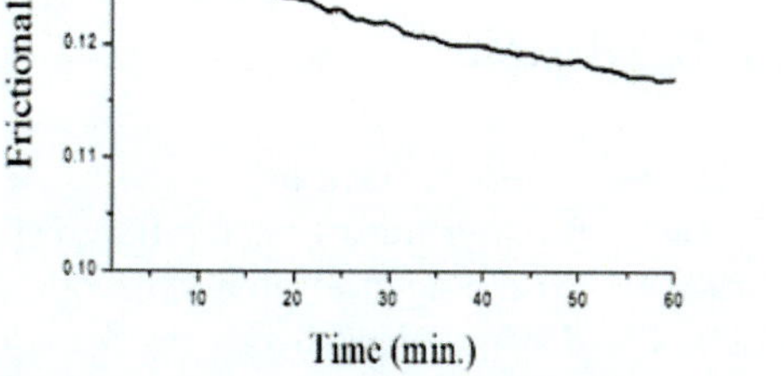

Fig. 6.12 Friction coefficient biolubricant

Table 6.3 The physicochemical properties of soybean oil, soybean oil-based biodiesel, and biolubricant

S. no.	Physical properties	Standard methods	Soybean oil	Soybean oil biodiesel	Soybean oil biolubricant
1	Appearance, and color	'Visual'	'Clear, and Pale light color'	'Clear and bright yellow color'	'Bright, golden-brown color'
2	Density @ 15°C	IS-1448 P-32	0.92 gm/cm^3	0.88 gm/cm^3	0.9 g/ml
3	Kinematic viscosity @ 40°C, mm^2/S	IS1448/ P-25	32.80 cSt	6.5 cSt	103 cSt
4	Kinematic viscosity @ 100°C, mm^2/S	IS1448/ P-25	–	–	15.8 cSt
5	Fire point °C	IS1448/ P-21	285	184	–
6	Flash point °C	IS1448/ P-21	254	170	182
7	Pour point °C	IS1448/ P-10	–	(–) 2	(–) 1
8	Viscosity index	IS-1448/ P-56	–	–	163

6.4 Conclusion

High production of soybean oil-based biolubricant was successfully done from the catalytic transesterification of FAME and trimethylolpropane, aided by calcined solid-base Ni–Al hydrotalcite heterogeneous catalyst synthesized by the coprecipitation method, and characterized by FT-IR, XRD, SEM, and TGA. This catalyst is used in the transesterification of soybean oil, and FAME. Characterization of the biolubricant using ^{1}H NMR and FT-IR techniques confirmed its structural configuration, and a gas chromatogram confirmed the composition necessary for the biolubricant's production. The high yield of the biolubricant was found at 5% w/w loading of catalyst and a 4:1 molar ratio of FAME to TMP. All physicochemical properties agree with ISO standards. The average friction coefficient of the biolubricant is 0.11 similar to that in the market.

Abbreviations

XRD	X-ray diffraction
FT-IR	Fourier-transformed infrared spectroscopy
SEM	Scanning electron microscopy
EDX	Energy dispersive X-ray

TGA	Thermogravimetric analysis
^{1}H NMR	Proton nuclear magnetic resonance spectroscopy
GC	Gas chromatography
HTs	Hydrotalcites
LDH	Layer double hydroxide
TMP	Trimethylolpropane
TMPTE	Trimethylolpropane triester
TFATE	Trimethylolpropane fatty acid triester
FAME	Fatty acid methyl ester

Acknowledgements The authors gratefully and thankfully acknowledge the CIF Dept., Jiwaji University, Gwalior, India, for XRD, DSC, FT-IR, and TGA facilities; Innovation Centre, Bundelkhand University, India, for gas chromatography analysis; SAIF, IIT Patna, India, for the ^{1}H NMR facility; IIT Roorkee, India, for SEM analysis; Hindustan Laboratories Regd., New Delhi, India, for physiochemical properties, and CART, IIT Delhi, India, for friction property of the biolubricant.

References

1. L. Lindemann, *Future Challenges of the Lubricants Industry* (FUCHS Capital Market Day, 2018
2. R.V. Sharma, A.K. Dalai, Synthesis of bio-lubricant from epoxy canola oil using sulfated Ti-SBA-15. Applied cat. B **142**, 604–614 (2013)
3. Y. Singh, A. Farooq, A. Raza, N.A. Mahmood, S. Jain, Sustainability of a nonedible vegetable oil based bio-lubricant for automotive applications: a review. Process Saf. Environ. Protect **111**, 701–713 (2017)
4. L.R. Rudnick, *Synthetics, mineral oils, and bio-based lubricants* (Boca Raton, CRC Press, Chemistry and Technology, 2020)
5. H.M. Mobarak, E. Niza Mohamad, H.H.Masjuki, M.A. Kalam, K.A.H. Al Mahmud, M. Habibullah, The prospects of biolubricants as alternatives in automotive applications. Renew. Sustain. Energy Rev. **33**, 34–43 (2014)
6. P. Nagendramma, S. Kaul, Development of ecofriendly/biodegradable lubricants: an overview. Renew. Sustain. Energy Rev. **16**, 64–74 (2012)
7. G. Gorla, S.M. Kour, K.V. Padmaja, M.S.L. Karuma, R.B.N. Prasad, Novel acylderivatives from karanja oil: alternative renewable lubricant base stocks. Ind. Eng. Chem. Res. **53**, 8685–8693 (2014)
8. M. Sarno, M. Iuliano, C. Cirillo, Optimized procedure for the preparation of an enzymatic nanocatalyst to produce a biolubricant from waste cooking oil. Chem. Eng. J. **3**, 377–384 (2019)
9. T. Romsdahl, A. Shirani, R.E. Minto, C. Zhang, E.B. Cahoon, K.D. Chapman, D. Berman, Nature-guided synthesis of advanced bio-lubricants. Sci. Rep. **9**, 1–11 (2019)
10. E.K. Heikal, Manufacturing of environment friendly biolubricants from vegetable oils. Egypt J Pet **26**(1), 53–59 (2017)
11. H.M. Mobarak, The prospects of biolubricants as alternatives in automotive applications. Renew. Sustain. Energy Rev. **33**, 34–43 (2014)
12. A.B. Kousaalya, S.D. Beyene, V. Gopal, B. Ayalew, S. Pilla, Green epoxy synthesized from Perilla frutescens: a study on epoxidation and oxirane cleavage kinetics of high-linolenic oil. Ind. Crop. Prod. **123**, 25–34 (2018)

13. P. Nagendramma, S. Kaul, Development of ecofriendly/biodegradable lubricants: an overview. Rene. Sust. Energ. Rev. **16**, 764–774 (2012)
14. L.A. Quinchia, M.A. Delgado, C. Valencia, J.M. Franco, C. Gallegos, Viscosity modification of different vegetable oils with EVA copolymer for lubricant applications. Ind. Crops Prod. **32**, 607–612 (2010)
15. J.K. Mannekote, S.V. Kailas, K. Venkatesh, N. Kathyayini, Environmentally friendly functional fluids from renewable and sustainable sources-a review. Renew. Sustain. Energy Rev. **81**, 1787–1801 (2018)
16. A.N. Suarez, M. Grahn, R. Pasaribu, R. Larsson, The influence of base oil polarity on the tribological performance of zinc dialkyl dithiophosphate additives. Tribol. Int. **43**, 2268–2278 (2010)
17. C.J. Reeves, The influence of fatty acids on tribological and thermal properties of natural oils as sustainable biolubricants. Tribol. Int. **90**, 123–134 (2015)
18. E. Durak, F. Karaosmanoglu, Using of cottonseed oil as an environmentally accepted lubricant additive. Energy Sources **26**(7), 611–625 (2004)
19. F. Zaccheria, M. Mariani, R. Psaro, P. Bondioli, N. Ravasio, Environmentally friendly lubricants through a zero waste process. Appl. Catal. B **181**, 581–586 (2016)
20. P.K. Sripada, R.V. Sharma, A.K. Dalai, Comparative study of tribological properties of trimethylolpropane-based biolubricants derived from methyl oleate and canola biodiesel Indus. Crop. Prod. **50**, 95–103 (2013)
21. J. McNutt, Q.S. He, Development of biolubricants from vegetable oils via chemical modification. J. Ind. Eng. Chem. **36**, 1–12 (2016)
22. D.Y.C. Leung, X. Wu, M.K.H. Leung, A review of biodiesel production using catalyzed transesterification. Appl. Energy **87**, 1083–1095 (2010)
23. Y.Y. Linko, X.Y. Wu, Biocatalytic production of useful esters by two forms of lipase from candida rugose. J. Chem. Tech. Biotechnol. **56**, 163–170 (1996)
24. E. Uosukainen, Y.Y. Linko, M. Lämsä, T. Tervakangas, P. Linko, Transesterification of trimethylolpropane and rapeseed oil methyl ester to environmentally acceptable lubricants. J. Am. Oil Chem. Soc. **75**, 1557–1563 (1998)
25. M.D. Serio, R. Tesser, M. Dimiccoli, F. Cammarota, M. Nastasi, E. Santacesaria, Synthesis of biodiesel via homogeneous lewis acid catalyst. J. Mol. Catal. A: Chem. **239**, 111–115 (2005)
26. M.H. Zong, Z.Q. Duan, W.Y. Lou, T.J. Smith, H. Wu, Preparation of a sugar catalyst and its use for highly efficient production of biodiesel. Green Chem. **9**, 434–437 (2007)
27. Z.Z. Yang, W.L. Xie, Y. Zuo, Application in transesterification of oil and fat via heterogeneous base catalysts. Cereal & Oil **7**, 13–16 (2006)
28. L.J. Gao, G.Y. Teng, G.M. Xiao, R.P. Wei, Biodiesel from palm oil via loading KF/Ca-Al hydrotalcite catalyst. Biomass Bioenergy **34**, 1283–1288 (2010)
29. X. Deng, Z. Fang, Y.H. Liu, C.L. Yu, Production of biodiesel from jatropha oil catalyzed by nanosized solid basic catalyst. Energy **36**, 777–784 (2011)
30. H. Liu, V.F. Su, F. Liu, C. Li, U.U. Solomon, Cinder supported K2CO3 as catalyst for biodiesel production. Appl. Catal. B: Environ. **106**, 550–558 (2011)
31. W.L. Xie, L.L. Zhao, Heterogeneous CaO–MoO3–SBA-15 catalysts for biodiesel production from soybean oil. Energy Convers. Manage. **79**, 34–42 (2014)
32. A. Kleinova, P. Fodran, L. Brncalova, Synthesis of biolubricants using sulfated zirconia catalysts. Biomass Bioenergy **32**, 366–371 (2008)
33. R. Jothiramalingam, M.K. Wang, Preparation of Ca/Zr mixed oxide catalysts through a birch-templating route for the synthesis of biodiesel via transesterification. Ind. Eng. Chem. Res. **48**, 6162–6172 (2009)
34. B. Djebarri, V.M. Gonzalez-Delacruz, D. Halliche, K. Bachari, A. Saadi, A. Caballero, ... & O. Cherifi, Promoting effect of Ce and Mg cations in Ni/Al catalysts prepared from hydrotalcites for the dry reforming of methane. React. Kinet., Mech. Catal. **111**(1), 259–275 (2014)
35. J. Guo, H. Lou, H. Zhao, D. Chai, X. Zheng, Dry reforming of methane over nickel catalysts supported on magnesium aluminate spinels. Appl. Catal. A **273**(1–2), 75–82 (2004)

36. L. Zhao, X. Li, Z. Qu, Q. Zhao, S. Liu, X. Hu, The NiAl mixed oxides: the relation between basicity and SO2 removal capacity. Sep. Purif. Technol. **80**(2), 345–350 (2011)
37. O.P. Ferreira, O.L. Alves, D.X. Gouveia, A.G. Souza Filho, J.A. de Paiva, J. Mendes Filho, Thermal decomposition and structural reconstruction effect on Mg–Fe-based hydrotalcite compounds. J. Solid State Chem. **177**(9), 3058–3069 (2004)
38. M. Hajar, F. Vahabzadeh, Artificial neural network modeling of biolubricant production using Novozym 435 and castor oil substrate. Ind. Crops Prod. **52**, 430–438 (2014)
39. M. Tariq, S. Ali, N. Khalid, Activity of homogeneous and heterogeneous catalysts, spectroscopic and chromatographic characterization of biodiesel: a review. Renew. Sustain. Energy Rev. **16**, 6303–6316 (2012)
40. J. Dweck, C.M.S. Sampaio, Analysis of the thermal decomposition of commercial vegetable oils in air by simultaneous TG/DTA. J. Therm. Anal. Calorim. **75**(2), 385–391 (2004)
41. K. Kamalakar, A.K. Rajak, R.B.N. Prasad, M.S.L. Karuna, Rubber seed oil-based biolubricant base stocks: a potential source for hydraulic oils. Ind. Crops Prod. **51**, 249–257 (2013)
42. K.V. Padmaja, B.V. Rao, R.K. Reddy, P.S. Bhaskar, A.K. Singh, R.B. Prasad, 10-Undecenoic acid-based polyol esters as potential lubricant base stocks. Ind. Crop. Prod. **35**(1), 237–240
43. E.N. Ferreira, T.B.M.G. Arruda, F.E.A. Rodrigues, D.T.D. Arruda, J.H. da Silva Júnior, D.L. Porto, N.M.P.S. Ricardo, Investigation of the thermal degradation of the biolubricant through TG-FTIR and characterization of the biodiesel–Pequi (Caryocar brasiliensis) as energetic raw material. Fuel **245**, 398–405 (2019)
44. P. Prajapati, S. Shrivastava, V. Sharma, P. Srivastava, V. Shankhwar, A. Sharma, S.K. Srivastava, D.D. Agarwal, Karanja seed shell ash: A sustainable green heterogeneous catalyst for biodiesel production. Results Eng. **18**, 101063 (2023)
45. R. Yunus, O.T. Lye, A. Fakhru'l-Razi, S. Basri, A simple capillary column GC method for analysis of palm oil-based polyol esters. J. Am. Oil. Chem. Soc. **79**(11), 1075–1080 (2002)
46. S. Shrivastava, P. Prajapati, P. Srivastava, A.P. Lodhi, D. Kumar, V. Sharma, S.K. Srivastava, D.D. Agarwal, Chemical transesterification of soybean oil as a feedstock for stable biodiesel and biolubricant production by using Zn Al hydrotalcites as a catalyst and perform tribological assessment. Ind. Crops Prod. **192**, 116002 (2023)
47. G.X. Chen, Z.S. Hu, J.X. Dong, L.G. Wang, Study on antiwear and reducing friction additive of nanometer cobalt hydroxide. Tribol. Lubr. Technol. **57**(4), 36 (2001)

Chapter 7
Pretreatment of Rice Husk with Acid and Alkali for Enhancement of Sugar Recovery and Techno-Economic Analysis of the Process

Anuradha and Muthu Kumar Sampath

Abstract The supply of fossil fuels is limited, and their usage is creating havoc on the ecosystem. There is a need for renewable energy sources that have a lower environmental effect. Biomass is becoming increasingly popular as a renewable energy resource for producing bioenergy. Using waste biomass as a fuel can be a viable option. Crop residues are generated in large amounts every year across the world, yet they are mostly underused. Waste disposal is becoming increasingly expensive, and in many locations, there is even a scarcity of land. As a result, there are several advantages to using waste biomass for the production of biobutanol. It can serve as a viable alternative fuel and appealing energy storage. The major objective of this research is to study different pretreatment approaches on rice husks and track the structural changes in the biomass with SEM and FTIR. This study also includes the techno-economic analysis of NaOH pretreatment on rice husk to estimate the production costs and capital investment for producing fermentable sugars from rice husk. Rice husk pretreated with 4% NaOH with 6 h of incubation yielded maximum reducing sugar.

Keywords Biofuels · Biobutanol · Lignocellulose · Rice husk

7.1 Introduction

Biobutanol is a possible gasoline substitute that can be mixed in considerable amounts with regular fuel according to recent research. Agricultural wastes are a viable candidate for butanol production and the leading alternative to petroleum-derived fuels. Due to their economic feasibility and environmental friendliness, biomass-derived fuels are gaining popularity. Second-generation biofuels do not compromise food

Anuradha · M. K. Sampath (✉)
Department of Bioengineering and Biotechnology, Birla Institute of Technology Mesra, Ranchi, Jharkhand 835215, India
e-mail: muthubio@bitmesra.ac.in

N. Gakkhar et al. (eds.), *Recent Advances in Bio-Energy Research*, Springer Proceedings in Energy, https://doi.org/10.1007/978-981-99-5758-3_7

security and are typically produced from agricultural leftovers that are generated as wastes during crop processing. As a result, biofuels generated from lignocellulose materials such as bio-alcohols (biobutanol and bioethanol) have emerged as potential fuel sources due to their environmental benefits and sustainability which might help reduce dependency on petroleum reserves [1]. Pre-treatment is a crucial technique for converting lignocellulose-based biomass for biofuel production [2]. The compact structure of LCBs generally breaks down during the pretreatment process and is further exposed to the cellulose fiber. As the composition of LCB components varies, the most convenient pretreatment process is determined by the type of biomass used. An ideal pretreatment process maximizes sugar yield from pretreated biomass while consuming the least amount of energy. Therefore, in this study, a comparison of alkali and acid pretreatment was performed on rice husks. Further, the suitable pretreatment approach that produced high reducing sugar was carried forward for enzymatic hydrolysis. SEM and FTIR analysis were also conducted to study the morphological and functional alterations of biomass after pretreatment.

Techno-economic analysis (TEA) is a method of simulating and designing processes based on empirical data in order to predict investment costs, operating expenditures, energy balances, and mass balances for a commercial-scale biorefinery. TEA is a useful tool for evaluating potential research objectives, identifying cost constraints early in the research process, and generating mass and energy statistics for life-cycle environmental assessments [3]. Recent research has resulted in the development of new tools and approaches that allow for faster iteration of prospective designs, more rigorous uncertainty analysis, and more accessibility through the use of open-source platforms. As a result, this study also contains a techno-economic analysis of the NaOH pretreatment of rice husk to calculate the production costs and capital investment for sugar production from biomass. In this study acid and alkali pretreatments were compared by varying the concentration of acid and alkali as well as the residence time. This study also includes the techno-economic analysis of alkali pretreatment that will provide a summary of production cost and capital investment to produce reducing sugar from biomass.

7.2 Material and Methods

7.2.1 Sample Collection

Sample (Rice husk) were collected from Birsa Agricultural University, Ranchi to the laboratory for pretreatment screening.

7.2.2 Preparation of Standard Glucose Calibration Curve

The DNS approach was used to create a standard glucose calibration curve. 0-1000ul of standard 1 mg/mL glucose solution was pipetted into a series of test tubes, and the volume was adjusted to 1 mL with water. Then 1 mL of DNS was added to all of the tubes, which were then immersed in boiled water for 10 min. As the test tubes cool down 9 mL of DW was added to dilute the color. The optical density was compared to that of a blank at 540 nm. A standard graph of optical density at 540 nm against the concentration of glucose was plotted. This was used to determine the reduced sugar produced by agro-waste.

7.2.3 Pretreatment Screening

Acid Pretreatment

15 g of rice husk was pretreated with sulfuric acid (H_2SO_4) at concentrations of 2, 4, and 6%. The samples were heated using a water bath at 80 °C for different retention times (1, 3, 6, and 24 h).

Alkaline Pretreatment

15 g of rice husk was pretreated with sodium hydroxide (NaOH) at concentrations of 2, 4, and 6%. The samples were heated using a water bath at 80 °C for different retention times (1, 3, 6, and 24 h).

Analysis of Reducing Sugar

The total reduced sugar was calculated using the DNS method. In the test tube, the sample (1 mL) was combined with 2 mL 3,5-dinitrosalicylic acid (DNS) reagent and submerged in a boiling water bath for 5 minutes. After cooling the sample to room temperature, the absorbance at 540 nm was measured.

7.2.4 Enzymatic Hydrolysis of Pretreated Rice Husk

3 g of NaOH pretreatment rice husk was dissolved in 100 mL of sodium acetate buffer and then 24 mg of cellulase from Trichoderma reesei was added to the lignocellulose biomass for hydrolysis. Further, the sample was incubated in a shaking incubator for 24 h.

7.2.5 SEM and FTIR Analysis

SEM was used to examine the changes in the morphological structure of rice husks before and after pretreatment. FT-IR spectroscopy was also used to detect the various functional groups present in raw and processed biomass. The spectra of the samples were taken in the 4000–500 cm^{-1} range.

7.2.6 Evaluation of Conversion Using SuperPro

The process model was developed using Super Pro Designer (Intelligen, Inc., Scotch Plains, NJ). The software generated a process flow sheet for the pretreatment and enzymatic hydrolysis of the rice husk.

7.3 Result and Discussion

7.3.1 Analysis of Reducing Sugar

In the study, rice husk samples were pretreated with alkali and acid i.e. NaOH & H_2SO_4 by varying two different parameters which were concentration of acid & alkali, and retention time of pretreatment. The total reduced sugar yield from each of the parameters is shown in Fig. 7.1a, b. The highest sugar yield was with 4% NaOH at a retention time of 6 h, which was 1.906 mg/ ml. Further retention time after 6 h at the same concentration gave no meaningful increase in sugar yield. Similarly, the effect of sulfuric acid (H_2SO_4) with different concentrations and retention time is shown in Fig. 7.1b. The highest sugar yield was obtained with 2% H_2SO_4 at a retention time of 3 h, which was 0.369 mg/ ml. Further, 4% NaOH pretreated biomass was carried forward for the enzymatic hydrolysis, and an increase in the reducing sugar was observed i.e. 3.916 mg/mL.

7.3.2 Morphological Analysis

Pretreatment of biomass changes the structural features of the lignocellulosic biomass [5] by breaking the cellulose-hemicellulose-lignin complex and converting the carbohydrate (cellulose) into fermentable sugar [6]. Scanning electron microscopy was done to observe the morphological alteration of rice husks. The morphological characteristics of untreated and pretreated rice husks with NaOH and H_2SO_4 that were studied can be seen in Fig. 7.2a–c. Figure 7.2a shows the well-organized and intact surface of untreated rice husks. Figure 7.2b shows biomass surface characteristics

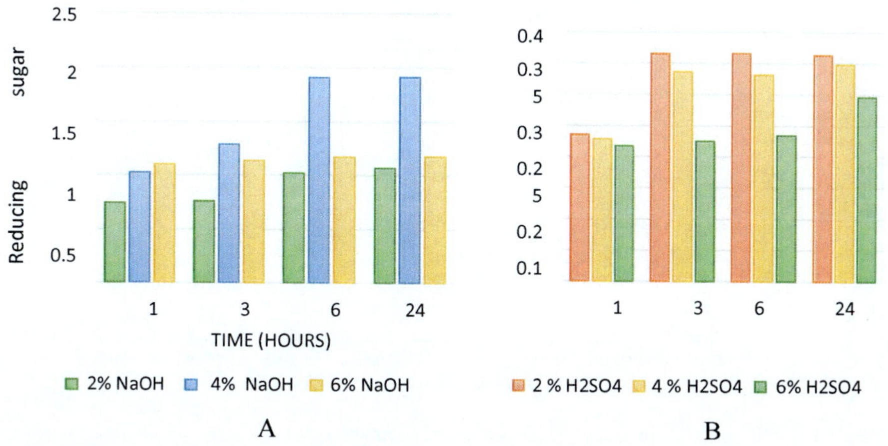

Fig. 7.1 Pretreatment with **a** NaOH **b** H_2SO_4

under treatment with 2% H_2SO_4. The surface coating of the biomass is removed and the fiber bundles which were tightly packed in the untreated biomass started to dismantle under the action of acid pretreatment. These changes can be attributed to some lignin removal [7], whereas, 4% NaOH pretreated sample exhibited rupture of rigid structure. The intrinsic fiber structure of the biomass was more exposed to NaOH pretreatment compared to H_2SO_4 pretreatment. The confirmation of lignocellulosic breakdown into its components was further analyzed with FTIR spectra.

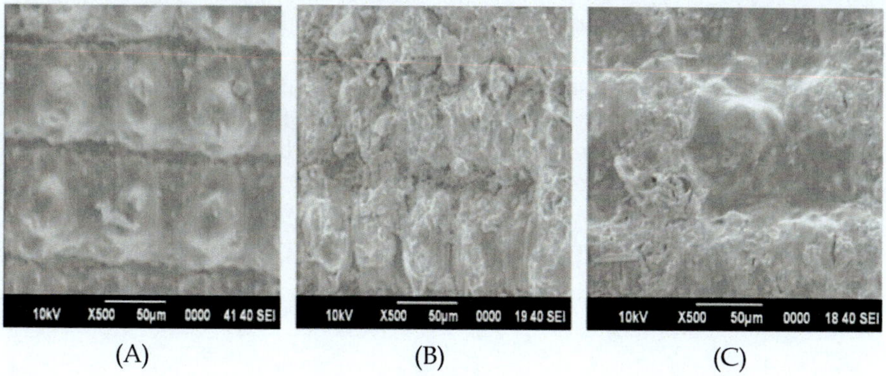

Fig. 7.2 SEM images of **a** raw biomass **b** 4% H_2SO_4 pretreated **c** NaOH pretreated

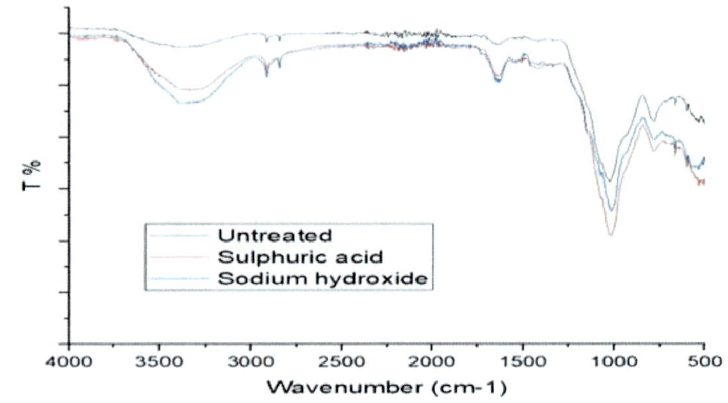

Fig. 7.3 FTIR spectroscopy of raw and pretreated biomass

7.3.3 FTIR Analysis

FTIR spectroscopy was used to determine the changed structure of lignocellulosic biomass. Figure 7.3 depicts the results. The FTIR spectra of raw, NaOH pretreatment, and H_2SO_4 pretreatment are shown in Fig. 7.3. Hydrogen bonded (O–H) stretching absorption is represented by a peak of 3329.89 cm^{-1}. The peaks at 2133.27 cm^{-1} and 2897.08 cm^{-1} show the alterations in the cellulose chain due to the chemical treatment [8]. Stretching at this specific location reveals the cellulose structure. The IR spectra at 1729.74 cm^{-1} were attributed to the $C = O$ acetyl group of hemicelluloses lignin or ester [9]. Pretreatment with H_2SO_4 results in a small change in lignin and hemicellulose content, but pretreatment with alkali (4% NaOH) for 6 h results in the removal of lignin. The absorption at 1512 cm^{-1} corresponds to aromatic stretching (C=C) from lignin [10]. Results show loss of aromatic functional group for NaOH pretreatment. No significant loss was observed with H_2SO_4 pretreatment. Removal of the C-O- C functional group of aryl alkyl ether in lignin also was observed in NaOH and H_2SO_4 pretreatment at an intensity of 1243 cm^{-1}. Therefore, chemical alteration in the composition of rice husk was studied with FTIR spectroscopy, and found that NaOH pretreatment is more effective than H_2SO_4 pretreatment for the delignification of rice husk.

7.3.4 Techno-Economic Analysis Using SUPERPRO Designer

The flowsheet generated by SuperPro Designer in Fig. 7.4 depicts the overall process of sugar recovery from rice husk. The biomass is held in the solid storage chamber (B/Storage), and then it is submitted to the pre-treatment section (P-2/R-101) via the S-103 stream, and water is supplied to the reactor via the S-104 stream (R-101).

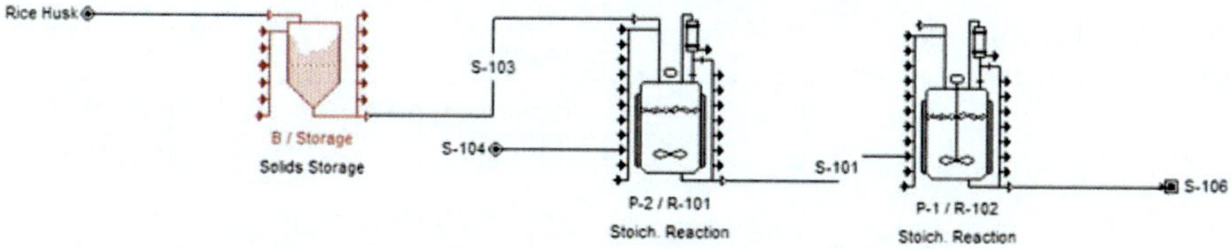

Fig. 7.4 A flow sheet for sugar recovery from rice husk using SuperPro Designer

Stream S-105 transports the processed biomass with 4% NaOH to the enzymatic hydrolysis chamber (P-1/R-102). Enzymatic hydrolysis takes place in the presence of cellulase enzyme. At the end of the process, the final recovered sugar is collected.

In the flowsheet, the component flow rates (kg/Batch), at the output destination and the overall balance of the components are represented in Fig. 7.5 where we can see that after pretreatment with NaOH the cellulose and hemicellulose content will be converted into hexose and pentose. An ideal pretreatment maximizes the sugar yield from the biomass that will be fermented and further processed for biofuel production. The entire cost analysis was performed in order to perform the design and economic analysis of the fermentable sugar manufacturing process from rice husk. The economic evaluation was also studied and a fixed capital estimate was generated which could help to calculate and evaluate the cost of construction and operation of a biobutanol plant [11]. In Table 7.1. We can see the total plant direct cost of different materials consist of equipment purchase cost, installation cost, process piping, instrumentation, insulation, etc. which is essential for a biobutanol plant. Therefore, this economic evaluation will help to evaluate the capital investment for the biobutanol plant.

COMPONENT	INITIAL	INPUT	OUTPUT	FINAL	IN-OUT
cellulose	0.00	40.00	2.05	0.00	37.95
hemicellulose	0.00	30.00	3.75	0.00	26.25
hexose	0.00	0.00	42.17	0.00	- 42.17
lignin	0.00	28.00	28.00	0.00	0.00
Nitrogen	0.12	0.00	0.00	0.12	0.00
Other solids	0.00	2.00	2.00	0.00	0.00
Oxygen	0.04	0.00	0.00	0.04	0.00
pentose	0.00	0.00	29.86	0.00	- 29.86
Sodium Hydroxid	0.00	12.29	12.29	0.00	0.00
Water	0.00	192.55	184.75	0.00	7.80
TOTAL	0.15	304.84	304.86	0.15	0.02
				Overall Error:	0.008%

Fig. 7.5 Overall component balance (kg/batch) for the recovery of sugar from rice husk using SuperPro Designer

Table 7.1 Fixed capital estimate summary

S. no.	Materials	Cost ($)
1	Equipment purchase cost	1,319,000
2	Installation	426,000
3	Process piping	462,000
4	Instrumentation	528,000
5	Insulation	40,000
6	Electrica	132,000
7	Buildings	593,000
8	Yard improvement	198,000
9	Auxiliary facilities	528,000
Total plant direct cost (TPDC)		4,224,000

7.4 Conclusions

In this investigation, acid and alkali pretreatment was used to study the most effective pretreatment for rice husks. The results showed that 4% NaOH effectively disintegrates the lignocellulosic biomasses with 6 h of incubation. Further SEM and FTIR also revealed that alkali pretreatment effectively eliminated the lignin from the biomass. Therefore, fermentable sugars produced by the lignocellulosic biomass waste such as rice husk can be employed as a suitable feedstock to produce biofuels.

References

1. M. Ali, M. Saleem, Z. Khan, I.A. Watson, The use of crop residues for biofuel production. Biopolymer-Based Materials, and Bioenergy Biopolym. Based Mater., Bioenergy, 369–395.https://doi.org/10.1016/b978-0-08-102426-3.00016-3
2. R. Sindhu, P. Binod, A. Pandey, Biological pretreatment of lignocellulosic biomass – an overview. Biores. Technol. **199**, 76–82 (2016). https://doi.org/10.1016/j.biortech.2015.08.030
3. C.D. Scown, N.R. Baral, M. Yang, N. Vora, T. Huntington, Techno-economic analysis for biofuels and bioproducts. Curr. Opin. Biotechnol. **67**, 58–64 (2021). https://doi.org/10.1016/j.copbio.2021.01.002
4. G.L. Miller, Use of dinitrosalicylic acid reagent for determination of reducing sugar. Anal. Chem. **31**(3), 426–428 (1959)
5. M.O.S. Dias, V.E. Adriano, A.N. Silvia, M.F. Rubens, R. Carlos, R.W.M. Maria, Production of bioethanol and other bio-based materials from sugarcane bagasse: integration to the conventional bioethanol production process. Chem. Eng. Res. Des. **87**, 1206–1216 (2009)
6. Z. Yan, L. Jihong, C. Sandra, C. Ting, J. Yan, Y. Menghui, Z. Lei, Z. Gang, Q. Panlun, L. Shizhong, Lignin relocation contributed to the alkaline pre-treatment efficiency of sweet sorghum bagasse. Fuel **158**, 152–158 (2015)
7. N.A. Latip, A.H. Sofian, M.F. Ali, S.N. Ismail, D.M.N.D. Idris, Structural and morphological studies on alkaline pre-treatment of oil palm empty fruit bunch (OPEFB) fiber for composite production. Mater. Today: Proc. **17**, 1105–1111 (2019). https://doi.org/10.1016/j.matpr.2019.06.529

8. Y. Cao, T. Huimin, Structural characterization of cellulose with enzymatic treatment. J. Mol. Struct. **705**, 189–193 (2004)
9. N. Zulkiple, M. Maskat, O. Hassan, Pretreatment of oil palm empty fruit fiber (OPEFB) with aqueous ammonia for high production of sugar. Procedia Chem. **18** (2016).https://doi.org/10.1016/j.proche.2016.01.024
10. N.A.M. Nor, W.A.W. Mustapha, O. Hassan, Deep eutectic solvent (DES) as a pretreatment for oil palm empty fruit bunch (OPEFB) in sugar production. Procedia Chem. **18**, 147–154 (2016). https://doi.org/10.1016/j.proche.2016.01.023
11. J. Peng, H. Xu, W. Wang, Y. Kong, Z. Su, B. Li, Techno-economic analysis of bioethanol preparation process via deep eutectic solvent pretreatment. Ind. Crops Prod. **172**, 114036 (2021). https://doi.org/10.1016/j.indcrop.2021.11403

Chapter 8
Investigation of Tribological Behavior of a Single Cylinder Diesel Engine Operated with Water-Added Palm Biodiesel–Diesel Fuel Blend

Sonu Kumar Patidar, Hifjur Raheman, and Neeraj Kumar

Abstract Recently, biodiesel has gained more attention as it is a promising substitute for non-renewable fossil fuels as well as it eases the current environmental challenges of higher vehicular emissions. However, a few studies have been reported higher NO_X emissions from biodiesel–diesel fuel blend-operated diesel engines as compared to conventional diesel. Various after and pre-treatment methods such as lean nitrous oxide trap, selective catalytic reduction, and exhaust gas recirculation can control the NO_X emissions but are expensive and affordable only in heavy-duty engines. Whereas, incorporation of water in the fuel could control the NO_X emissions in a cost-effective manner. However, such fuels need to be tested in the engine for a prolonged period to ensure their sustainable future and long-term adaptability. Therefore, the present study was aimed to investigate the tribological behavior of a 10-kW water-cooled, 4-stroke DI diesel engine operated with a water-added palm biodiesel–diesel blend (WB20) and pure diesel. For this, soot deposition formation and crankcase oil defilement characteristics have been investigated by running the engine for 100 h with each fuel. WB20 fuel showed 1.14% higher average BTE, whereas 7.04 and 13.9% lesser average NO_X and HC emissions respectively, as compared to diesel. Similarly, WB20 fuel showed 16.94 and 14.45% lesser amount of soot deposition over the cylinder head and piston top respectively, as compared to the diesel–operated engine after 100-h long run. Similarly, WB20 operated engine showed 0.48–28.57% lesser concentration of various engine wear metals (i.e., Cu, Zn, Fe, Mg, and Mn) in the lubricating oil during 100 h long run test as compared to the diesel-operated engine.

Keywords Tribological behavior · Soot deposition · Wear metals

---

S. K. Patidar (✉) · H. Raheman
Agricultural and Food Engineering Department, Indian Institute of Technology, Kharagpur, India
e-mail: sonupatidariitkgp@gmail.com

N. Kumar
ICAR - Indian Institute of Wheat and Barley Research, Karnal, India

N. Gakkhar et al. (eds.), *Recent Advances in Bio-Energy Research*, Springer Proceedings in Energy, https://doi.org/10.1007/978-981-99-5758-3_8

8.1 Introduction

Presently, India ranks 3rd in primary energy consumption with a global share of 5.8% of total consumption. More than 90% of this share is mainly oriented from the requirement of fossil fuels such as coal, gasoline, and natural gas [1]. Petroleum fuels are facing serious challenges such as strict emission norms, non-renewable nature, and fluctuating prices. Diesel engines are used as a potential energy supplier for transportation, construction, and agriculture sectors such as irrigation and electricity generation. Diesel fuel consumption is ever-increasing as the number of diesel engines is increasing year by year. To address such issues, well-organized and compact research solutions need to be promoted, invented, and imposed by the policy formers related to the harmful air pollutants raised from fossil fuel burning.

Biodiesel and its derivatives are such set of promising alternative fuels and have the potential to replace diesel partially or completely from the existing diesel engines [2]. Biodiesel has similar physicochemical properties as diesel and provides enhanced brake power and decreased harmful exhaust emission as compared to diesel except NO_X emissions [3, 4]. The NO_X emissions are mainly depending on the higher in-cylinder temperature and the oxygen content of the fuel.

Various Pre and after-treatment strategies have been adopted to lower the NO_X emissions from biofuel-operated diesel engines. After-treatment strategies such as selective catalytic reduction (SCR) and lean NO_X trap offer more than 90% conversion efficiency but these are cost-ineffective devices which makes them suitable for heavy-duty vehicular engines [5]. On the other hand, fuel pre-treatment methods are simple to employ and do not require major engine modifications to address the higher NO_X problem from biodiesel-fueled engines. The addition of water in the biodiesel–diesel blend could lower the in-cylinder temperature during combustion which results in lower NO_X emission [6–8]. Additionally, the subsidiary atomization of water droplets present in the fuel could improve air–fuel mixing and helps in improving the energy efficiency of the engine [9]. However, these conclusions are based on short-duration engine testing and require to be derived based on long-run engine tests. The higher viscosity of water-emulsified biodiesel as compared to diesel could lead to indigent atomization, and coking of fuel spraying needle and filter. Moreover, the low volatility of biodiesel blends may lead to the dilution of crankcase oil.

A few attempts have been made in the recent past to ensure the adaptability of the biodiesel–diesel blends in the diesel engine based on long-run engine testing varying from 100 to 512 h. Raheman et al. [10] conducted a 100-h long-run test using Mahua and Simarouba oil mixture biodiesel–diesel blend (B10) in a diesel engine and reported average of 21% lesser soot deposits over the cylinder head, piston top, and injector tip with B10 operated engine compared to diesel engine due to better oxidation ability of the B10 blend. In addition, they reported 11–50% lesser heavy metal concentration (Al, Cu, Zn, Mg, Pb, and Fe) in the lubricating oil with B10 blend over diesel. Similarly, Chourasia et al. [11] conducted a 512-h long engine endurance test and reported 63% and 52.7% greater wear of the internal surface of the combustion cylinder and piston outer surface respectively, operated with B20A4 (biodiesel, di-ethyl ether, and diesel in 20:4:76 v/v respectively) as compared to diesel.

The presence of H_2O molecules in the fuel may lead to the corrosion of engine components in the fuel-burning chamber and abrasive metal debris may contaminate and reduce lube oil service life and could lead to the ultimate failure of an engine. No such attempt has been made in the past to conduct a long-run endurance test to ensure the adaptability of water-emulsified biodiesel–diesel fuel blend in the diesel engine by measuring the carbon deposits over engine components and heavy wear metal addition in lube oil for quality analysis. Therefore, in the present study, the comprehensive behavior of a diesel engine was investigated in terms of combustion, performance, emission as well as soot deposition formation and lube oil analysis based on a 100-h long run test when operated with a water-emulsified biodiesel–diesel fuel blend and compared with those of diesel.

8.2 Material and Methods

8.2.1 Preparation of Stable Fuel Blends

Preparation of biodiesel

In the present investigation, raw palm oil (RPO) having free fatty acid (FFA) content of 0.80% was procured from the local market of Kharagpur, India to produce biodiesel. As the raw oil had lesser FFA content, a base transesterification reaction was performed with the raw palm oil to produce biodiesel (B100). The process parameters for the reaction were 25% methanol/RPO (v/v), KOH (initial acid value (%) + 3.5) gram per liter of RPO, and 60 °C reaction temperature with a 30 min reaction process [12].

8.2.2 Preparation of Stable Water–Biodiesel Emulsion

After the production of biodiesel (B100), a biodiesel–diesel fuel blend (B20) was prepared by mixing 20% v/v of the B100 blend and 80% v/v of diesel. To produce a stable water-emulsified blend, two commercially available surfactants (i.e., Tween 80 and Span 80) were used as suggested by several past studies. In the present investigation, the water-emulsified biodiesel (WB20) was prepared with 1% v/v surfactants maintaining HLB value as 5 (0.93% v/v Span 80 + 0.7% v/v Tween 80), 10% v/v water and 89% v/v B20 blend using a homogenizer rotating at speed of 2500 rpm for 15 min. After the preparation, the prepared fuel sample was kept under observation to investigate any separation of the water layer and it was found that there was no

Fig. 8.1 Prepared fuel blends

water and creamy layer separation in the WB20 fuel sample up to 48 h. The prepared fuel blends, i.e., WB20, B20, B100, and diesel are shown in Fig. 8.1.

8.2.3 Measurement of Fuel Properties of Prepared Fuel Blend

Various fuel processing parameters such as air–fuel mixing, spray formation patterns, burning behavior, performance, and exhaust emissions characteristics are mainly depends on the fuel properties. Also, any developed or modified fuel should have undergone the measurement of fuel properties to ensure its adaptability. Hence, in the present investigation, various fuel properties such as calorific value, density, flash point, acid value, and kinematic viscosity were measured by following the latest ASTM standards.

8.2.4 Investigation of Combustion, Performance, and Exhaust Emission Characteristics

The experimental investigation of combustion, performance, and exhaust emissions was performed in a 10-kW single-cylinder, water-cooled, constant speed, DI diesel engine. For this reason, preliminary tests were performed in the engine. The test engine setup specifications are given in Table 8.1, and a schematic view of the experimental setup is shown in Fig. 8.2. The maximum engine brake power using diesel as a reference fuel was obtained at a load of 68 $N - m$ and observed as 9.80 kW at 1400 ± 100 engine rpm. Therefore, the rated load or 100% load was taken as 68 $N - m$ and accordingly, the intermediate loads (25, 50, and 75%) were computed. Both the fuels (WB20 and diesel) were tested separately at different loads by operating the

Table 8.1 Engine specifications

	Particulars	Details
Engine	Make	Chetak, India
	Model	SDM-14
	Number of cylinders	1
	Bore × stroke (mm)	114.3 × 116
	Stokes	4-stroke
	Highest power (kW)	10
	Engine rated speed (rpm)	1400
	Compression ratio	17.5:1
	Fuel injection time	24° before top dead center

engine for 2.5 h with each fuel. The various measurement instruments/sensors were installed on the engine test rig such as piezoelectric pressure transducer for pressure measurement, eddy current dynamometer for load application, rotary encoder for crank angle measurement, and proximity sensor for measuring the engine rpm etc. The operational signals from these sensors were received and processed in the data acquisition system to find out ignition delay, brake power and brake-specific fuel consumption. Whereas an Indus flue gas analyzer was used to investigate the NO_X and hydrocarbon emissions at the exhaust tailpipe. Later on, these parameters were compared for both the fuels (i.e., test fuel (WB20) and ref. fuel (diesel).

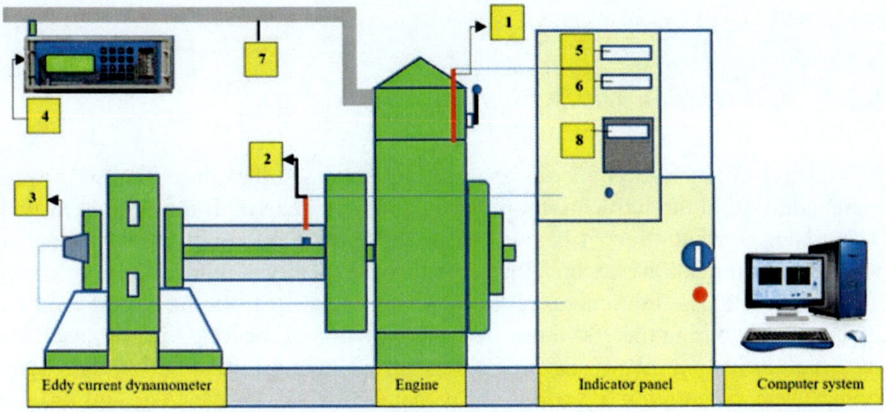

Fig. 8.2 Schematic view of engine test setup. 1. Piezoelectric pressure transducer, 2. Proximity sensor, 3. Rotary encoder, 4. Exhaust gas analyzer, 5. Load indicator, 6. Engine rpm indicator, 7. Exhaust pipe, 8. Brake power indicator

Table 8.2 Loading pattern for 100-h test

Load (% of rated load)	Running time (min)
100	93.75 (11.72 min warm-up)
50	93.75
100	23.45
No-load	11.72
100	70.31
50	82
Total	375 min or 6.25 h

8.2.5 Investigation of Tribological Behavior of the Engine

Investigation of soot deposition formation

The assessment of soot deposition was performed by running the engine for a prolonged period of 100 h with each of the fuels (WB20 and Diesel). Initially, the engine was dismantled, and the engine components (piston top and cylinder head) were cleaned before the long run. The 100-h run was completed in 16 test cycles with 6.25 h each. The loading pattern for each test cycle was decided as per IS: 10000-Part: 9 [13] as given in Table 8.2. After the completion of the 100-h test with each fuel, the images of the piston top and cylinder head were captured for quality analysis. Similarly, deposited soot particles were carefully scraped using a wooden scraper from the engine components and weighed for quantitative assessment.

8.2.6 Lubrication Oil Defilement Analysis

The quality of lubrication oil was analyzed by obtaining the concentration of wear metal addition in the lubrication oil during and after the 100-h run with each fuel. The higher concentration of heavy wear metal in lube oil could lead to increased viscosity of lubrication oil which further reduce the quality of lubrication and service life of lube oil. The lubrication oil sample was collected from the engine sump at 25-h intervals during the 100-h test. After completion of the long run with each test fuel, the lubricating oil was drained out and fresh lube oil was filled in the engine sump. The concentration of various heavy metals (i.e., Fe, Pb, Mn, Mg, and Zn) was obtained using atomic absorption spectroscopy (AAS) by performing the dry ash technique [10, 14]. The lube oil sample was collected in a 100-ml glass container and thoroughly mixed at 50 °C for 1 h using a water bath shaker. Approximately 10 g of mixed lube oil sample was then transferred to a dried silica crucible. The lube oil sample was then dried for 1 h at 120 °C using a hot plate. Thereafter, the sample was placed in a muffle furnace at a temperature of 450 and 650 °C for a time period of 4 and 2 h, respectively. The persisting ash in crucibles was then dissolved in 1.5 ml of

Table 8.3 Fuel properties of different test fuels used in the present investigation

Fuel type	Density (kg/m^3)	Viscosity (cSt)	Acid value (mg KOH/g)	Calorific value (MJ/kg)	Flash point (°C)
WB20	850	3.9	0.33	38.90	75
Diesel	824	2.55	0.13	42.63	58
IS 15607:2016 for biodiesel B100	860–900	3.5–5.0	<0.5	–	> 101
IS 1460:2017 for diesel	820–860	2.0–5.0	<0.2	–	< 66
ASTM methods	D 4052-22	D 446-12	D 5555-14	D 240-19	D 93–20

HCl. Thereafter, 100 ml of deionized water was added to the solution and subjected to metal detection using AAS.

8.3 Results and Discussion

8.3.1 Fuel Properties

Fuel properties of WB20 and diesel were measured and compared as mentioned in Sect. 2.2. The various fuel properties are mentioned in Table 8.3. All the fuel properties for both WB20 and diesel fuel were found within the limits prescribed by the updated Indian standards for diesel and biodiesel.

8.3.2 Combustion, Performance, and Exhaust Emission Characteristics

Ignition delay (ID)

In diesel engines, investigation of chemical delay has a great impact as it displays the rate of fuel burning and ability of the fuel to form NO_X emissions during combustion. The variation of ID with engine load and fuel type is portrayed in Fig. 3a.

Effect of load on ID

From the figure, it can be concluded as the engine load increased, ID was reduced for both the test fuels. At greater loads, the elevated cylinder temperature favors the vaporization of the fuel and causes a decrement in ignition delay.

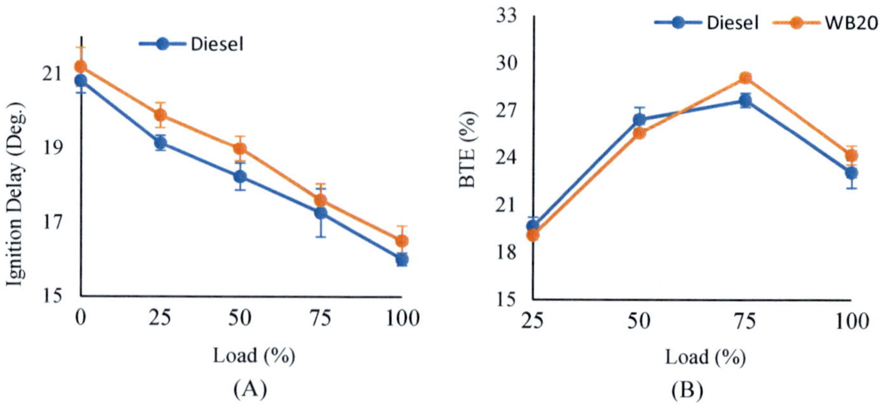

Fig. 8.3 Variation of ID and BTE with engine load and fuel type

Effect of fuel type on ID

The water-emulsified biodiesel WB20 showed higher chemical delay as compared to diesel at all the tested loads. The WB20 fuel absorbed a fraction of heat to vaporize its fuel-borne H_2O during the burning process, hence fending the initiation of combustion favoring to higher ignition delay.

8.3.3 Brake Thermal Efficiency (BTE)

Brake thermal efficiency is an indication of the transformation of chemical energy into mechanical work. The variation of BTE with engine load and fuel type is portrayed in Fig. 3b.

Effect of load on BTE

Figure 3b displays that the BTE improved with the engine loading for both the tested fuels up to semi highest engine loading (i.e., up to 75% engine load). The brake power increased at a higher rate compared to fuel consumption rate leading to increased BTE for loading up to 75% load. At full load, the losses of injected fuel were dominating factors and subsequently reduced the BTE.

Effect of fuel type on BTE

AT the initial loads (up to 50%), diesel showed higher BTE as compared to WB20 as it has a higher heating value than the WB20 fuel which ensured lesser fuel consumption of diesel as compared to WB20. Beyond 50% load, WB20 showed higher BTE as compared to pure diesel. It was due to subsidiary atomization provided by a micro-explosion of the H_2O droplets at the elevated temperature of the combustion chamber for WB20. The enhanced fuel mixing favored the combustion efficiency of WB20 leading to increased BTE as compared to diesel.

8.3.4 NO_X Emissions

The elevated combustion temperature, presence of reactive O_2, and availability of reaction time are the main factors for NO_X formation in the exhaust emission. The behavior of NO_X formation with engine loading and fuel type is portrayed in Fig. 4a.

Effect of load on NO_X

With the increase in load NO_X emissions were found to be increased. As the load increased, the combustion temperature was increased due to increased fuel consumption at higher loads which liberated more energy in the combustion chamber leading to the higher in-cylinder temperature at higher loads. The elevated temperature contributed more to the formation of NO_X at higher engine loads as compared to low engine loads.

Effect of fuel type on NO_X

Water-emulsified biodiesel (WB20) fuel showed lesser NO_X concentration as compared to diesel. The energy absorption phenomenon by water droplets decreased the in-cylinder temperature in the case of WB20 fuel. The enhanced air–fuel mixing in WB20 mended the combustion efficiency and the extra oxygen of WB20 was taken for the combustion of fuel resulting in reduced NO_X emission formation [15, 16]. Both reduced temperature and lesser oxygen availability helped WB20 to produce lesser NO_X emissions.

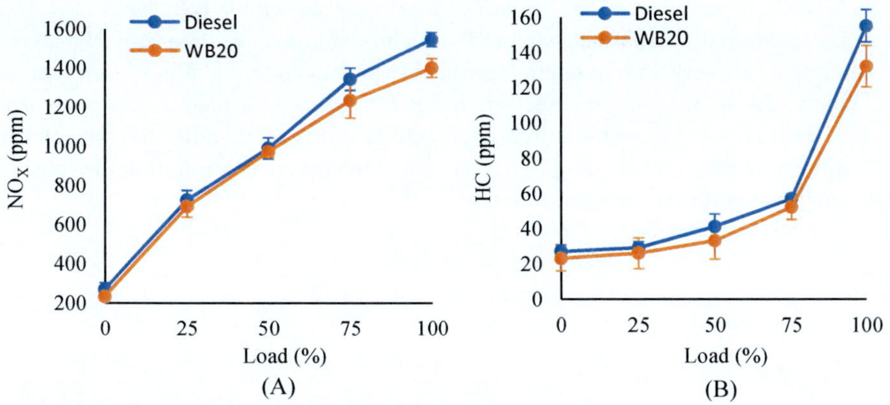

Fig. 8.4 Variation of NO_X and HC with engine load and fuel type

8.3.5 Emissions of Hydrocarbon (HC)

The hydrocarbons were mainly emitted due to the partial combustion of fuel in the combustion chamber. The behavior of HC formation with engine load and fuel type is portrayed in Fig. 4b.

Effect of load on HC

As the load increased HC emissions were increased for the fuels as the higher amount of fuel was injected at a higher load. The higher quantity of fuel requires extra oxygen for complete burning and due to this unburnt fuel remained in the combustion chamber leading to higher HC at higher loads.

Effect of fuel type on HC

The water-emulsified fuel showed lesser HC emissions as compared to diesel owing to its higher combustion efficiency as discussed in earlier section of ignition delay. The micro-explosion phenomenon and presence of extra oxygen in WB20 fuel contributed to more conversion of fuel particles into combustion products resulting in lesser HC emissions over diesel.

8.3.6 Tribological Behavior of the Engine

The carbon deposits over the cylinder head and piston crown were measured quantitatively and qualitatively and shown in Fig. 8.5. The water-emulsified fuel showed lesser soot deposits after 100 h long run of the engine as it burnt more effectively in the engine as discussed in the section of BTE. The lesser amount of soot deposits with WB20 fuel was also supported by the result obtained in the HC emissions section as the lesser concentration of HC emissions was observed with WB20 fuel.

The addition of various heavy metals in the lubrication oil was measured by taking out samples at every 25 h of engine run and are portrayed in Fig. 8.6. It is clear from the figure that all the wear metals were lesser for water-emulsified fuel as compared to diesel fuel. The vegetable oil-based WB20 fuel provided additional lubricity to the engine components leading to a lesser heavy metal concentration in the lube oil as compared to diesel-operated engine.

Fig. 8.5 Soot deposition formation over cylinder head and piston top

8.4 Conclusions

Based on the experiments conducted, it can be concluded that water emulsified biodiesel showed improved performance and reduced emission level as compared to diesel-operated engine. WB20 showed 5.14% higher BTE at 75% engine load, whereas 8.77–19.71% lesser HC, 1.71–13.80% lesser NO_X emissions as compared to diesel. The developed fuel WB20 needs no engine modifications as the engine behaved normally during the 100-h long run tests. WB20 showed 16.94 and 14.45% lesser carbon deposits over cylinder head and piston top, respectively. Therefore, the developed water-added biodiesel–diesel fuel blend (WB20) can be recommended as potential replacement of the diesel.

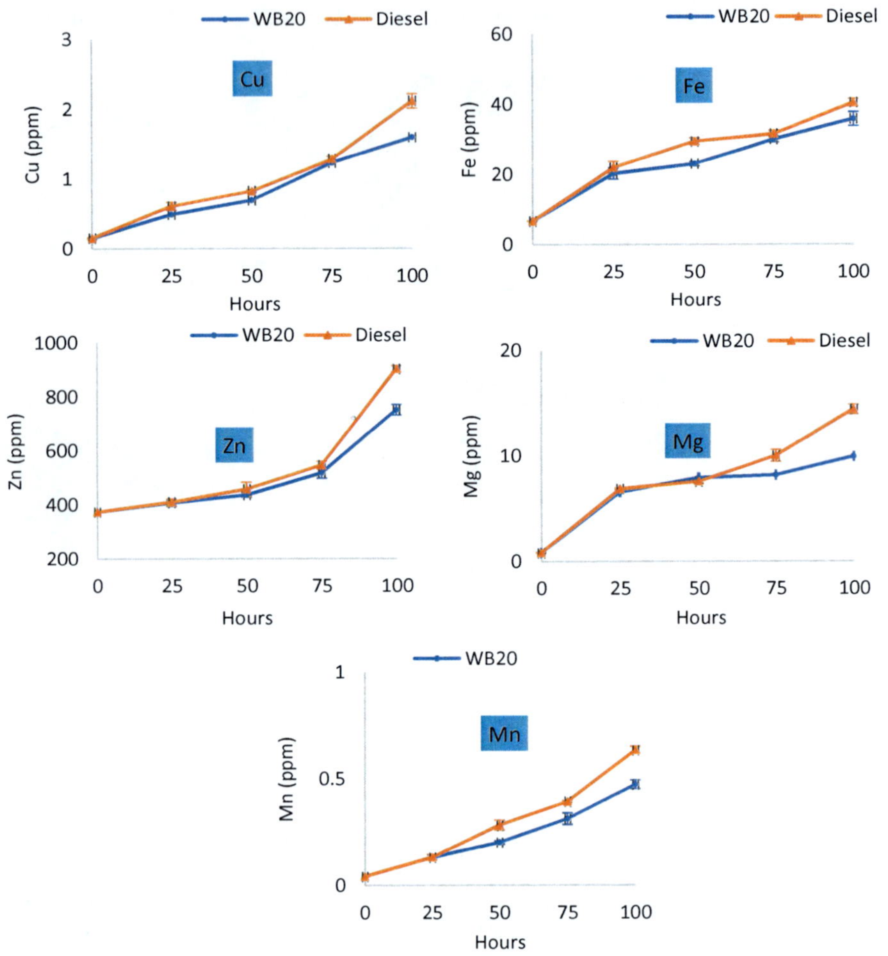

Fig. 8.6 Concentration of various heavy metal in lubricating oil of the engine

References

1. *Statistical Review of World Energy*, 69th edn (2020). https://www.bp.com/content/dam/bp/businesssites/en/global/corporate/pdfs/energyeconomics/statistical-review/bp-stats-review-2020-full-report.pdf. Accessed 10 Sep 2020
2. H. Raheman, S.V. Ghadge, Performance of compression ignition engine with mahua (Madhuca indica) biodiesel. Fuel **86**(16), 2568–2573 (2007)
3. A.K. Agarwal, Biofuels (alcohols and biodiesel) applications as fuel for internal combustion engines. Progress in Energy Combustion Science **33**(3), 233–271 (2007)
4. M.N. Nabi, M.G. Rasul, Influence of second-generation biodiesel on engine performance, emissions, energy and exergy parameters. Energy Convers. Manage. **169**, 326–333 (2018)

5. N. Kumar, H. Raheman, Thermal and environmental performance of CI engine using CeO_2 nanoparticles as additive in water–diesel–biodiesel fuel blend. Int. J. Environ. Sci. Technol. **31**, 1–8 (2021)
6. H. Raheman, S. Kumari, Combustion characteristics and emissions of a compression ignition engine using emulsified jatropha biodiesel blend. Biosys. Eng. **123**, 29–39 (2014)
7. Z. Chen, X. Wang, Y. Pei, C. Zhang, M. Xiao, J. He, Experimental investigation of the performance and emissions of diesel engines by a novel emulsified diesel fuel. Energy Convers. Manage. **95**, 334–341 (2015)
8. N. Kumar, H. Raheman, Combustion, performance, and emission characteristics of diesel engine with nanocera added water emulsified Mahua biodiesel blend. Environ. Prog. Sustain. Energy **40**(4), e13572 (2021)
9. N. Kumar, R. Machavaram, H. Raheman, Combustion behavior of diesel engine operated with water emulsified diesel. J. Sci. Res. Rep. **8**, 1–8 (2017)
10. H. Raheman, P.C. Jena, S.S. Jadav, Performance of a diesel engine with blends of biodiesel (from a mixture of oils) and high-speed diesel. Int. J. Energy Environ. Eng. **4**(1), 1 (2013)
11. S. Chourasia, P.D. Patel, A. Lakdawala, R.N. Patel, Study on tribological behavior of biodiesel-Diethyl ether (B20A4) blend for long run test on compression ignition engine. Fuel **230**, 64–77 (2018)
12. N. Kumar, H. Raheman, Characterization of nano-oxide added water emulsified biodiesel blend prepared with optimal emulsifying parameters. Renew. Energy **145**, 308–317 (2020)
13. Indian Standard code 10000-9: Methods of tests for internal combustion engines. Part - 9: Endurance tests (1980)
14. S.K. Patidar, H. Raheman, Performance and durability analysis of a single-cylinder direct injection diesel engine operated with water emulsified biodiesel-diesel fuel blend. Fuel **1**(273), 117779 (2020)
15. N. Kumar, H. Raheman, Production, characterization and utilization of second generation biodiesel blend in diesel engine using water and nanoparticles as additives. Fuel **15**(308), 122063 (2022)
16. O. Armas, R. Ballesteros, F.J. Martos, J.R. Agudelo, Characterization of light duty diesel engine pollutant emissions using water-emulsified fuel. Fuel **84**(7–8), 1011–1018 (2005)

Chapter 9
Catalytic Pyrolysis of Mixed Plastic Waste Using Synthesized Composite Catalyst

Prathiba Rex

Abstract Thermocatalytic pyrolysis was used to degrade two types of plastic waste, such as polystyrene waste (PSW) and polypropylene waste (PPW) and their mixtures. The product yield (in wt%) of oil, gas and residue was obtained, and the oil properties were determined. Three types of catalysts were used: kaolin, hydrochloric acid-treated bentonite and a composite catalyst, Ni:Al:Mg (equal ratio). The composite catalyst was prepared by using the precipitation method. For the pyrolysis of single and mixed plastic waste, all three catalysts were used. It was found that acid-treated bentonite and composite catalysts slightly increased the oil yield of mixed plastic waste. The product yield of the synthesized Ni:Al:Mg catalyst was 86.43 wt% of oil yield, 13.43 wt% of gas yield and 0.2 wt% of solid residue at a ratio of 10:1 wt/w (polymer:catalyst). It was noted that the addition of catalyst reduced the reaction temperature from 482 to 475 °C. The oil properties were compared to the diesel values of the Bharat Stage IV standard. It is highlighted that the addition of polypropylene waste to polystyrene waste significantly reduced the aromatic content of the oil.

Keywords Pyrolysis · Polystyrene · Polypropylene · Catalyst · Density · Viscosity

9.1 Introduction

Plastic has a major role in our daily lives, which has increased the production rate of plastics. It is durable, flexible, thermally stable and energy efficient. Plastics' rapid growth leads to an increase in plastic waste in landfills. Almost 300 million tonnes of plastic are produced annually and widely used. The majority of plastic waste is buried or dumped in the ocean, with very little recycled [1]. Plastics are non-biodegradable and take approximately 100 years to degrade, which causes serious degradation to

P. Rex (✉)
Institute of Agricultural Engineering, Saveetha School of Engineering, SIMATS, Chennai 602 105, India
e-mail: prathireeta@gmail.com

N. Gakkhar et al. (eds.), *Recent Advances in Bio-Energy Research*, Springer Proceedings in Energy, https://doi.org/10.1007/978-981-99-5758-3_9

the environment [2]. Hence, to mitigate plastic pollution, environmentally sound methods should be developed. Nowadays, the recovery of energy from plastic waste is an efficient method to reduce the plastic waste load on treatment facilities, and it is also economically viable [3]. Thermoplastics are reusable and can be converted into liquid form; some examples are polyethylene, polypropylene, polystyrene and low-density polyethylene. Thermoset's plastics have a cross-linked structure and are irreversible, e.g. epoxy resin, melamine formaldehyde, polyester and polyurethane. Thermoplastics can be used to reuse, recycle and reduce, but thermoset plastics cannot be recycled due to their irreversibility. There are different methods to reduce plastic waste, such as landfilling, incineration, gasification and hydrogenation. Landfilling is suitable for biodegradable waste; plastics are non-biodegradable and cause air pollution. They release harmful gases such as CO_2 and CH_4, leach heavy metals into the ground and generate toxic chemicals [4]. Incineration is an exothermic process in which the waste materials are burned at very high temperatures to generate heat for electricity and release flue gas, CO_2 and a stream. Generation of air pollutants such as nitrogen (NO_x), sulphur dioxide (SO_2) and dioxins requires waste disposal [5]. Gasification is a partial combustion process that converts waste into syngas ($CO + H_2$), CO_2, H_2O, CH_4 and hydrocarbons. The disadvantage of this process is that it produces a lot of tar and generates a lot of H_2S and HCl. Hydrogenation is a process to convert the feedstock (plastic) to syncrude oil under high pressure (100 bar) in the presence of hydrogen, and it produces pollutants like HCl and residues [6].

Pyrolysis is a process that decomposes materials at high temperatures in the absence of air to convert complex molecules into simpler ones and produce solids, liquids and gases. This process depends on the operating conditions such as pyrolysis temperature, catalyst, heating rate, composition of feedstock and residence time. It is a promising technique to reduce plastic waste [7]. Low-density polyethylene (LDPE) was used for pyrolysis to produce jet fuel and hydrogen. The condensed pyrolyzed vapours were passed through magnesium oxide and biomass-derived activated carbon, and it was noted that the catalyst was more efficient and cost-effective [6]. The influence of heating rate and vacuum condition on the composition of the fuel product was studied in the pyrolysis of polypropylene waste. From the study, it was noted that light and heavy fractions of hydrocarbons were obtained. At 525 °C, the oil yield was 93 wt% and the heating value was 46 MJ/kg [8]. Sivagami K [9] studied the catalytic pyrolysis of various types of polyolefin and multilayer plastics such as polyethylene terephthalate (PET), metalized polyethylene terephthalate (MET/PET), biaxially oriented polypropylene (BOPP), metalized biaxially oriented polypropylene layers (MET/BOPP) and combined PET and PE with polyolefin. 65–70% of the oil yield with a calorific value of 45.24 kJ/g was obtained in BOPP.

Merve Soganciogl u [10] investigated the quantity and energy recovery potential of waste plastics pyrolysis liquid products. Pyrolysis temperature (300–700 °C), oil yield, unwashed, pre-washed and heat value were examined for plastics like polyethylene terephthalate (PET), polystyrene, high-density polyethylene (HDPE), low-density polyethylene and polypropylene. HDPE has a higher yield and a wide range of hydrocarbons from C_{10} to C_{40}. The heat values of HDPE and PP increased in prewashing; no effect was observed in LDPE, PET or PS. Anuar Sharuddin [11] investigated the energy recovery from plastic waste pyrolysis: A study on non-recycled plastics (NRP) as a true measure of plastic waste. With different weight percentages of PS, a fixed weight mixture of plastics (HDPE, 26.2%, LDPE, 31.1%), PP, 8.2% and PS, 13.0%) is used. The maximum amount of oil was obtained at 100 wt.% of PS; it was understood that oil yield increased with an increase in PS in waste mixtures.

From the literature, it reveals that there are limited studies for the pyrolysis of plastic mixtures using composite catalyst hence this study was conducted to experiment the different catalysts that can be used for plastic and its mixtures. The product oil was also ascertained to determine its usage as an alternative for fuel production.

9.2 Experimental Methods

9.2.1 Materials

Polystyrene waste is the rulers, and polypropylene waste is the plastic syringes; they are shredded, washed, and dried. Kaolin, bentonite and dried magnesium sulphate were procured from Sisco Research Laboratories (SRL). Nickel sulphate hexahydrate was purchased from Sigma Aldrich. All the chemicals used are of analytical grade, and the solution was prepared using distilled water.

9.2.2 Preparation of Acid Bentonite Catalyst

100 g of bentonite were mixed with 2 L of 1 M HCl and stirred continuously for 24 h at room temperature. The mixture was centrifuged and washed with distilled water until a neutral pH was reached, and then filtered and dried at 105 °C for 24 h [12]. Figure 9.1 shows the bentonite catalysts before acid treatment and after acid treatment.

9.2.3 Preparation of Composite Catalyst

A composite catalyst was prepared by using the precipitation method shown in Fig. 9.2. Ni:Al:Mg (1:1:1). The catalyst was made up of equal parts 1 M $NiSO_4 \cdot 6H_2O$, $Al(SO_4)_3 \cdot 18H_2O$ and $MgSO_4$. They were mixed together, and 1 molar of NH_4OH was added until the pH value reached 8.3. For 12 h, the precipitate was stirred at 350 rpm at 40 °C. Then, the mixture was cooled to room temperature, washed with distilled water and centrifuged to remove unreacted molecules. The mixture was then dried at 105 °C for 24 h and calcined at 750 °C for 3 h [13].

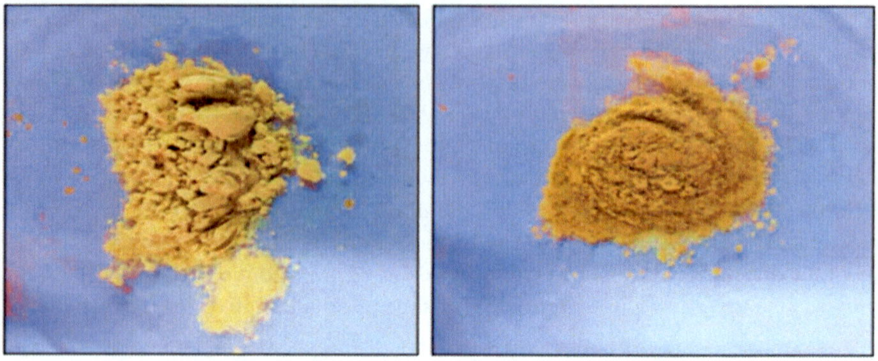

(a) (b)

Fig. 9.1 **a** Bentonite **b** acid-treated bentonite

(a) (b)

Fig. 9.2 **a** Composite catalyst before calcination **b** composite catalyst after calcination

9.2.4 Recyclability Experiments

Xylene was used to regenerate the catalyst. To remove oil, xylene acts as a clearing agent. The catalyst was soaked in 30 ml of xylene for 10 min after use. The catalyst was filtered and thoroughly washed with distilled water after 10 min to remove any residual xylene. The catalyst was then dried in an oven and calcined at 750 °C before being named "RC" (recycled catalyst) [14].

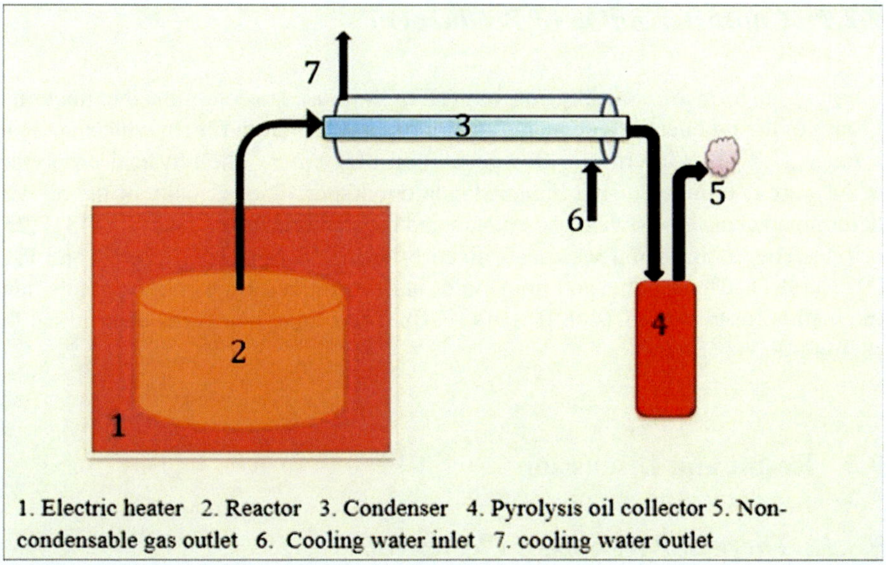

Fig. 9.3 Pyrolysis experimental setup

9.2.5 Pyrolysis Experiments

Figure 9.3 depicts the pyrolysis experiment of plastics. An electrical heater was used as a heating source where the reactor is heated at a higher temperature to decompose the plastic, and an outlet at the top of the reactor was connected to the condenser where the cooling water was circulated, which condensed the volatile compounds. The condensed liquid was collected in a pyrolysis oil collector and stored. Residue was collected at the bottom of the reactor.

9.2.6 Characterization of Catalyst

The surface area properties of the catalyst were analysed using a porosimeter. In order to determine the total acidity, catalyst was mixed with 0.1 N of NaOH (1:40 w/v) and stirred for 3 h. Two drops of phenolphthalein indicator were added to the mixture and then titrated with 0.1 N of HCl solution [15]. Equation 9.1 gives the total acidity equation, where V_1 and N_1 are the volume and normality of NaOH, V_2 and N_2 are the volume and normality of HCl, and M1 is the mass of catalyst in g.

$$\text{Total acidity, (mmol/g)} = \frac{v_1 N_1 - v_2 N_2}{M_1} \tag{9.1}$$

9.2.7 *Characterization of Product Oil*

Fourier transform infrared spectroscopy (FTIR) was used to determine the functional groups of the product oil. The manufacturer of FTIR is Perkin Elmer, which can scan a range of 4000–450 cm^{-1} with a resolution of 1 cm^{-1}. The physical properties of oil were determined using Indian standard methods. The viscosity of the oil was determined using an Ostwald viscometer, and the standard method was IS:448 [P:25] (1976). The aniline point was determined by using IS:1448 [P:3] (2007) and ISO 2977 (1997). Cloud point and pour point are determined by using apparatus, and the method used was IS:1448 [P:10] (1970). The smoke point was determined by IS:1448 [P:31] (1968).

9.3 Result and Discussion

9.3.1 *Thermal Pyrolysis of Plastic Waste*

50 g of samples were used for each pyrolysis process. From Table 9.1, it can be seen that the thermal pyrolysis of polystyrene waste produced 85.76 wt% of oil yield. The addition of PSW to polypropylene waste increased the oil yield and reduced the gas yield. This is due to the reduction of long-chain polymer molecules, which has resulted in an increase in oil yield [16]. It was also noted that the production of non-condensable gases will be higher in the pyrolysis of PPW. The reaction temperature of PPW pyrolysis was 496 °C, and in the literature, it was studied that an increase in the reaction temperature would promote the production of gases. It is clear that there is a cross-interaction between the reaction intermediates and the volatiles in mixed plastic of PSW + PPW; additionally, cofeeding of PSW + PPW increases the gas yield while decreasing the oil yield [17].

Table 9.1 Yield of thermal pyrolysis

Plastic	Wt. (g)	Temperature (°C)	Time (min)	Oil yield (wt%)	Residue yield (wt%)	Gas yield (wt%)
PSW	100	468	66	85.76	0.77	13.47
PPW	100	496	74	74.3	0.01	25.69
PSW + PPW	50geach	482	67	81.3	0.9	17.8

Table 9.2 Characteristics of catalyst

Catalyst	Surface area (m^2/g)	Pore size (nm)	Total acidity (mmol/g)
Kaolin	46.8	21.34	0.010
Acid-treated bentonite	128.13	37.80	0.112
Ni:Al:Mg	133.76	29.35	0.135

9.3.2 Catalytic Pyrolysis

In catalytic pyrolysis, the ratio was maintained at 10:1 (polymer:catalyst) for all the experiments. The use of kaolin catalyst has slightly increased the oil yield of PSW and its mixtures. A slight reduction in gas yield was also noted with the use of kaolin as a catalyst. This enunciates that the addition of kaolin catalyst will enhance the oil production since the catalyst has a well-formed porous structure and acid values for the degradation reaction to occur, which is shown in Table 9.2 [18]. Product yield of catalytic pyrolysis is given in Table 9.3. It should be noted that the composite catalyst Ni:Al:Mg showed a good increase in oil yield for plastic mixtures. This can be due to a reduction in reaction temperature from 482 to 475 °C, since a lower reaction temperature will increase the production of oil. The residue yield of catalytic pyrolysis shows a slight increase compared to thermal pyrolysis; the reason will be the clogging of pores after multiple reactions take place on the surface of the catalyst. In general, all three catalysts have shown a good increase in oil yield and a slight increase in residue content. Figure 9.4 shows the pyrolytic oil obtained from different types of catalysts. Recyclability experiments were performed to ensure that the catalyst could withstand multiple experimental runs. Recyclability experiments were carried out in accordance with Sect. 2.4. Kaolin recycled catalyst was able to retain an oil yield of 87.35, 2.16 wt% of residue and 10.49 wt% of gas yield. It was observed that for the first two recyclable runs, the kaolin catalyst retained its oil yield, and thereafter, gas yield started to increase. This increase in gas yield can be due to the weakening of pores and active sites [14].

9.3.3 FTIR Analysis of Product Oil

Fourier transform infrared spectroscopy is a technique for determining a sample's chemical composition and physical state. It is used to identify the unknown components and also quantify the functional group of the sample using a standard reference. Figure 9.5 shows the peak at 3695, 3070, 3075 and 3017 cm^{-1} that corresponds to O–H stretching [19]. The peaks at 2840 and 2912 cm^{-1} contain C–H stretching that implies the presence of an alkane group. Be consistent with N=N=N stretching, which contains the azide group, at 2156 cm^{-1}. 1741 and 1729 cm^{-1} correspond to CO stretching. A peak of 1372 cm^{-1} resembles S–O stretching, which represents the sulfonate group. The peak at 1202 cm^{-1} corresponds to C–O stretching,

Table 9.3 Yield of catalytic pyrolysis

Plastic	Catalyst	Ratio	Temperature °C	Time min	Oil yield (wt)	Residue yield (wt%)	Gas yield (wt%)
PSW	Kaolin	10:1	472	69	89.28	1.5	9.22
PPW			506	78	78.96	0.12	20.92
PSW + PPW			491	69	82.86	0.26	16.88
PSW	Acid-treated bentonite	10:1	466	71	87.22	0.16	12.62
PPW			490	77	77.67	0.05	22.28
PSW + PPW			479	71	86.94	0.19	12.87
PSW	Ni:Al: Mg	10:1	469	72	86.69	1.33	11.98
PPW			503	80	77.48	0.15	22.37
PSW + PPW			475	70	86.37	0.2	13.43

(a) **(b)** **(c)**

Fig. 9.4 Pyrolytic oil obtained from **a** PSW + PPW + Kaolin **b** PSW + PPW + acid treated bentonite **c** PSW + PPW + Ni:Al:Mg

which contains the ester group. 985 cm1 is associated with C–C bending, indicating the presence of an alkene group. 884 cm1 and 770 cm^{-1} correlate with C–H bending. Properties of oil describe the oil quality, ignition quality, aromaticity, ignition delay period, heavy/lighter fraction, engine performance, suitability of oil at cold temperatures and flow of oil.

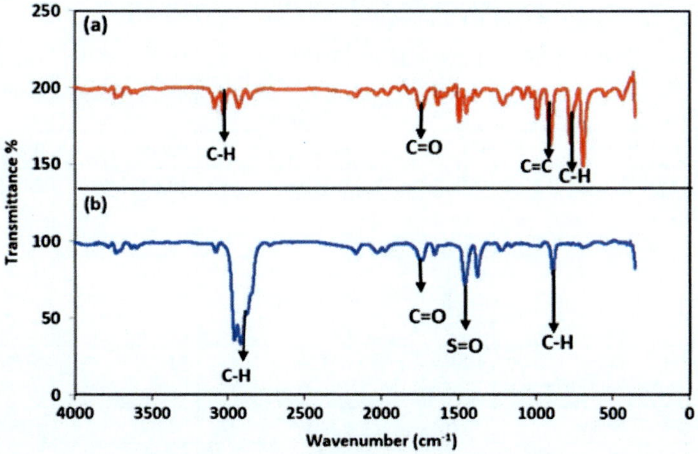

Fig. 9.5 FTIR analysis of **a** PSW product oil and **b** PPW product oil

9.3.4 Properties of Oil

Properties of oil describe the oil quality, ignition quality, aromaticity, ignition delay period, heavy/lighter fraction, engine performance, suitability of oil at cold temperatures and flow of oil. Table 9.4 shows the properties of oil, in which the density of PPW pyrolysis oil from both the thermal and catalytic processes have a lower value than diesel, which implies the presence of lighter hydrocarbons. Lower kinematic viscosity pyrolysis oils than diesel provide better fuel atomization. Aniline points greater than 200 are paraffinic, lower than 150 are aromatic, and in between these two temperatures are naphthene and olefins, indicating that PSW pyrolysis oil has a higher aromatic content than PPW pyrolysis oil [20]. PPW oil has a higher API gravity than diesel, whereas PSW oil has a lower API gravity than diesel. The diesel index provides the ignition quality of oil; if the diesel index value is high, it implies the presence of a higher quantity of paraffin, and a lower value shows a higher aromatic content. Diesel with an index greater than 50 burns oil completely and is more expensive. The cetane number of PPW is higher than diesel, which implies good fuel atomization. The cloud point and pour point for the oils are lower, which indicates a lower operating condition. The flash and fire points of pyrolysis oil are lower than standards, which gives better ignition quality.

Table 9.4 Oil properties of mixed plastic waste

Properties	PSW + PPW	PSW + PPW + Kaolin	PSW + PPW + acid treated catalyst	PSW + PPW + Ni:Al:Mg	Diesel (Bharat stage IV 2017)
Density (g/ml)	0.884	0.862	0.872	0.870	0.815–0.845
Kinematic viscosity (cSt) @ 40 °C	1.102	0.980	1.103	1.064	2–4.5
Specific gravity @ 15 °C	0.929	0.847	0.868	0.861	0.84
Aniline point (°F)	98.6	109.4	113	111.2	>120
API gravity	20.81	35.52	31.61	32.82	30–40
Diesel index	20.52	38.85	35.72	36.49	<55
Cetane number	24.36	37.20	35.0	35.54	51
Cloud point (°C)	−5	−5	−5	2	–
Pour point (°C)	<−10	<−10	<−10	−1	3
Flash point (°C)	30	30	30	30	66
Fire point (°C)	34	34	34	34	–

9.4 Conclusion

Pyrolysis of plastic wastes such as polystyrene (PSW), polypropylene (PPW) and mixtures of PSW and PPW was conducted with thermal and catalytic (kaolin, acid-treated bentonite and Ni:Al:Mg) processes. Three types of catalysts were used, including natural clay kaolin and HCl acid-treated bentonite clay. Among these three catalysts, kaolin provided a better oil yield with a lower gas yield. The oil yield of PSW in both thermal and catalytic pyrolysis was higher than that of PPW and PSW + PPW. The properties of pyrolysis oil were discovered to be slightly similar to diesel values. The properties of PSW pyrolysis oils imply a larger aromatic content, whereas PPW contains a higher paraffinic content. Thus, this study highlights the use of plastic waste to produce oils, which can be used for fuel production.

References

1. D.K. Ratnasari, M.A. Nahil, P.T. Williams, Catalytic pyrolysis of waste plastics using staged catalysis for production of gasoline range hydrocarbon oils. J. Anal. Appl. Pyrolysis (2017). https://doi.org/10.1016/j.jaap.2016.12.027
2. T. Sangpatch, N. Supakata, V. Kanokkantapong, B. Jongsomjit, Fuel oil generated from the cogon grass-derived Al–Si (Imperata cylindrica (L.) Beauv) catalysed pyrolysis of waste plastics. Heliyon **5**, e02324 (2019). https://doi.org/10.1016/j.heliyon.2019.e02324
3. S.L. Wu, J.H. Kuo, M.Y. Wey, Thermal degradation of waste plastics in a two-stage pyrolysis-catalysis reactor over core-shell type catalyst. J. Anal. Appl. Pyrolysis **142**, 104641 (2019). https://doi.org/10.1016/j.jaap.2019.104641
4. R. Prathiba, M. Shruthi, L.R. Miranda, Pyrolysis of polystyrene waste in the presence of activated carbon in conventional and microwave heating using modified thermocouple. Waste Manag. **76**, 528–536 (2018). https://doi.org/10.1016/j.wasman.2018.03.029
5. S.M. Al-Salem, Feedstock and optimal operation for plastics to fuel conversion in pyrolysis (Elsevier Inc., 2018)
6. N. Premalatha, R. Prathiba, M. Angelo, et al., Pyrolysis of polypropylene waste using sulfonated carbon catalyst synthesized from sugarcane bagasse. J. Mater. Cycles Waste Manag. (2021). https://doi.org/10.1007/s10163-021-01188-6
7. P. Rex, V. Ganesan, V. Sivashankar, S. Tajudeen, Prospective review for development of sustainable catalyst and absorbents from biomass and application on plastic waste pyrolysis. Int. J. Environ. Sci. Technol. (2022). https://doi.org/10.1007/s13762-022-04292-8
8. G.K. Parku, F.X. Collard, J.F. Görgens, Pyrolysis of waste polypropylene plastics for energy recovery: influence of heating rate and vacuum conditions on composition of fuel product. Fuel Process Technol. **209**, 36–38 (2020). https://doi.org/10.1016/j.fuproc.2020.106522
9. K. Sivagami, G. Divyapriya, R. Selvaraj et al., Catalytic pyrolysis of polyolefin and multilayer packaging based waste plastics: a pilot scale study. Process. Saf. Environ. Prot. **149**, 497–506 (2021). https://doi.org/10.1016/j.psep.2020.10.038
10. M. Sogancioglu, G. Ahmetli, E. Yel, A comparative study on waste plastics pyrolysis liquid products quantity and energy recovery potential. Energy Procedia **118**, 221–226 (2017). https://doi.org/10.1016/j.egypro.2017.07.020
11. S.D. Anuar Sharuddin, F. Abnisa, W.M.A. Wan Daud, M.K. Aroua, Energy recovery from pyrolysis of plastic waste: Study on non-recycled plastics (NRP) data as the real measure of plastic waste. Energy Convers Manag **148**, 925–934 (2017). https://doi.org/10.1016/j.enconman.2017.06.046
12. F.M.T. Luna, J.A. Cecilia, R.M.A. Saboya et al., Natural and modified montmorillonite clays as catalysts for synthesis of biolubricants. Materials (Basel) **11**, 6–9 (2018). https://doi.org/10.3390/ma11091764
13. C. Wu, P.T. Williams, Investigation of coke formation on Ni-Mg-Al catalyst for hydrogen production from the catalytic steam pyrolysis-gasification of polypropylene. Appl. Catal. B Environ. **96**, 198–207 (2010). https://doi.org/10.1016/j.apcatb.2010.02.022
14. P. Rex, L.R. Miranda, Catalytic activity of acid-treated biomass for the degradation of expanded polystyrene waste. Environ. Sci. Pollut. Res. **27**, 438–455 (2020). https://doi.org/10.1007/s11356-019-07034-5
15. K. Lee, Journal of analytical and applied pyrolysis effects of the types of zeolites on catalytic upgrading of pyrolysis wax oil. J. Anal. Appl. Pyrolysis **94**, 209–214 (2012). https://doi.org/10.1016/j.jaap.2011.12.015
16. D. Li, S. Lei, P. Wang et al., Study on the pyrolysis behaviors of mixed waste plastics. Renew Energy **173**, 662–674 (2021). https://doi.org/10.1016/j.renene.2021.04.035
17. D.V. Suriapparao, R. Vinu, Resource recovery from synthetic polymers via microwave pyrolysis using different susceptors. J. Anal. Appl. Pyrolysis **113**, 701–712 (2015). https://doi.org/10.1016/j.jaap.2015.04.021

18. W. Luo, Q. Hu, Z.y. Fan, et al., The effect of different particle sizes and HCl-modified kaolin on catalytic pyrolysis characteristics of reworked polypropylene plastics. Energy **213**, 119080(2020). https://doi.org/10.1016/j.energy.2020.119080
19. E. Weidemann, W. Buss, M. Edo, et al., Correction to: Influence of pyrolysis temperature and production unit on formation of selected PAHs, oxy-PAHs, N-PACs, PCDDs, and PCDFs in biochar—a screening study. Environ. Sci. Pollut. Res. **25**(4), 3933–3940 (2018), https://doi.org/10.1007/s11356. Environ. Sci. Pollut. Res. **25**, 3941–3942. https://doi.org/10.1007/s11356-017-0804-6
20. M.S.N. Awang, N.W.M. Zulkifli, M.M. Abbas et al., Effect of diesel-palm biodiesel fuel with plastic pyrolysis oil and waste cooking biodiesel on tribological characteristics of lubricating oil. Alexandria Eng J **61**, 7221–7231 (2022). https://doi.org/10.1016/j.aej.2021.12.062

Chapter 10
Utilization of Karanja Seed Shells as a Sustainable Heterogeneous Catalyst for Biodiesel Production

Pooja Prajapati, Sakshi Shrivastava, Varsha Sharma, Priyanka Srivastava, Virendra Shankhwar, Arun Sharma, S. K. Srivastava, and D. D. Agarwal

Abstract Development of inexpensive catalysts for biodiesel production has long been favored by researchers. In the present study, the ash from burned discarded Karanja seed shells (KSS) has been found to be one of the most economically and environmentally sustainable materials as a heterogeneous base catalysts for biodiesel production. The catalytic activity of Karanja seed shells ash as a heterogeneous catalyst in biodiesel production was studied using soybean oil methanolysis. Various methods characterized catalysts, such as XRD, WD-XRF, SEM–EDX, FT-IR, BET, and TGA. Soybean oil biodiesel was converted under the following experimental conditions, as determined by ^{1}H NMR, FT-IR, and GC–MS: a catalyst dose of 2 wt%, methanol to oil molar ratio of 10:1, a reaction temperature of 65 °C, a reaction time of 60 min. The reuse of a catalyst was also evaluated, keeping a high biodiesel yield for up to four cycles.

Keywords Karanja seed shell · Transesterification · Heterogeneous catalyst · Soybean oil · Biodiesel

10.1 Introduction

Consumption of fossil fuels, which have been depleted due to rapid scientific and technological progress particularly industrialization is increasing rapidly, as is the energy demand. As a result, researchers are looking for green and clean fuels as viable sources of alternative energy because of current environmental concerns.

P. Prajapati (✉) · S. Shrivastava · V. Sharma · A. Sharma · S. K. Srivastava · D. D. Agarwal
School of Studies in Chemistry, Jiwaji University, Gwalior 474001, India
e-mail: poojakrg92@gmail.com

P. Srivastava
Coventry University, Coventry, UK

V. Shankhwar
Central Instrumentation Facilities, Jiwaji University, Gwalior 474001, India

N. Gakkhar et al. (eds.), *Recent Advances in Bio-Energy Research*, Springer Proceedings in Energy, https://doi.org/10.1007/978-981-99-5758-3_10

Carbon dioxide, sulfide, and nitrogen oxides are among the dangerous gases emitted by traditional fossil fuel combustion. These pollutants have a variety of negative health and environmental effects. Clean and environmentally friendly fuels are being researched as a result of the above issues [1–7]. Biodiesel, being a renewable fuel, adapts as a great alternative to fossil fuels because it is environmentally friendly [8–10]. Biodiesel provides several benefits due to its high cetane number, high flash point, excellent lubricity, low viscosity, non-toxicity, biodegradability, and produces low greenhouse gases. Additionally, without requiring any modification, it can be used straight in diesel engines [11, 12].

Biodiesel, like conventional diesel, is derived chemically from the triglycerides of vegetable or animal oils and alcohol. Biodiesel production depends on either homogeneous or heterogeneous acidic or basic catalysts [13–16]. The catalyst is essential to the production of biodiesel. Typical transesterification methods use homogeneous basic catalysts, such as NaOH or KOH, which are poisonous, corrosive, and chemically manufactured [17]. Furthermore, homogenous catalysts drive production costs, produce wastewater, reduce biodiesel yield, and cannot be recycled. While low-cost, solid heterogeneous catalysts with high catalytic activity may accelerate the production of green biodiesel. In this context, waste utilization improves solid waste disposal and pollution reduction while being cost-efficient. The use of heterogeneous catalysts derived from biomass waste is becoming increasingly common to produce biodiesel.

Waste biomass ash catalysts that can be used to produce biodiesel have been evaluated in numerous studies, such as the use of banana peel ash, orange peel, pineapple leaves ash and sugarcane leaves ash, Brassica nigra plant ash, wheat bran ash, coconut husk ash, rice husk ash, palm oil mill boiler ash, biomass bottom ash and fly ash, Torrey ash, waste ginger leaves ash, moringa leaves ash, wood ash, Tamarindus indica fruit ash, walnut shell ash, hazelnut shell ash and acai seed ash among others [18–34]. These catalysts are typically produced by drying and burning biomass (at temperatures ranging from 350 to 900 °C). Milling, impregnation of active compounds, drying, and calcinations are all part of their preparation. Although incorporating the active element into the ash enhances the stability of these catalysts, the production process seems to be more difficult and requires the use of solvents. As a result, in the current work, simple combustion was used to obtain ash from discarded Karanja seed shells.

Karanja seed shells, environmentally and economically are a suitable choice for catalyst development due to their high alkaline (basic) content. Karanja, also known as Pongamia Pinnata, can easily grow on the edges of roadways, rivers, and agricultural boundaries with no maintenance. It is a medium-sized tree that develops in four to five years and is abundantly available in India. It is resistant to heat, drought, salinity, and cold. From the Himalayan foothills to Kanyakumari, it can be found in India and many other regions [35]. The Karanja tree plant has a variety of uses notably it is used in the production of soap, lamp fuel, finishing and tanning of leather, veterinary medication, and other products that are used to cure humans and animals. The oil of Karanja seed is utilized as a feedstock in the production of biodiesel. But karanja seed shells, which are elliptical, 2–3 cm broad, 3–6 cm long have a thick walls and contain a single seed, are discarded [35]. However in this research we used discarded Karanja seed shells as catalyst because they are rich in alkaline elements and act as a viable, long-lasting precursors to produce biodiesel. Particularly in rural India, the Karanja (Pongamia pinnata) tree's seed shells are a common agricultural waste.

As a consequence, the feasibility of KSS ash as a green heterogeneous catalyst for biodiesel production utilizing soybean oil has been investigated in the present work. The catalyst was analysed to illustrate how the structure, elemental composition, and morphology impacted the production of biodiesel. Investigations were conducted into how reaction conditions affected the transesterification reaction. In addition, four reuse cycles were performed to assess the catalyst's reusability. Thus, it has a promising possibility for biodiesel production. As far as we are aware, no one has discussed KSS ash as a heterogeneous catalyst for the production of biodiesel.

10.2 Materials and Methods

10.2.1 Materials

For the catalyst preparation, Karanja seed shells were collected from the Jiwaji University campus in Gwalior, Madhya Pradesh, India. A market in Gwalior, Madhya Pradesh, India, was visited to obtain soybean oil to test the proposed catalyst's catalytic performance. The methanol was acquired from Merck and was of HPLC quality (purity of 99%). Rankem's distilled water was used throughout the research. All chemicals were utilized without any purification.

10.2.2 Catalyst Preparation

The Karanja seed shells were collected and sun-dried for 10 days after being rinsed many times with distilled water to remove impurities. The Karanja seed shells were crushed and burned in the air to produce an ash catalyst. In a muffle furnace, the catalyst was calcined for 4 h at 250–850 °C. To keep the catalyst from coming into contact with air, it was placed in a desiccator.

10.2.3 Catalyst Characterization

The crystalline phases of the calcined KSS ash catalyst is investigated by an X-ray powder diffractometer (XRD) on a 5th-generation Rigaku (Model No. Mini flex 600). X-ray fluorescence (WD-XRF, PANalytical spectrometer, AxiosMAX, The Netherlands) was employed to determine the elemental composition of the catalyst.

Scanning electron microscopy (SEM) and energy dispersive X-ray (EDX) is used to measure the surface morphology and elemental distribution of the catalyst on the Carl Zeiss Ultra Plus model. A Fourier transform infrared spectrometer (FT-IR; serial number 105627) from Perkin Elmer was used to record the functional groups of the calcined KSS ash catalyst. The BET (Brunauer—Emmett—Teller) method was used to determine surface area, porosity, and pore diameter in N_2 gas using the BELSORP MAX equipment. To ascertain the thermal decomposition of the catalyst, a thermogravimetric instrument (Shimadzu TGA50 series), was used. And the basicity of the KSS ash catalyst was evaluated by the Hammet indicator titration method using benzoic acid [15].

10.2.4 Transesterification Procedure

There were three-necked 250 mL round-bottom flasks with refluxing condensers and temperature-regulated magnetic stirrer used for transesterification. At room temperature, the magnetic stirrer was used to swirl the round bottom flask for 10 min to ensure that the catalyst and methanol were homogeneously mixed. Then the soybean oil was poured into a round-bottom flask. At 65 °C and steady stirring, the transesterification reaction takes 60 min. After cooling, the mixture was transported to a three-layer separation funnel for further processing. Top to bottom, the layers were dominated by FAME (fatty acid methyl ester), followed by glycerin and catalyst. The top fraction's unreacted methanol was evaporated at a lower pressure. Before drying, the liquid was properly washed to remove any contaminants and then used to make soybean oil methyl esters.

10.2.5 Biodiesel Analysis

The produced biodiesel was tested using ^{1}H NMR (Nuclear magnetic resonance spectroscopy), GC–MS (Gas chromatography-mass spectrometry), FT-IR, and ASTM (American society for testing and materials) standards. In the presence of $CDCl_3$, a solvent, a JEOL ECZ500R/S1 (500 MHz) spectrometer recorded ^{1}H NMR spectra of the soybean oil and formed biodiesel. To check the conversion of triglycerides into Fatty Acid Methyl Esters (FAME), GC–MS (Clarus*680 GC, Clarus*SQ8C MS) was used. Functional groups were studied using the Perkin Elmer's FT-IR spectrophotometer (Serial No. 105627).

10.3 Results and Discussion

10.3.1 Catalysts Characterization

Powder XRD Analysis

X-ray diffraction spectrum of calcined KSS ash is given in Fig. 10.1. Various potassium (KCl, K_2O, and K_2CO_3) and oxide (CaO and SiO_2) compounds responsible for catalytic activity in the catalyst were found. Potassium was found in the catalyst in the forms of KCl, K_2CO_3, and K_2O, and was determined to be a key component. According to the XRD pattern, the peaks at 2θ values of 28.298, 40.427, 50.088, 58.601, 66.297, and 73.594 are attributable to KCl (JCPDS file no 41-1476). This discovery was made in a similar experiment by Vadery and colleagues, who found that 2θ values for KCl in coconut husk were identical [26]. K_2CO_3 (JCPDS file no 87-0730) is attributed to the peaks at 26.278, 29.734, 31.27, and 41.722, whereas K_2O (JCPDS file no 77–2176) is assigned to the peaks at 27.87, 38.792, 46.78, and 48.10. The presence of CaO was identified at $2\theta = 32.670$, 37.109, and 54.05 (JCPDS file no 82-1691) and the peaks at 21.283 and 43.296 confirm the existence of SiO_2 in the catalyst (JCPDS file no 81-0069). According to the XRD data, the catalyst contains a number of basic oxides and carbonates of K, Ca, and Si.

WD-XRF Analysis

The inorganic component of the calcined KSS ash catalyst was assessed using WD-XRF analysis, and the findings are shown K (51.199%), Ca (12.915%), P (9.686%), Cl (9.117%), Mg (4.088%), Na (0.560%), Al (0.324%), S (7.413%), and other elements were in abundance in the catalyst. Some transition metal oxides also coexisted with these metal oxides in trace amounts. From the XRF results we can conclude that K,

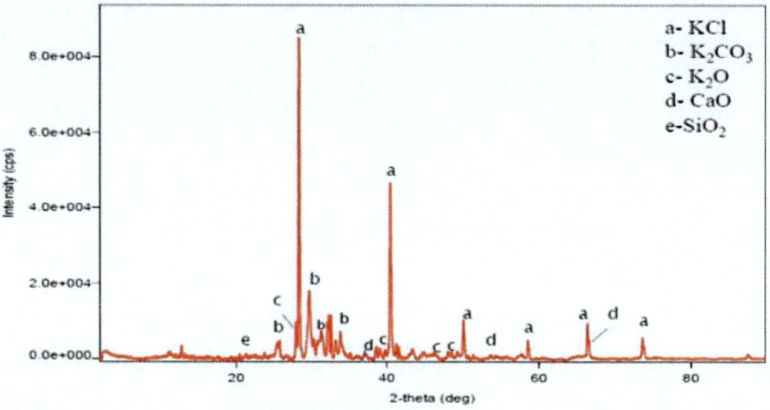

Fig. 10.1 XRD pattern of calcined KSS ash catalyst

is the primary basic metal accountable for catalytic activity in the transesterification reaction to produce the biodiesel [20], as proven by XRD analysis.

FT-IR Analysis

As depicted in the Fig. 10.2, in the calcined KSS catalyst OH groups are attributed to 3139 cm^{-1}, whereas CO stretching frequencies at 1646 and 1373 cm^{-1} are ascribed to CO stretching frequencies in the form of K_2CO_3 [40, 41], and CO bending frequency in carbonates was evident at 1117 cm^{-1}, which confirmed that K_2CO_3 was present [24, 37, 39]. The signal at 844 cm^{-1} may likely show the presence of the $CO_3{}^{2-}$ group [42]. Si–O–Si stretching band of SiO_2 is represented by a peak at 1041 cm^{-1}, while OH bending vibrations of water molecules adsorbing on the catalyst are represented by peaks at 616 cm^{-1}. The 532 cm^{-1} signal is caused by bond stretching vibrations of K–O and CaO, indicating the presence of these components in the catalyst. All of the peaks found in the calcined KSS ash catalyst are observed in the calcined ashes of banana peel, *B. nigra* leaves, cocoa pod husk, plantain peel, and *C. papaya* stem [19, 24, 36–38]. When K and Ca oxides were present, their catalytic activity was improved, as was demonstrated in this study.

SEM analysis. Figure 10.3a–c, reveals surface structure and morphology of the prepared KSS catalyst. Porous surface morphology and aggregation of large surface area particles suggest that the KSS-calcined catalyst has a large surface area. This micrograph displays the calcined catalyst's fibrous structure and spongy texture.

EDX analysis. An EDX test was performed to determine the quantitative element composition of the catalyst, which revealed that it includes various alkali and alkaline earth metals, including K (65.88%), Ca (25.55%), Si (3.34%), Mg (3.22%), Na (4.34%), Al (0.38%), and Ti (0.21%), as shown in Fig. 10.4. According to EDX

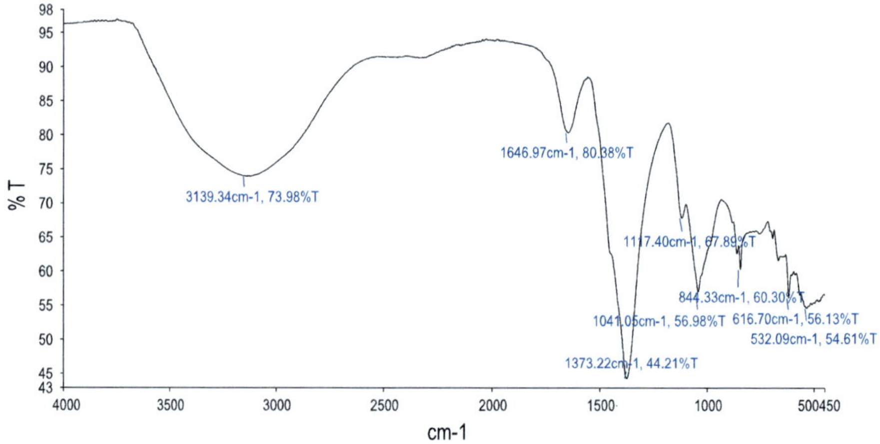

Fig. 10.2 FTIR pattern of calcined KSS ash catalyst

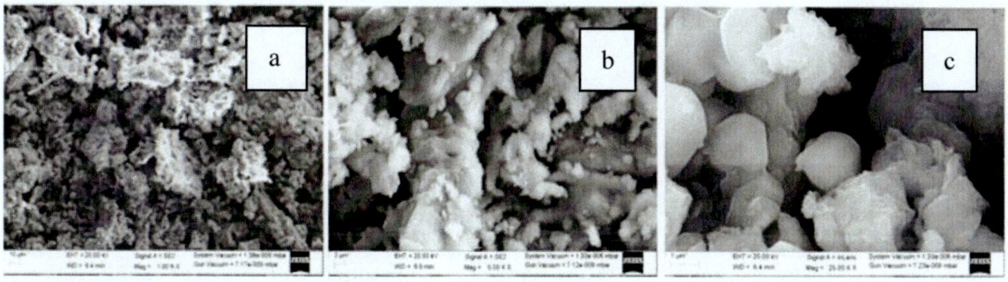

Fig. 10.3 SEM images (**a**–**c**) of calcined KSS ash catalyst

analysis, K is the most prevalent metal, followed by Ca. The outcomes of EDX and WD-XRF are indeed very similar.

TGA analysis. Thermal gravimetric analysis (TGA) was used to investigate the weight loss percentage of the calcined KSS ash catalyst. Figure 10.5, depicts the relationship between weight loss percentage and temperature as a whole. The weight loss of 22% upto 200 °C is due to KSS catalyst decomposition in an N_2 environment. The carbonaceous material of the calcined KSS ash catalyst undergoes oxidation with the generation of CO_2, CO, and other gases, resulting in a further drop in catalyst mass above 200 °C [20].

BET analysis. Surface area and pore structure had an impact on catalytic activity in the transesterification reaction. KSS has a total specific surface area of 4.2454 $m^2\ g^{-1}$, and pore volume of 0.0056302 $cm^3\ g^{-1}$. The catalyst's N_2 adsorption–desorption isotherm exhibited the Type-IV isotherm (Fig. 10.6), that is particularly closely followed by mesoporous material. A mesoporous material can increase the rate of reaction by dispersing the reactants throughout its pores. On the other hand,

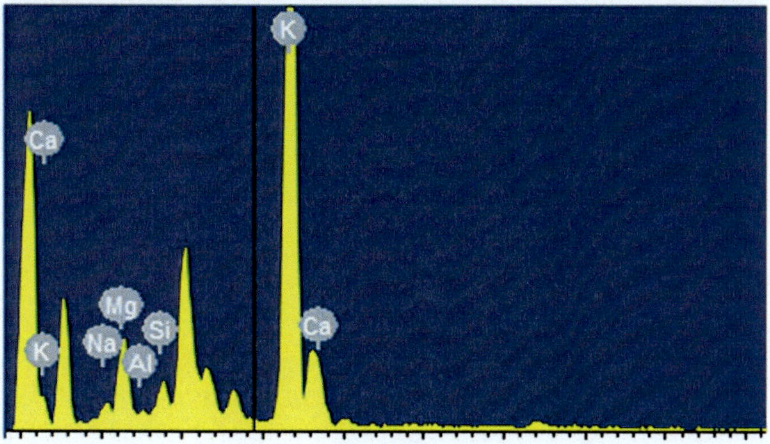

Fig. 10.4 EDX spectra of calcined KSS ash catalyst

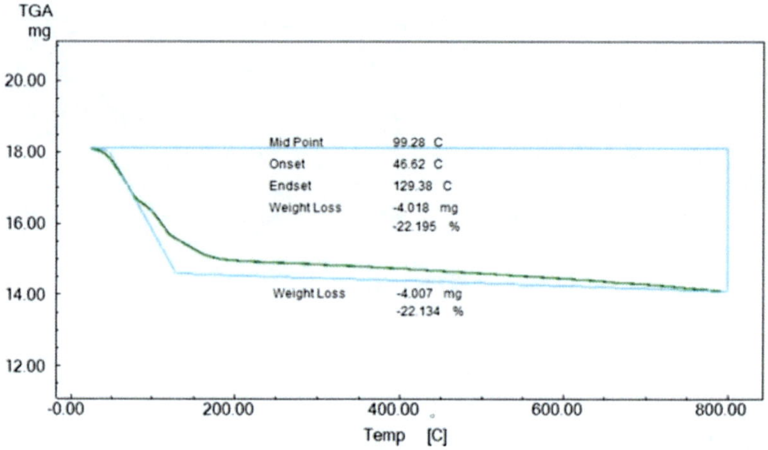

Fig. 10.5 TGA thermogram of calcined KSS ash catalyst

microporous materials have a lower reaction rate than mesoporous materials because reaction occurs at the pores' entrances [39, 48]. As a conclusion, this mesoporous calcined KSS ash catalyst has the potential to considerably improve reaction rate while biodiesel synthesis.

Basic strength analysis. For the Hammett indicator test to determine basicity, the results were based on color variation [43]. Transesterification can be carried out using this catalyst due to its high basic strength, as shown by the Hammett indicator test, which shows it to be in the $11.5 < H < 15$ range in strength. Its basic strength is comparable to walnut shell ash, and calcined KSS ash resembles walnut shell ash in terms of constituents, with minor differences in element percentages [31]. Increasing the basic strength of metal carbonate salts after calcination by heating them to active

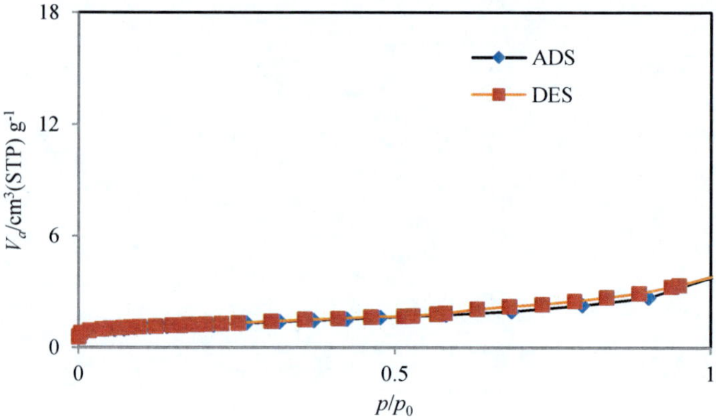

Fig. 10.6 N_2 adsorption–desorption isotherm Calcined KSS ash catalyst

metal oxide is possible [44]. Metal oxides in groups II and I have a higher basicity in terms of transesterification activity [45]. Active methoxy species can be formed when protons from MeOH are removed by strong bases [46]. The basicity of a catalyst has been shown to boost transesterification activity.

10.3.2 Catalytic Activity of Catalyst

Effect of calcinations temperature. Catalytic activity increases when an uncalcined catalyst is activated at 650 °C, but it decreases slightly when it is further calcined at 850 °C (Fig. 10.7). The catalysts carbonate (CO_3^{2-}) concentration dropped when the calcination temperature was raised to 850 °C; it could be linked to the partial or complete breakdown of K_2CO_3 to K_2O at higher temperatures (850 °C). Compared to K_2O, K_2CO_3 is more basic because of its weak acid and strong base nature, as well as the decrease in carbonate at higher calcinations temperatures [47].

Effect of catalyst loading. Transesterification reaction rate and biodiesel yields can be affected by catalyst loading. Figure 10.8 shows the results of the catalyst loading (0.5, 1, 1.5, 2, 2.5, 3, and 4) in a biodiesel production process using a KSS catalyst calcined at 650 °C with a 10:1 methanol to oil ratio at 65 °C. According to the findings, the KSS catalyst loading of 2 wt% resulted in the best results, yielding 96% biodiesel in a reaction time of 60 min compared to the other catalyst loads. When the catalyst loading was increased from 0.5 to 2wt% the reaction time was decreased from 180 to 60 min with a marginal rise in biodiesel yield from 55 to 96%. Moreover, despite increasing the catalyst loading by 4 wt% while maintaining the same reaction conditions, no noticeable improvement/boost in biodiesel yield and reaction time was observed. It is also possible that a further increase in catalyst loading may make the three-phase solution more viscous, restricting mass transfer between the phases and saponification side reactions can be produced, lowering oil conversion for base-catalyzed reactions. Accordingly, KSS catalyst concentrations of 2 wt% were found

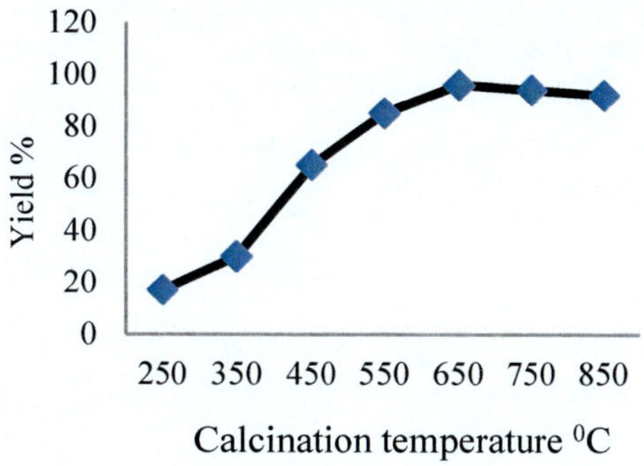

Fig. 10.7 Effect of calcination temperature on biodiesel yield

to be the best experimental conditions for soybean oil transesterification to biodiesel [47].

Effect of methanol to oil ratio. Methanol to oil ratio effects on biodiesel production was also investigated at 65 °C with an optimized reaction time 60 min and catalyst dose (2 wt%). Figure 10.9 demonstrates that raising the methanol-to-oil ratio from 3:1 to 15:1 increases biodiesel output by almost 96%. When the methanol to oil ratio is increased to 10:1, it does not reduce the reaction time further without a considerable increase in biodiesel production. This could be because transesterification is a reversible process that can be catalyzed both forward and backward by a base; however, due to high concentrations of methanol in the reaction mixture, reversible reactions occur, formulating monoglycerides and diglycerides, resulting in low oil to biodiesel conversion [20].

Reusability of the catalyst. Catalyst reusability research is required to establish the efficiency of utilizing a catalyst to minimize production costs. The KSS catalyst was tested for reusability under ideal conditions, which included a methanol to oil ratio of 10:1, 2 wt% catalysts loading, and a reaction time of 60 min at 65 °C temperatures. After the transesterification reaction, the product was centrifuged, cleaned, and dried overnight in an oven at 100 °C. After 4 h of calcination at 650 °C, the catalyst was

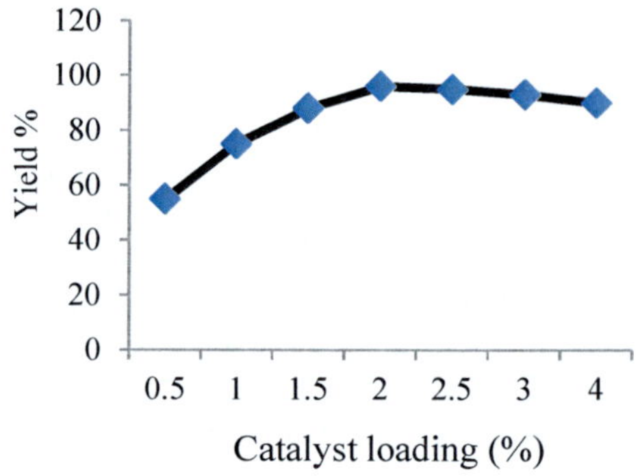

Fig. 10.8 Effect of catalyst loading on biodiesel yield

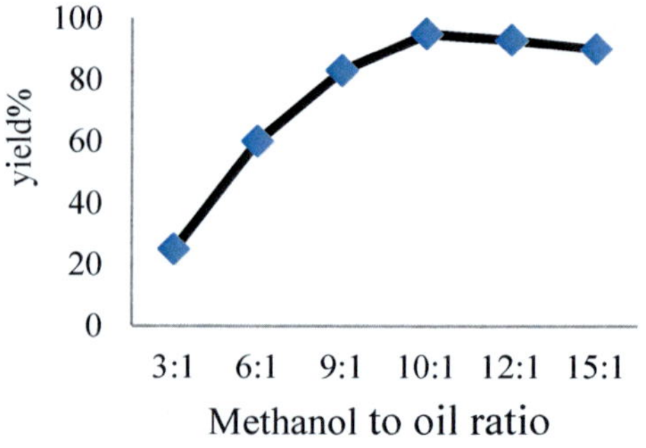

Fig. 10.9 Effect of methanol to oil ratio on biodiesel yield

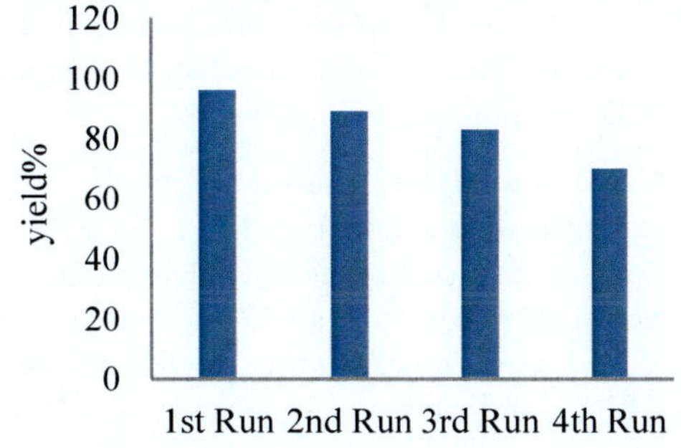

Fig. 10.10 Reusability of the KSS catalyst

reactivated. Then, using fresh reactants, a similar reaction was carried out. The experimental results, shown in Fig. 10.10, demonstrated that the yield dropped to about 70% after 4th run. As a result, biodiesel yields were found to have decreased significantly, possibly due to triglyceride contamination and pores clogged by glycerol and triglycerides on the catalyst surface. The catalyst loading was changed after each usage to achieve optimal reaction conditions. However, because of leaching during the washing phase, there was some catalyst loss in each cycle. This could explain the steady decline in yield after each catalytic cycle. Using SEM–EDX analysis of the reused catalyst, we discovered a significant decrease in K (59.79%) and Ca (13.89%) concentration, which resulted in a low biodiesel yield [20, 48].

10.3.3 Biodiesel Characterization

^{1}H NMR analysis. ^{1}H NMR spectroscopy validated the FAME (soybean oil biodiesel) conversion, as shown in Fig. 10.11(1) and (2).

In the ^{1}H NMR spectra of soybean oil biodiesel, the exclusion of glyceridic protons and other hydrocarbon protons in the region of 4.0–5.4 ppm, and the inclusion of a strong peak at 3.66 ppm (singlet of methoxy proton characteristics) and at 2.31 ppm

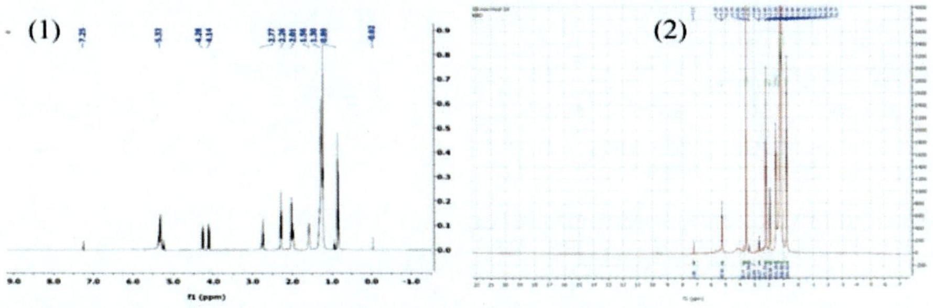

Fig. 10.11. ^{1}HNMR of (1) soybean oil and (2) soybean oil biodiesel

(a triplet of –CH_2 proton characteristics) [20], confirm the synthesis of biodiesel after transesterification of soybean oil to fatty acid methyl ester (FAME) by the calcined KSS ash catalyst [Fig. 10.11(2)].

GC–MS analysis. Analysis of biodiesel's chemical composition utilizing GC–MS testing is depicted in Fig. 10.12. The GC spectra reveals the five different methyl esters of fatty acids found in the prepared soybean oil biodiesel were methyl linoleate (Rt-21.93 min, C18:2, 43.23%), methyl oleate (Rt-20.38 min, C18:1, 7.57%), methyl palmitoleate (Rt-22.03 min, C16:0, 6.05%), methyl stearate (Rt-20.84 min, C18:0, 5.87%), and methyl nonadecan (Rt-23.18 min C20:0, 3.72%).

FT-IR analysis. Soybean oil's C=O stretching vibration at 1744 cm^{-1} is replaced by the methyl esters in the biodiesel's 1739 cm^{-1} peak, indicating that the transesterification process has gone through an intermediate step (Fig. 10.13). As a result of C–H stretching vibrations, soybean oil has peaks at 2923 cm^{-1} and 2854 cm^{-1}, while biodiesel has peaks at 2924 cm^{-1} and 2854 cm^{-1}. At 1463 and 1377 cm^{-1}, the CH_3 bending vibrations can be seen in soybean oil; at 1456 and 1436 cm^{-1} in biodiesel. The IR peaks at 1196 cm^{-1} and 1170 cm^{-1} for biodiesel and 1160 cm^{-1} and 1098 cm^{-1} for soybean oil demonstrate triglyceride and ester molecule C–O stretching bands, respectively. Methyl ester molecules emitted an infrared signal at 722 cm^{-1} and 723 cm^{-1} due to the rocking of fatty acid chains [24].

Physico-chemical properties. Soybean oil biodiesel's physicochemical qualities have been studied using established techniques, as shown in the Table 10.1. The properties of the biodiesel fuel made from soybean oil are well within acceptable ranges.

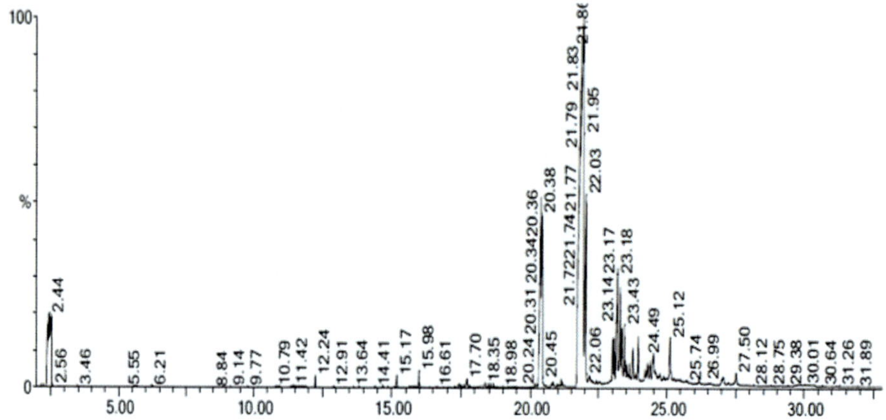

Fig. 10.12 GC spectra of soybean oil biodiesel

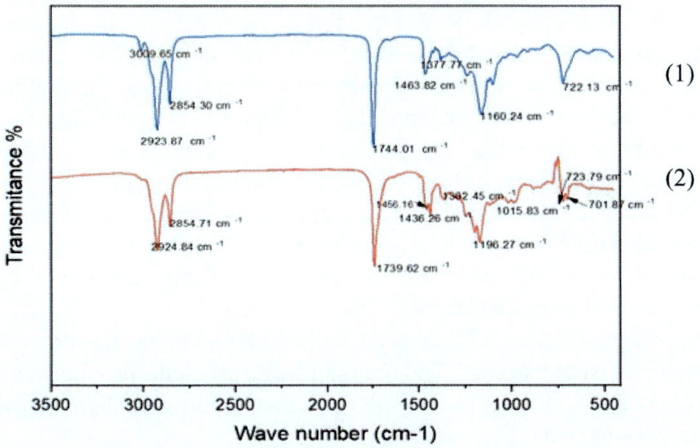

Fig. 10.13 FTIR spectra of (1) soybean oil (2) soybean oil biodiesel

Table 10.1 Physico-chemical properties of produced soybean oil biodiesel

S. no	Physical properties	Standard methods	Soybean oil biodiesel	Biodiesel standards
1	Appearance/color	Visual	Clear and bright yellow	–
2	Density @15 °C	–	0.8795 gm/cm^3	0.86–0.90 g/cm^3
3	Kinematic viscosity @40 °C, mm^2/S	IS1448/P-25	6.52 cSt	2.5–6.0 cSt
4	Flash point °C (PMCC)	IS1448/P-21	170	120 min
5	Fire point °C (PMCC)	IS1448/P-21	184	–
6	Cloud point °C	IS1448/P-10	(+) 1	–
7	Pour point °C	IS1448/P-10	(−) 2	–
8	Cetane index	D 4737	48.5	51
9	Oxidation stability @110 °C	EN 14,112	4 h	Minimum 3 h

10.4 Conclusions

We have emphasized in this work, the Karanja seed shells (KSS) of the Karanja tree, which is a widely available as agricultural waste, can be used as a green heterogeneous catalyst for the synthesis of biodiesel that is affordable, sustainable, biodegradable, and environmentally benign.

- To provide an efficacy of KSS ash catalyst for soybean oil methanolysis, the Karanja seed shell was calcined at 650 °C. According to XRF, EDS, and XRD analysis, the catalyst shows the presence of several alkalies and alkaline earth metals. K and Ca were the most prevalent elements in the ash's elemental composition, which suggests that the prepared catalyst can used as a heterogeneous catalyst and increased its catalytic activity to produce biodiesel.
- The KSS ash catalyst demonstrated a yield of 96% production of biodiesel from soybean oil in 60 min at 65 °C. The catalyst was reused up to the 4th cycle which yielded a drop to 70% after the 4th run. The fuel properties of the produced biodiesel also meet biodiesel standards.
- Consequently, utilizing KSS catalyst will result in simple reaction procedures, low-cost catalyst preparation, and minimal waste problems for biodiesel synthesis on a wide scale. Additionally, this will helps in lowering greenhouse gas emissions and waste management for wealth.

Acknowledgements The authors gratefully acknowledge the CIF, Jiwaji University, Gwalior (M.P.), India for providing XRD, FT-IR, and TGA facilities; the Innovation Centre, Bundelkhand University (U.P.), India for GC-MS analysis; the SAIF, IIT Patna, India for the ^{1}H NMR facility; the CSIF, University of Calicut, India for providing BET facility; IIT Roorkee, India for SEM-EDX analysis; SICART Anand, India for WD-XRF facility; and Hindustan Laboratories Regd., New Delhi, India for testing biodiesel properties.

References

1. V. Kavitha, V. Geetha, P.J. Jacqueline, Production of biodiesel from dairy waste scum using eggshell waste. Process Saf. Environ. Prot. **125**, 279–287 (2019)
2. W. A. G. Pessoa Junior, M. L. Takeno, F. X. Nobre, I.S.C. Barros, E. P. Silva, et al., Application of water treatment sludge as a low-cost and eco-friendly catalyst in the biodiesel production via fatty acids esterification: Process optimization. Energy **213**, 118824 (2020)
3. Asadi, O.N. Kadijani, M.H. Doranehgard, M.V. Bozorg, Q. Xiong, M.S. Shadloo et al., Numerical study on the application of biodiesel and bioethanol in a multiple injection diesel engine. Renew. Energy **150**, 1019–1029 (2020)
4. E. Karbasi, J. Karimi-Sabet, J. Mohammadi-Rovshandeh, M. Ali Moosavian, H. Ahadi, Y. Amini, Experimental and numerical study of air-gap membrane distillation (AGMD): novel AGMD module for Oxygen-18 stable isotope enrichment. Chem. Eng. J. **322**, 667–678 (2017)
5. H. Lu, P. Tian, L. He, Evaluating the global potential of aquifer thermal energy storage and determining the potential worldwide hotspots driven by socio-economic, geo-hydrologic and climatic conditions. Renew. Sustain. Energy Rev. **112**, 788–796 (2019)
6. Li. He, Y. Chen, H. Zhao, P. Tian, Y. Xue, L. Chen, Game-based analysis of energy-water nexus for identifying environmental impacts during Shale gas operations under stochastic input. Sci. Total Environ. **627**, 1585–1601 (2018)
7. Y. Zheng, M.S. Shadloo, H. Nasiri, A. Maleki, A. Karimipour, I. Tlili, Prediction of viscosity of biodiesel blends using various artificial model and comparison with empirical correlations. Renew. Energy **153**, 1296–1306 (2020)
8. C. Zapata, P. Nieuwenhuis, Exploring innovation in the automotive industry: new technologies for cleaner cars. J. Clean. Prod. **18**, 14–20 (2010)

9. M. Mohadesi, B. Aghel, M. Maleki, A. Ansari, Study of the transesterification of waste cooking oil for the production of biodiesel in a microreactor pilot: the effect of acetone as the co-solvent. Fuel **273**, 117736 (2020). https://doi.org/10.1016/j.fuel.2020.117736
10. I. Ambat, V. Srivastava, M. Sillanp, Recent advancement in biodiesel production methodologies using various feedstock: a review. Renew. Sustain. Energy Rev. **90**, 356–369 (2018)
11. R. Shan, L. Lu, Y. Shi, H. Yuan, J. Shi, Catalysts from renewable resources for biodiesel production. Energy Convers. Manag. **178**, 277–289 (2018)
12. F.A.S. Mota, J.T. Costa Filho, G.A. Barreto, The Nile tilapia viscera oil extraction for biodiesel production in Brazil: an economic analysis. Renew. Sustain. Energy Rev. **108**, 1–10 (2019). https://doi.org/10.1016/j.rser.2019.03.035
13. A.F. Abomohra, M. Elsayed, S. Esakkimuthu, M. El-Sheekh, D. Hanelt, Potential of fat, oil and grease (FOG) for biodiesel production: a critical review on the recent progress and future perspectives. Prog. Energy Combust. Sci **81**, 100868 (2020)
14. K.Y. Wong, J.H. Ng, C.T. Chong, S.S. Lam, W.T. Chong, Biodiesel process intensification through catalytic enhancement and emerging reactor designs: a critical review. Renew. Sustain. Energy Rev. **116**, 109399 (2019). https://doi.org/10.1016/j.rser.2019.109399
15. T. Roy, S. Sahani, Y.C. Sharma, Study on kinetics-thermodynamics and environmental parameter of biodiesel production from waste cooking oil and castor oil using potassium modified ceria oxide catalyst. J. Clean. Prod. **247**, 119166 (2020)
16. V.B. Veljkovi, M.O. Biberd, I.B. Bankovi, I.G. Djalovi, M.B. Tasi, Z.B. Nje et al., Biodiesel production from corn oil: a review. Renew. Sustain. Energy Rev. **91**, 531–548 (2018)
17. A. Zheng, S.J. Huang, S.B. Liu, F. Deng, Acid properties of solid acid catalysts characterized by solid-state ^{31}P NMR of adsorbed phosphorous probe molecules. Phys. Chem. Chem. Phys. **13**, 14889–14901 (2011)
18. M.R. Miladinovic, J.B. Krstic, M.V. Zdujic et al., Transesterification of used cooking oil catalyzed by hazelnut shell ash. Renew. Energy **183**, 103–113 (2022)
19. E. Betiku, A.M. Akintunde, T.V. Ojumu, Banana peels as a biobase catalyst for fatty acid methyl esters production using Napoleon's plume (*Bauhinia monandra*) seed oil: a process parameters optimization study. Energy **103**, 797–806 (2016)
20. G. Pathak, D. Das, K. Rajkumari, L. Rokhum, Exploiting waste: towards a sustainable production of biodiesel using: Musa acuminata peel ash as a heterogeneous catalyst. Green Chem. **20**, 2365–2373 (2018). https://doi.org/10.1039/c8gc00071a
21. B. Changmai, P. Sudarsanam, L. Rokhum, Biodiesel production using a renewable mesoporous solid catalyst. Ind. Crop. Prod. **145**, 111911 (2020)
22. S. Barros, W.A.G. Pessoa Junior, M.L. Takeno, F.X. Nobre, W. Pinheiro, L. Manzato, S. Iglauer, F.A. Freitas, Pineapple (*Anan ascomosus*) leaves ash as a solid base catalyst for biodiesel synthesis. Bioresour. Technol. **312**, 123569 (2020)
23. A. Arumugam, P. Sankaranarayanan, Biodiesel production and parameter optimization: an approach to utilize residual ash from sugarcane leaf, a novel heterogeneous catalyst, from Calophylluminophyllum oil. Renew. Energy **153**, 1272–1282 (2020)
24. B. Nath, B. Das, P. Kalita, S. Basumatary, Waste to value addition: utilization of waste *Brassica nigra* plant derived novel green heterogeneous base catalyst for effective synthesis of biodiesel. J. Clean. Prod. **239**, 118112 (2019). https://doi.org/10.1016/j.jclepro.2019.118112
25. A. Gouran, B. Aghel, F. Nasirmanesh, Biodiesel production from waste cooking oil using wheat bran ash as a sustainable biomass. Fuel **295**, 120542 (2021)
26. V. Vadery, B.N. Narayanan, R.M. Ramakrishnan, S.K. Cherikkallinmel, S. Sugunan, D.P. Narayanan, S. Sasidharan, Room temperature production of jatropha biodiesel over coconut husk ash. Energy **70**, 588–594 (2014). https://doi.org/10.1016/j.energy.2014.04.045
27. G.Y. Chen, R. Shan, J.F. Shi, B.B. Yan, Transesterification of palm oil to biodiesel using rice husk ash-based catalysts. Fuel Process. Technol. **133**, 8–13 (2015)
28. W.W.S. Ho, H.K. Ng, S. Gan, Development and characterisation of novel heterogeneous palm oil mill boiler ash-based catalysts for biodiesel production. Bioresour. Technol. **125**, 158–164 (2012). https://doi.org/10.1016/j.biortech.2012.08.099

29. T. Maneerung, S. Kawi, C.H. Wang, Biomass gasification bottom ash as a source of CaO catalyst for biodiesel production via transesterification of palm oil. Energy Convers. Manag. **92**, 234–243 (2015). https://doi.org/10.1016/j.enconman.2014.12.057
30. E.M. Vargas, L. Ospina, M.C. Neves, L.A.C. Tarelho, M.I. Nunes, Optimization of FAME production from blends of waste cooking oil and refined palm oil using biomass fly ash as a catalyst. Renew. Energy **163**, 1637–1647 (2021)
31. B.K. Uprety, W. Chaiwong, C. Ewelike, S.K. Rakshit, Biodiesel production using heterogeneous catalysts including wood ash and the importance of enhancing by product glycerol purity. Energy Convers. Manag. **115**, 191–199 (2016)
32. C.S. Nabora, C.K. Kingondu, T.T. Kivevele, Tamarindus Indica fruit shell ash: a low cost and effective catalyst for biodiesel production from Parinaricuratellifolia seeds oil. SN Appl. Sci. **1**, 253 (2019). https://doi.org/10.1007/s42452-019-0256-3
33. M.R. Miladinovi, M.V. Zduji, D.N. Veljovi, J.B. Krsti, I.B. Bankovi, V.B. Veljkovi, O.S. Stamenkovi, Valorization of walnut shell ash as a catalyst for biodiesel production. Renew. Energy **147**, 1033–1043 (2020). https://doi.org/10.1016/j.renene.2019.09.056
34. E.K.L. Mares, M.A. Gonçalves, P.T.S. Da Luz, G.N. Da Rocha Filho, J.R. Zamian, L.R.V. Da Conceiç, Acai seed ash as a novel basic heterogeneous catalyst for biodiesel synthesis: optimization of the biodiesel production process, Fuel **299**, 120887 (2021)
35. R.L. Patel, C.D. Sankhavara, Biodiesel production from Karanja oil and its use in diesel engine: a review. Renew. Sustain. Energy Rev. **71**, 464–474 (2016)
36. E. Betiku, A.O. Etim, O. Pereao, T.V. Ojumu, Two-step conversion of neem (Azadirachtaindica) seed oil into fatty methyl esters using a heterogeneous biomass-based catalyst: an example of cocoa pod husk. Energy Fuels **31**, 6182–6193 (2017)
37. A.O. Etim, E. Betiku, S.O. Ajala, P.J. Olaniyi, T.V. Ojumu, Potential of ripe plantain fruit peels as an ecofriendly catalyst for biodiesel synthesis: optimization by artificial neural network integrated with genetic algorithm. Sustainability **10**, 707 (2018)
38. M. Gohain, K. Laskar, A.K. Paul, N. Daimary, M. Maharana, I.K. Goswami, A. Hazarika, U. Bora, D. Deka, Carica papaya stem: a source of versatile heterogeneous catalyst for biodiesel production and C–C bond formation. Renew. Energy **147**, 541–555 (2020)
39. E. Betiku, A.A. Okeleye, N.B. Ishola, A.S. Osunleke, T.V. Ojumu, Development of a novel mesoporous biocatalyst derived from kola nut pod husk for conversion of Kariya seed oil to methyl esters: a case of synthesis, modeling and optimization studies. Catal. Lett. **149**, 1772–1787 (2019)
40. M. Sharma, A.A. Khan, S. Puri, D. Tuli, Wood ash as a potential heterogeneous catalyst for biodiesel synthesis. Biomass Bioenergy **41**, 94–106 (2012)
41. C. Zhao, P. Lv, L. Yang, S. Xing, W. Luo, Z. Wang, Biodiesel synthesis over biochar-based catalyst from biomass waste pomelo peel. Energy Convers. Manag. **160**, 477–485 (2018)
42. H. Li, F. Liu, X. Ma, Z. Wu, Y. Li, L. Zhang, S. Zhou, Y. Helian, Catalytic performance of strontium oxide supported by MIL-100 (Fe) derivate as transesterification catalyst for biodiesel production. Energy Convers. Manag. **180**, 401–410 (2019)
43. E.A. Olatundun, O.O. Borokini, E. Betiku, Cocoa pod husk-plantain peel blend as a novel green heterogeneous catalyst for renewable and sustainable Honne oil biodiesel synthesis: a case of biowastes-to-wealth. Renew. Energy **166**, 163–175 (2020)
44. S.H. Teo, Y.H. Taufiq-Yap, U. Rashid, A. Islam, Hydrothermal effect on synthesis, characterization and catalytic properties of calcium methoxide for biodiesel production from crude *Jatropha curcas*. RSC Adv. **5**, 4266–4276 (2015)
45. E.W. Thiele, Relation between catalytic activity and size of particle. Ind. Eng. Chem. **31**, 916–920 (1939)
46. A.F. Lee, K. Wilson, Recent developments in heterogeneous catalysis for the sustainable production of biodiesel. Catal. Today. **242**, 3–18 (2015)

47. C.X.A. Da Silva, V.L.C. Gonalves, C.J.A. Mota, Water-tolerant zeolite catalyst for the acetalisation of glycerol. Green Chem. **11**, 38–41 (2009). https://doi.org/10.1039/b813564a
48. B. Nath, P. Kalita, B. Das, S. Basumatary, Highly efficient renewable heterogeneous base catalyst derived from waste Sesamumindicum plant for synthesis of biodiesel. Renew. Energy **151**, 295–310 (2020)

Chapter 11
Combined Effect of Preheating and Addition of CeO_2 Nanoparticles in Biodiesel Blend (B20) on Combustion, Emission, and Performance of a 10-kW Diesel Engine

Ankush Halba and Hifjur Raheman

Abstract The objective of the present study was to address several concerns associated with biodiesel, such as its higher viscosity and density, while simultaneously improving the diesel engine's efficiency and environmental performance. For this purpose, the combined effect of preheating and inclusion of cerium oxide nanoparticles (CeO_2) to a biodiesel blend is investigated in a 10-kW diesel engine by analyzing its combustion, performance, and emission characteristics. The biodiesel (B100) was produced via transesterification of Palm oil in the presence of potassium hydroxide and methanol. Subsequently, the fuel blends B20 (80% v/v diesel and 20% v/v palm biodiesel) and NB20 (B20 + 60 ppm CeO_2 nanoparticles) were prepared for conducting experiments. A separate external fuel preheating setup was developed for supplying preheated fuel directly to the engine at a specific temperature for the fuel blends HB100 (B100 at 70 °C), HB20 (B20 at 55 °C), and NHB20 (NB20 at 55 °C). The six fuels (B20, HB20, NHB20, B100, HB100, and diesel) were investigated in a diesel engine at various engine loads (0, 25, 50, 75, and 100%) for the present study. The findings of the present investigation revealed that NHB20 showed nearly identical performance and emission characteristics to diesel, with the highest brake thermal efficiency of 26.98% at 75% engine load compared to diesel (27.17%), among the other tested fuels.

Keywords Biodiesel · Combustion · Engine performance · Exhaust emissions · Preheating · Nanoparticles

A. Halba (✉)
Hydro and Renewable Energy Department, Indian Institute of Technology Roorkee, Roorkee 247667, India
e-mail: ankush_h@hre.iitr.ac.in

A. Halba · H. Raheman
Agricultural and Food Engineering Department, Indian Institute of Technology Kharagpur, Kharagpur 721302, India

N. Gakkhar et al. (eds.), *Recent Advances in Bio-Energy Research*, Springer Proceedings in Energy, https://doi.org/10.1007/978-981-99-5758-3_11

11.1 Introduction

Non-polluting alternatives would take decades to replace fossil-based energy carriers and combustion engines, such as diesel engines, emphasizing the importance of renewable, ecologically friendly energy sources like biodiesel in mitigating the vulnerabilities faced during the transition period [1]. However, on the other hand, biodiesel has several flaws, such as higher viscosity, fuel economy, density, and lower calorific value over diesel fuel, which restricts it from being used in diesel engines directly [2]. Injecting preheated biodiesel into the engine cylinder appears to be an effective way to deal with the issues of using biodiesel. For instance, Kodate et al. [3] wrapped heating cords around the fuel supply line to inject the Karanja biodiesel (KB100) at 95 °C preheating temperature in a diesel engine which improved the engine's brake thermal efficiency (BTE) by 9.1% and reduced HC (10.6%) and CO (8.1%) emissions compared to non-heated KB100. However, preheating of KB100 resulted in an increment (5.1%) in NO_x emission. Several authors [4, 5] also reported similar trends of findings; however, they utilized waste heat of engine exhaust to preheat biodiesel and were only able to achieve temperatures of up to 70 °C. In addition, it was difficult to heat biodiesel uniformly and consistently with waste heat. It has been observed that preheating biodiesel alone was unable to address all of the biodiesel's shortcomings.

Recent research has concentrated more on fuel formulation strategies, such as adding metal oxide nanoparticles such as CeO_2, MgO, and TiO_2 to biodiesel blends, which has emerged as a promising new biodiesel ingredient to enhance engine efficiency and minimize exhaust emissions [2]. Hoseini et al. [6] investigated diesel engine performance using graphene oxide (GO) nanoparticles in an Ailanthus altissima oil biodiesel blend (B10 and B20). They found that GO addition improved the performance parameters but hiked the NO_x emissions. Kumar et al. [7] studied the effect of adding CeO_2 nanoparticles to the Mahua biodiesel blend (MB20) on a 3-kW diesel engine. Their findings revealed that the CeO_2 additive reduced NO_x, CO, and HC emissions and enhanced engine performance. Nonetheless, increased concentrations of nanoparticles in biodiesel blends may increase the risk of knocking and instability during operation. Therefore, the quantity and selection of metal oxide nanoparticles are critical because each has different physical, chemical, electrical, and optical properties.

To the best of the authors' knowledge, no research has been found to study the combined effect of preheating and adding CeO_2 nanoparticles in biodiesel blends on a diesel engine's combustion, performance, and emission characteristics. Moreover, it is evident from the literature review that researchers have not focused on developing an external fuel preheating arrangement to provide consistent and uniform heating during diesel engine operation and before the operation. Therefore, in the present research, the combined effect of adding the cerium oxide nanoparticles to the preheated biodiesel blend on the engine's combustion, performance, and emission characteristics was studied. In addition, an external fuel preheating setup was developed to facilitate the fuel's uniform and constant heating.

11.2 Materials and Methods

11.2.1 Development of an External Fuel Preheating System

The fuel preheating system was developed and fabricated at IIT Kharagpur and illustrated in Fig. 11.1. This system consists of a fuel tank (capacity: 5.84 L), a heater (2 kW) for heating the fuel, a K-type thermocouple sensor for measuring the temperature inside the fuel tank, a stirrer powered by a DC motor (12 V, 200 rpm) to facilitate homogeneous heating of fuel inside the fuel tank, and a temperature controller (Make: Multispan India, Model: UTC 2131) for controlling the temperature inside the fuel tank.

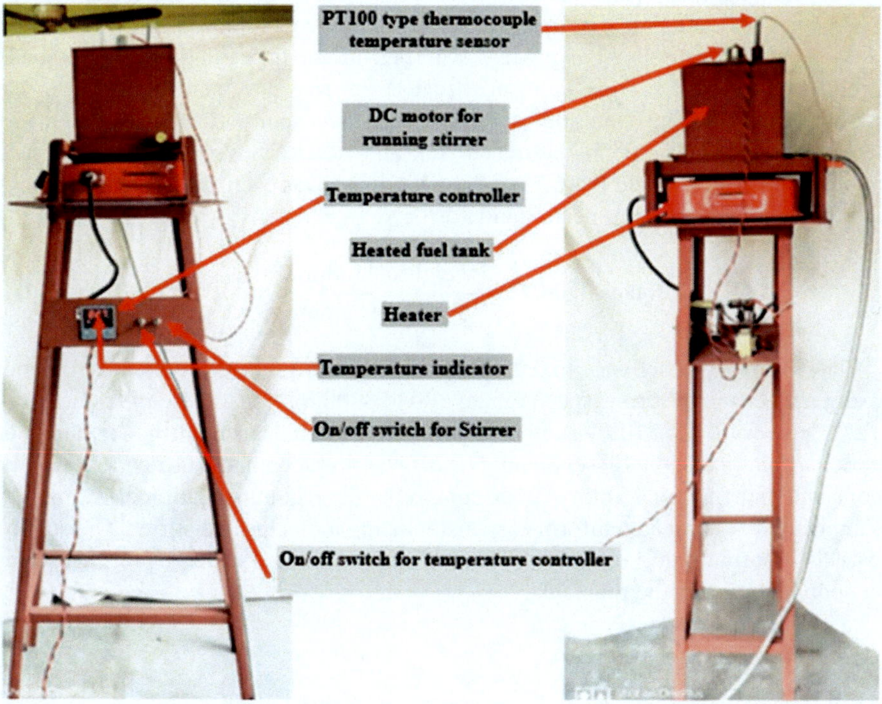

Fig. 11.1 External fuel preheating system

11.2.2 Selection of Cerium Nano-Oxide, Biodiesel Production, and Preparation of Biodiesel Blends

Leach et al. [8] reported that cerium oxide (CeO_2) acted as a three-way catalyst (TWC) for the conversion of hydrocarbons (HCs), carbon monoxide (CO), and oxides of Nitrogen (NO_x) exhaust emissions into H_2O, CO_2, and N_2 simultaneously. According to Kumar and Raheman [9], CeO_2 exhibited in both + 3 and + 4 valence states, making the conversion of Ce_2O_3 to CeO_2 easier. Kang et al. [10] reported that cerium nano-oxide induces high catalytic behavior due to its higher surface-to-volume ratio, resulting in an enhanced combustion reaction that might improve engine performance. Therefore, CeO_2 was selected for the present investigation. The CeO_2 nanoparticles (density: 7.13 gm/ml; surface area: 30 mm^2/g (BET); appearance: pale yellow or white powder; size: 30–50 nm (BET) and molecular weight: 172.11 g/mol) were procured from Sigma Aldrich Corporation for the present study.

A base transesterification process was performed in a biodiesel processor to produce the biodiesel from Palm oil because the free fatty acid of Palm oil was found to be 0.8%. The operating conditions were the response time (0.5 h), methanol to oil ratio (1:4 w/v), temperature (60 °C), and KOH catalyst (3.5 g + final acid value). The average yield (97.6%) of Palm biodiesel was estimated using Eq. (11.1), which complied with European Standards (EN: 14,214).

$$\text{Biodieselyield} = \frac{(\text{Weightofbiodiesel})}{(\text{WeightofPalmoil})} \times 100 \tag{11.1}$$

The Palm biodiesel was mixed with diesel fuel to prepare the biodiesel blends using a mechanical stirrer (speed range: 300 to 8000 rpm; make: Remi, model: RQ-127A, power AC/DC: 100W) with a shearing type stirring mechanism. Cerium oxide nanoparticles added biodiesel blend (NB20) was prepared in the heated fuel tank by constantly stirring for 15 min at 2500 rpm as per the recommendation of Kumar and Raheman [11] to create uniform suspension using a mechanical stirrer. The stirring was also continued during testing by the DC motor-operated stirrer to prevent settling or sedimentation in the fuel tank.

11.2.3 Fuel Properties, Experimental Setup, and Test Methods

The fuel properties flash point (ASTM D93-02), calorific value (ASTM D240-02), viscosity (ASTM D446-07), acid value (ASTM D5555-95), and density (ASTM D4052-96) were determined following the ASTM standards and given in Table 11.1. They were compared with the Indian diesel (IS 1460-2017) and Indian biodiesel (IS 15607:2016) standards.

Table 11.1 Standard methods followed for measuring fuel properties

Property	Unit	Method
Viscosity at 40 °C	cSt	Cannon Fenske viscometer
Density at 15 °C	Kg/ m^3	Relative density bottle
Calorific value	MJ/Kg	Bomb calorimeter
Flash point	°C	Pensky Martin closed cup
Acid value	mg KOH/g	Standard titrimetry

The schematic of the experimental setup to carry out the experiments is shown in Fig. 11.2. The engine testing was conducted on a 10-kW, four-stroke, rated speed (1500 rpm), direct injection (DI), single-cylinder, naturally aspirated, water-cooled Chetak diesel engine. The 4-way valve was used to connect the developed external fuel preheating system to the existing experimental system. The four-way valve can direct the fuel supply (either heated or unheated) to the engine through the glass pipe.

The engine load was varied with the help of the eddy current dynamometer. The performance tests of the diesel engine for each test fuel were performed according to IS: 10,000 (Part VIII), and fuel consumption was measured according to IS: 10,000 (Part IV). The engine was operated for 30 min to achieve thermal stabilization after applying each load for each fuel; after that, different readings were taken at 5 min intervals, and all the experiments were replicated thrice. The engine speed and crank angle were measured using the rotary encoder (Model E50S8). A piezoelectric pressure transducer (Make: Kistler, Model-SN14) was used to measure in-cylinder pressure. Cylinder gas pressure, engine rpm, and crankshaft position were used to evaluate the combustion parameters, such as heat release rate (HRR) and ignition delay (ID). Fuel consumption, engine rpm, and engine torque were used to evaluate performance parameters such as brake thermal efficiency (BTE) and brake specific fuel consumption (BSFC). Oxides of nitrogen (NO_x), hydrocarbon (HC), and carbon monoxide

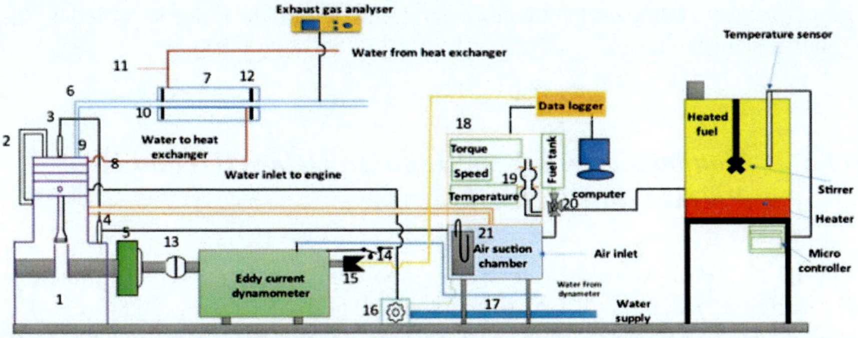

1. Diesel engine 2. Inlet manifold 3. Fuel injector 4. Fuel pump 5. Fly wheel 6. Exhaust manifold 7. Heat exchanger 8. Temperature of water coming back from engine 9. Temperature of exhaust gases leaving the engine 10. Temperature of exhaust gas entering the heat exchanger 11. Temperature of water leaving the heat exchanger 12. Temperature of exhaust gases leaving the heat exchanger 13. Universal coupling 14. Pressureswitch 15. Rotary encoder 16. Water flow meter 17. Temperature of water engine the engine 18. Control panel 19. Glass bulb of 25 ml capacity fitted with IR sensors20. Four-waycontrolvalve 21. U Tube manometer

Fig. 11.2 Schematic diagram of the experimental setup

Table 11.2 Measured fuel properties

Property	Diesel	B100	B20	NB20	HB100 (70 °C)	HB20 (55 °C)	NHB20 (55 °C)	IS1460: 2017 for diesel	IS15607: 2016 for biodiesel
Viscosity (cSt)	2.46	4.18	3.11	3.14	2.40	2.38	2.41	2.0–5.0	3.5–5.0
Density (kg/m^3)	824.40	872.00	837.60	837.80	841.50	825.80	826.10	820–860	860–900
Calorific value (MJ/kg)	42.63	37.32	41.22	41.26	37.36	41.24	41.28	–	–
Flash point (°C)	59.00	148.00	72.00	70.00	127.00	64.00	62.00	>66	>101
Acid value (mgKOH/g)	0.15	0.36	0.24	0.25	0.38	0.25	0.26	<0.2	<0.5

(CO) emissions were measured using a flue gas analyzer (Make: Indus Scientific Pvt. Ltd., Model-FGA53X).

11.3 Results and Discussion

11.3.1 Fuel Properties

The fuel properties of B100, B20, NB20, HB100, HB20, and NHB20 were determined following the ASTM standards and are summarized in Table 11.2. The fuel properties were compared with diesel and found within the range specified by Indian Standards for biodiesel.

11.3.2 Combustion Analysis [Ignition Delay (ID) and Heat Release Rate (HRR)]

The ID and HRR were measured in degree crank angle (°CA) and joule energy per degree crank angle (J/°CA), respectively. The ID for diesel (20.86–15.90°CA), B20 (20.67–15.45°CA), HB20 (20.10–15.11°CA), NHB20 (20.02–14.87°CA), B100 (19.11–13.39°CA), and HB100 (18.98–12.38°CA) was obtained at varying engine load (0–100%) and presented Fig. 11.3. The ignition delay for all test fuels was reduced with the rise in engine load. Compared to diesel, the peak HRR for B20,

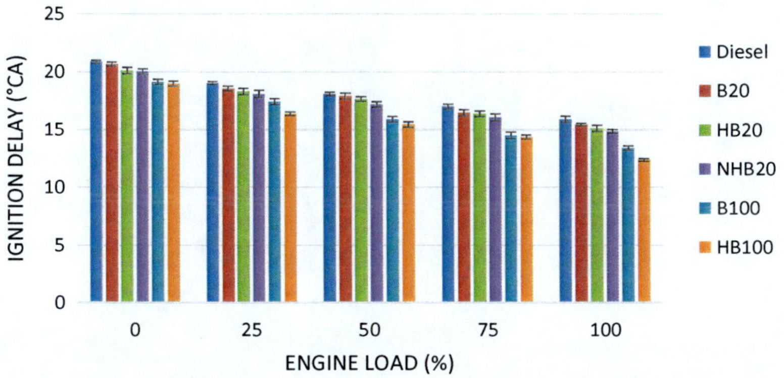

Fig. 11.3 Variation of ID for test fuels at different engine loads

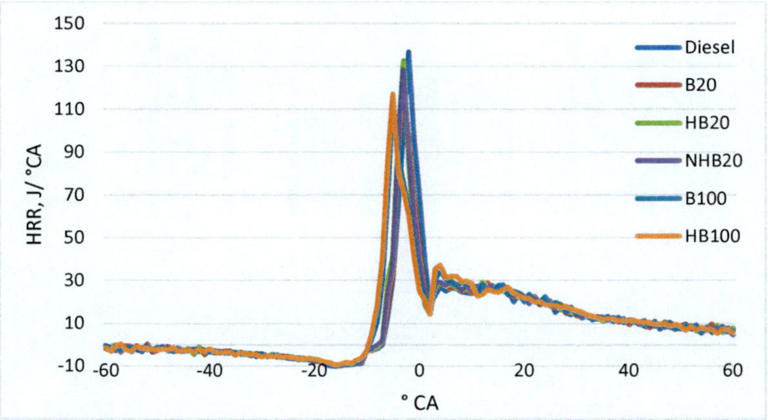

Fig. 11.4 Variation of HRR with the crank angle for test fuels at 75% engine load

HB20, NHB20, B100, and HB100 was found to be lowered by 11.35–16.28%, 3.81–7.71%, 6.23–11.80%, 18.51–36.90%, and 14.55–29.02% for all tested loads for B20, HB20, NHB20, B100, and HB100, respectively. The peak HRR was discovered for all six tested fuels at 75% engine load and is shown in Fig. 11.4.

11.3.3 Performance Analysis [Brake Thermal Efficiency (BTE) and Brake Specific Fuel Consumption (BSFC)]

Engine load and fuel type impacted the BTE and BSFC of the engine, as shown in Figs. 11.5 and 11.6. BTE obtained at 75% engine load was found to be 27.17%, 26.43%, 26.74%, 26.98%, 25.47%, and 25.74%, respectively, for diesel, B20, HB20,

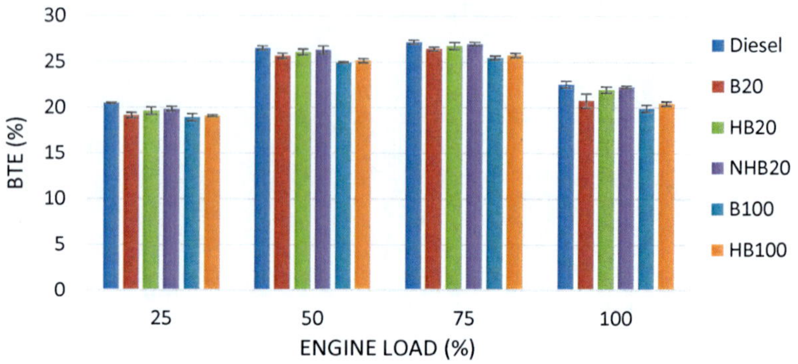

Fig. 11.5 Variation of BTE of test fuels at different engine loads

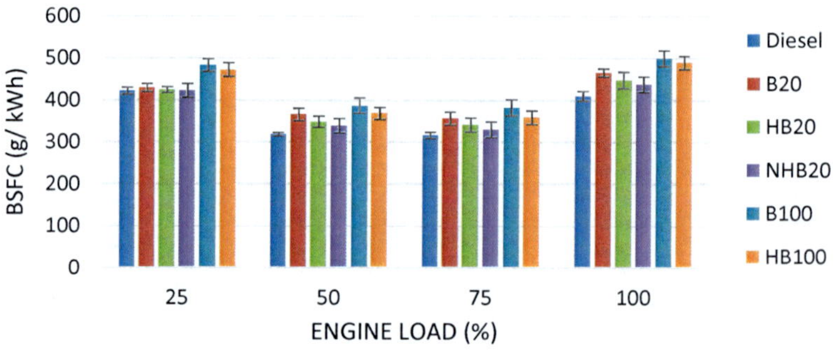

Fig. 11.6 Variation of BSFC of test fuels at different engine loads

NHB20, B100, and HB100. The BSFC was found to be minimum at 75% of engine load for all test fuels and was 315.78, 356.12, 340.64, 329.22, 381.90, and 358.26 g/kWh for diesel, B20, HB20, NHB20, B100, and HB100, respectively.

11.3.4 Exhaust Emissions

The formation of CO and HC emission significantly hiked at maximum engine load (100%), as shown in Figs. 11.7 and 11.8. The CO emission at full load was 2.218 g/kWh, 2.179 g/kWh, 2.107 g/kWh, 1.984 g/kWh, 1.836 g/kWh, and 1.674 g/kWh for diesel, B20, HB20, NHB20, B100, and HB100, respectively. The maximum HC emission for diesel, B20, HB20, NHB20, B100, and HB100 was obtained at 0.344 g/kWh, 0.320 g/kWh, 0.294 g/kWh, 0.278 g/kWh, 0.272 g/kWh, and 0.252 g/kWh, respectively at full engine load as compared to 0.060 g/kWh, 0.046 g/kWh, 0.036 g/kWh, 0.030 g/kWh, 0.028 g/kWh, and 0.024 g/kWh at no-load condition. Both engine load and fuel type affected the NO_x emissions, as evident from Fig. 11.9. The NO_x

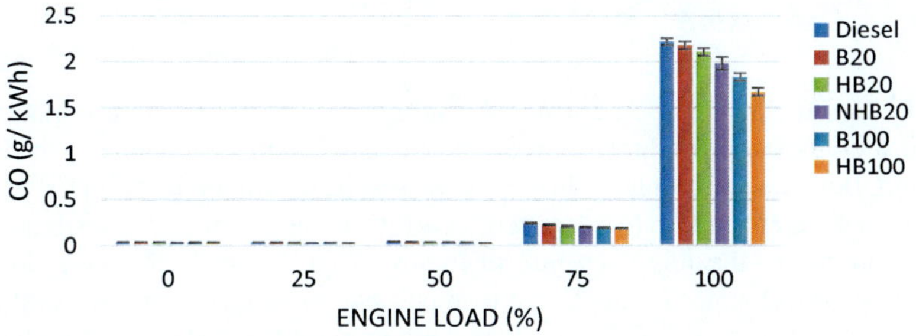

Fig. 11.7 Variation of CO emission of test fuels at different engine loads

emission for diesel, B20, HB20, NHB20, B100, and HB100 were measured in the range of 3.82–9.20 g/kWh, 4.30–9.38 g/kWh, 4.74–9.72 g/kWh, 4.56–9.46 g/kWh, 4.80–9.77 g/kWh, and 5.38–9.91 g/kWh, respectively from zero to maximum load (100%).

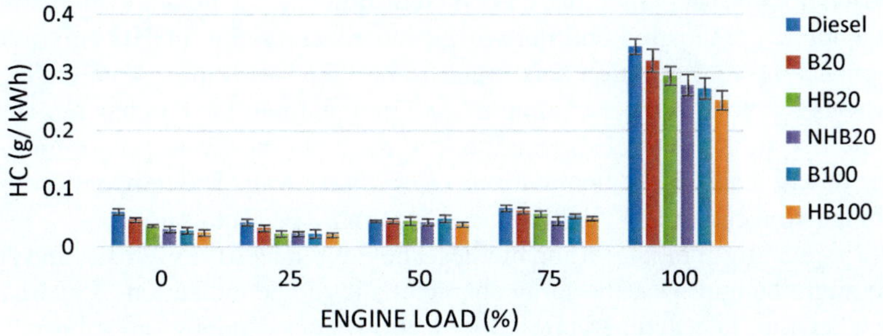

Fig. 11.8 Variation of HC emission of test fuels at different engine loads

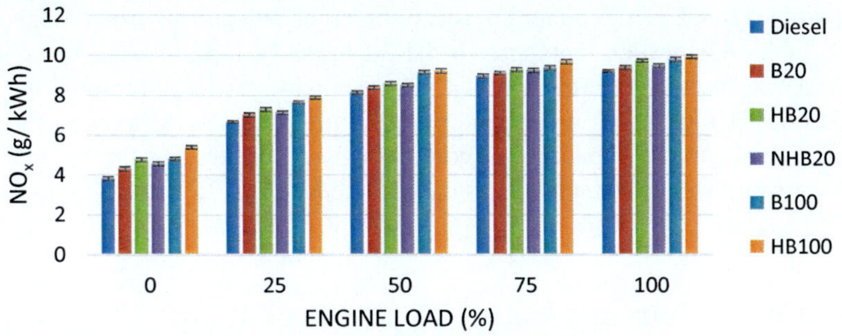

Fig. 11.9 Variation of NO_x emission of test fuels at different engine loads

11.3.5 Discussion

The BTE and BSFC were improved by up to 75% of the maximum load for all tested fuels because the brake power rate increased quickly compared to the fuel consumption rate and reduction in friction losses. At maximum load, BTE and BSFC showed a different pattern for all tested fuels due to the higher fuel consumption rate corresponding to the increased rate of brake power. Up to 75% engine load, the formation of CO emission did not vary much due to sufficient oxygen availability in the cylinder. However, CO emissions were initially slightly reduced with engine load due to better oxidation of hydrocarbons at higher cylinder temperatures. The maximum CO emission was obtained at full engine load conditions. An extra amount of fuel was injected at higher loads, making the mixture rich and supporting incomplete combustion. Due to this, the abrupt increment of CO emission was observed at maximum load. The HC emissions for engine loadings up to 25% were found to be minimum due to ample oxygen and elevated cylinder temperature, leading to better oxidation of hydrocarbons. HC emissions have risen significantly with an increase in load above 25% due to a larger quantity of fuel injected, resulting in a lower air–fuel ratio (rich mixture) at higher loads. Oxygen insufficiency at higher loads contributed to incomplete combustion and thus hiked the concentration of HC emission. The NO_x emissions concentration was found to be hiked with the rise in engine load, which was due to the larger amount of fuel pumped into the combustion chamber, with the rise in the load of the engine leading to an increased proportion of heat energy released and the temperature of the cylinder wall. The temperature rises of the combustion chamber at higher loads led to more NO_x emissions.

The preheating of biodiesel and biodiesel blends reduced the viscosity and density, which might be improved the spray characteristics and atomization of fuel particles at a higher fuel inlet temperature. Due to this, better evaporation occurred, and a homogenous mixture was produced, which helped enhance the combustion reaction, resulting in better performance, better fuel combustion, and reduced emission except for NO_x compared to non-unheated biodiesel blends. The addition of cerium oxide nanoparticles in the HB20 blend improved the efficiency and environmental performance, including the NO_x, compared to HB20 due to the high catalytic behavior and the higher surface-to-volume ratio of CeO_2 nanoparticles that promoted the better combustion of fuel, resulting in better engine performance. Moreover, the CeO_2 nanoparticles' specific capability of continuously absorbing and liberation oxygen acted as a three-way catalyst to simultaneously reduce CO, HC, and NO_x emissions.

11.4 Conclusions and Future Recommendations

The present study aimed to reveal the combined effect of preheating and adding cerium oxide nanoparticles to the biodiesel blend B20 on the combustion, emission, and performance characteristics of a 10-kW stationary diesel engine. Additionally, the present investigation provided a comparative analysis of six tested fuels and compared all biodiesel blends to the reference fuel (diesel). Compared to diesel, the BTE of B20, HB20, NHB20, B100, and HB100 decreased by 2.72–7.93%, 1.58–4.15%, 0.69–3.07%, 5.99–11.56%, and 5.20–9.26%, respectively, while the BSFC increased by 1.81–14.73%, 0.67–9.26%, 0.22–6.74%, 14.44–21.82%, and 12.00–19.39%. CO emissions for B20, HB20, NHB20, B100, and HB100 were found to be reduced by 1.76–12.17%, 5.02–14.78%, 10.57–19.13%, 8.24–22.60%, and 9.27–30.43%, respectively, compared to diesel, whereas HC emissions of these fuels were reduced by 6.97%, 14.53%, 19.18%, 20.93%, and 26.74%, respectively at full load condition. NO_x emissions were found to be raised by 1.62–2.69%, 3.55–24.34%, 2.95–19.47%, 4.44–25.73%, and 7.79–41.04% for B20, HB20, NHB20, B100, and HB100 over diesel at varying loads.

It may be possible that using the various types of biodiesel and nanoparticles could perform better than diesel fuel. Future research on life cycle analysis and techno-economic analysis would be beneficial in commercializing NHB20. A study based on durability analysis and non-regulated emissions is also advised to see the multi-faceted effects of NHB20 on diesel engines. The study to improve the storage ability of the nanoparticles added biodiesel blend should also be carried out to increase their feasibility in a diesel engine. Furthermore, SEM and TEM analysis research should be performed to learn more about nanoparticles' structural and chemical characterization.

References

1. G. Subbiah, Y. Devarajan, R. Mishra, J. Thangaraja, Experimental research on waste and inedible feedstock as a partial alternate fuel: environmental protection and energy-saving initiative. Biomass Convers. Biorefinery 1–13 (2022)
2. M. Dhanarasu, K.A. Ramesh Kumar, P. Maadeswaran, Recent trends in role of nanoadditives with diesel-biodiesel blend on performance, combustion and emission in diesel engine: a review. Int. J. Thermophys. **43**(11), 1–23 (2022)
3. S.V. Kodate, A.K. Yadav, G.N. Kumar, Combustion, performance and emission analysis of preheated KOME biodiesel as an alternate fuel for a diesel engine. J. Therm. Anal. Calorim. **141**(6), 2335–2345 (2020)
4. M.W. Mekonen, N. Sahoo, Effect of fuel preheating with blended fuels and exhaust gas recirculation on diesel engine operating parameters. Renew. Energy Focus **26**, 58–70 (2018)
5. M. Walle Mekonen, N. Sahoo, Combined effects of fuel injection pressure and injector nozzle holes on the performance of preheated palm oil methyl ester used in a diesel engine. Biofuels **11**(1), 19–35 (2020)
6. S.S. Hoseini, G. Najafi, B. Ghobadian, R. Mamat, M.T. Ebadi, T. Yusaf, Novel environmentally friendly fuel: the effects of nanographene oxide additives on the performance and emission characteristics of diesel engines fuelled with *Ailanthus altissima* biodiesel. Renew. Energy **125**, 283–294 (2018)
7. M.V. Kumar, A.V. Babu, P.R. Kumar, Influence of metal-based cerium oxide nanoparticle additive on performance, combustion, and emissions with biodiesel in diesel engine. Environ. Sci. Pollut. Res. **26**(8), 7651–7664 (2019)

8. F.C. Leach, M. Davy, B. Terry, Combustion and emissions from cerium oxide nanoparticle dosed diesel fuel in a high speed diesel research engine under low temperature combustion (LTC) conditions. Fuel **288**, 119636 (2021)
9. N. Kumar, H. Raheman, Characterization of nano-oxide added water emulsified biodiesel blend prepared with optimal emulsifying parameters. Renew. Energy **145**, 308–317 (2020)
10. W. Kang, D.O. Ozgur, A. Varma, Solution combustion synthesis of high surface area CeO_2 nanopowders for catalytic applications: reaction mechanism and properties. ACS Appl. Nano Mater. **1**(2), 675–685 (2018)
11. N. Kumar, H. Raheman, Production, characterization and utilization of second generation biodiesel blend in diesel engine using water and nanoparticles as additives. Fuel **308**, 122063 (2022)

Chapter 12
Techno–Economic Analysis of Protease Enzyme Production and its Biofuel Application

Aparupa Das, Anuradha, and Muthu Kumar Sampath

Abstract The urge for sustainable production of biofuel is presently an active enormous concern in research and discussion so to protect future energy needs. Due to extensive growth in world population and so are the worldwide energy demands with exhausting non-renewable energy source. Hence, in such framework there is a requirement for sustainable development by generating renewable energy sources at convenient cost along with the least drawback. Enzymes play a crucial role in biofuel production because it aids to minimize the utilization of synthetic process, and they may show a positive influence on environment with better biofuel condition. Utilization of protease enzyme in biofuel industry as they act as nitrogen supplier in the yeast fermentation and also permits for rapid kinetics. Protease enzyme helps to hydrolyze the protein content which is present in the lignocellulosic biomass which exhibit the maximal activity in several climatic states. In the biofuel industries, distinct types of protease are initiated which remove the distinct proteolysis step as well as lower the production cost, expand the performance of biofuel production. Utilization of protease under optimum condition will enhance biofuel output. In this work, Techno–Economic analysis (TEA) is performed for one similar method generating protease production from sugar through industrial scale to acquire economic viability. The change of leading machinery to commercialization product is thought to be highly cost effective, time consumption, hazardous attempt. Therefore, balance on advance process and study should be formulated on Techno–Economic analysis. Because of absence of materials on the required process design, limitation, and relevant theories in the initial phase of process advancement, TEA is bounded to contributing a rough evaluation of the funding for a theoretical biofuel production.

Keywords Biofuel · Non–renewable energy · Renewable energy · Enzyme · Protease · Nitrogen supplier · Techno- economic analysis (TEA)

A. Das · Anuradha · M. K. Sampath (✉)
Department of Bioengineering and Biotechnology, Birla Institute of Technology Mesra, Ranchi, Jharkhand 835215, India
e-mail: muthubio@bitmesra.ac.in

N. Gakkhar et al. (eds.), *Recent Advances in Bio-Energy Research*, Springer Proceedings in Energy, https://doi.org/10.1007/978-981-99-5758-3_12

12.1 Introduction

In laboratory, revolution at a small segment by a way of enzyme introduction have immense impact when turned out in industrial methods, which is displayed in the current scenario for commercial biofuel production. With continuous discovery, enzymes such as lipase, protease, and cellulase have often established itself to be the most crucial component in the biofuel production [1]. Biofuel which are acquired from renewable organic matter are termed as advanced biofuel where protease is used to contribute free amino acid to yeast, increases fermentation as well as decrease the by-products (glycerol) discharged into the environment since glycerol possess high oxygen content which causes fuel degradation. Another advantage of protease in fermentation by breaking the complex of starch-glucose molecules, thus assembling the starch more convenient for hydrolysis through amylase [2]. Protease are also utilized for hydrolyzing the oleosins which is a protein found in the corn seed feedstock. Large-scale research has been done to exhibit that particular protease hydrolyses oleosin and ensuing in the release of more oil. Lately, extracting oil from the dried-crushed ethanol plant has started generating a high-valued secondary product for utilization in biodiesel production along with animal feed [3].

Low carbon biofuel is crucial to the quick decarbonization of shipping especially aviation [4]. The protease enzyme also possesses the ability to utilize crude feedstock in the absence of segregation of fatty acids which are generally found in crude feedstock. Conventionally, protease aid in the conversion of biomass into biofuel [5]. The various microbial strain reported for high protease productions are *Bacillus cereus KM05, Bacillus clausii, and Micrococcus* sp [6].

Techno–economic analysis (TEA) is a method for performing process design, along with simulation, evaluating capital expenditures, mass and energy balances, operating cost for commercial biofuel production [7]. TEA, as simulation point of view, is extremely dependent on model theory that could generate error and notable mistake [8]. It is functional to conduct and organize R&D stage by utilizing lab-scale data to outline and examine a conceptual scaled-up process. The latest research has directed the barrier of biofuel and its product TEA in the complication along with innovation of the structure to be modeled, the access for evaluating and describing uncertainty, along with the advancement of facilitated model [9].

There are several enzymes such as protease, amylase, esterase, and many other alcohol dehydrogenases have been used in biofuel production. But the cost of developing the process with the enzymes is normally high [10]. The aim of the paper is to simulate the industrial scale production of alkaline protease enzyme for biofuel application as well as to evaluate its economic performance. This will surely help to understand whether the developed lab scale microbial-based process is feasible and economical at the industrial scale.

12.2 Process Description

SuperPro Designer v9.5 was selected to enhance model and simulate the biofuel production and diversity of this process with reference to technical and economic framework. The process flow diagram of simulation modeling with respect to evaluation mechanism executed. For this study, *Bacillus clausii* has been used and through the process simulation protein estimation is analyzed. The process flow diagram of the biofuel production was separated into three vital parts: Fermentation section, Primary Recovery section, and Downstream segment.

12.2.1 Fermentation Section

Initially, fermentation media is prepared in a stainless-steel tank (V-101) and sterilized in a continuous sterilizer (ST-101). A compressor (G-101) and an absolute air filter (AF-101) provide sterile air to the fermenter (FR-101). The operation requires three vital medium mixtures: Glucose, ammonium hydroxide along with salts which constitute a mixture of mineral with other elements required for microbial growth.

Other elements above the ones defined as component of mixtures are Yeast extract, Biomass. However, all these three compounds Glucose, NH_4OH, and salt are mixed with water in a distinct blending tank as to acquire suitable stock solution.

12.2.2 Primary Recovery Section

First step in this section is cell harvesting to decrease the proportion of the broth along with discharge extracellular impurities, and it is executed by membrane microfilter (MF-101) in which the broth is concentrated into two-folds. Further step is cell disruption accomplished in a homogenizer (HG-101) under a high pressure. Following homogenization, a disk-stack centrifuge (DS-101) is utilized to eliminate mostly all the cell waste particles, later with the dead-end polishing filter (DE-101) that discharge leftover cell waste particle. Lastly, the resultant protein solution is concentrated into two-fold through an ultrafilter.

12.2.3 Downstream Segment

The primary step is the removal of microbial cell out of the extracellular solution carrying the enzyme of interest, be usually attained by centrifugation or cloth filtration. Two filters are generally utilized, namely, microfilter and ultrafilter for the purification and concentration of the biofuel product. The microfilter serves as polishing filter releasing residual cells along with remaining cell debris which are not eliminated through the dead-end filtration.

Finally, the ion exchange chromatography (C-101) is used to purify the product stream.

12.3 Process Scheduling

The designing of ultimate product is thoroughly based on biofuel characteristics and its studied application. Through the equipment occupancy option, process schedule is visualized and will create Equipment Occupancy Chart (EOC). EOC is an analytical tool for considering scheduling border along with conducting appropriate adjustment so as to minimize the cycle time based on method and increase in product.

The next section of process schedule is introduced by the Operation Gantt Chart, and it exhibits full scheduling feature for either/or one multiple batches.

12.4 Results and Discussion

The process needs a main fermenter and around the equivalent quantity of cell broth is introduced in the direction of the termination of the microbial culture. The cell stock is agitated in the storage tank also then disintegrates in a pressure homogenizer. Following, the cell waste produced is separated by working with six dish -stack throughput along with the residual debris is separated by microfiltration. Lastly, the enzyme blend is concentrated and preserved in a specific buffer practicing with diafiltration network so that the concentrated enzyme blend is generated per cycle as shown in Fig. 12.1.

Additionally, the other factors to be held into consideration at the time of batch process is the recognition of prolonged process as this process enhances the limiting step which regulates the entire time of the batch process [11]. Examining the restrictive in the size of the accessible reactors, it was essential to utilize hydrolysis process in alignment to establish the processing of the entire input stream penetrating the steps of the process.

12.4.1 Material Balance

The outcome of protease concentration at the termination of the fermentation section along with the result of the protease production are shown in Table 12.1. The use of glucose in an entire process was 6,143.000 kg/batch on a 100 m^3. The equipment occupancy chart for the process was as shown in Fig. 12.2. The bulk materials (kg/batch) utilized in protease production were as shown in Fig. 12.3.

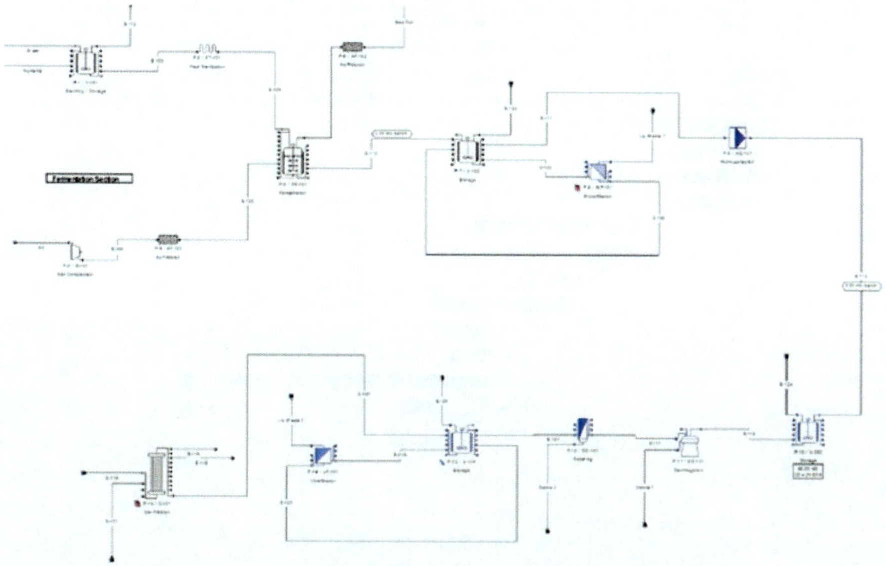

Fig. 12.1 Flowsheet diagram of alkaline protease production Material Balance

Table 12.1 Final component balance of an entire batch process (kg/batch)

Component	Input	Output	Component	Input	Output
Biomass	0.000	12.744	Process water	73,398.119	76,500.092
CO_2	0.000	4,238.806	Protease	0.000	120.543
Debris	0.000	723.255	Proteins	0.000	1,084.883
Glucose	6,143.000	122.860	Salts	1,026.000	230.136
Nitrogen	36,989.523	37,176.948	Sodium chloride	2,246.941	2,246.941
Nucleic acids	0.000	482.170	Sodium hydroxide	282.107	282.107
Oxygen	11,229.302	8,337.219	Water	14,111.112	14,111.112
Phosphoric acid	422.915	422.915	WFI	792,235.481	792,235.481
			Total	938,084.500	938,328.213

12.4.2 Economic Evaluation

Direct fixed capital cost

The direct fixed capital cost of the plant is derived from the equipment purchase cost (EPC) with subsequent recognized engineered procedure. With the help of the simulation tool, an equipment was evaluated based upon the M&E balances. The

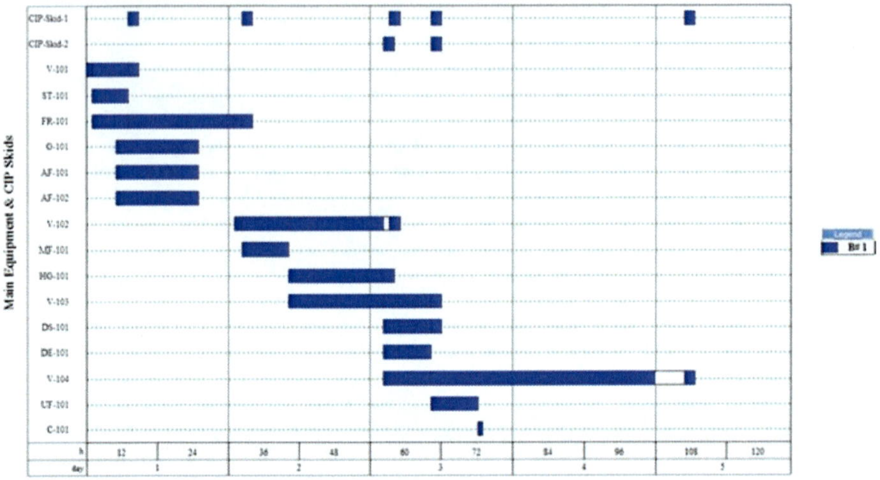

Fig. 12.2 Equipment occupancy chart

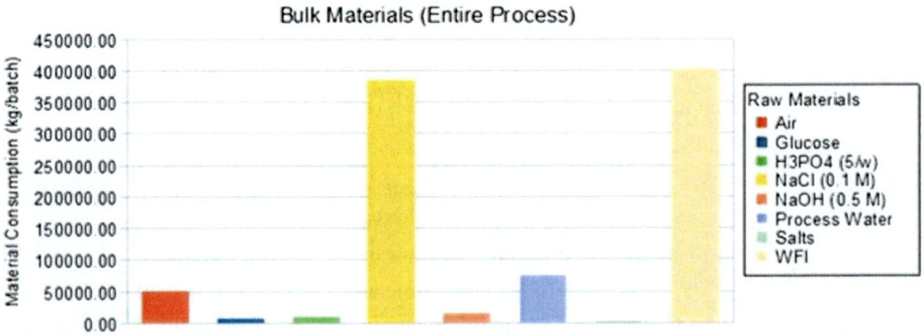

Fig. 12.3 Bulk materials (kg/batch) utilized in protease production

EPC was evaluated based upon data from retailer, literature reviews along with the SuperPro Designer equipment cost database unit [12]. The analysis of the direct fixed capital cost and with their distinct factors utilized for its evaluation is shown in Table 12.2.

Table 12.2 Summary of direct fixed capital cost

Total plant direct cost (TPDC) (prices in USD)	
Equipment purchase cost	409,416,000
Installation	60,267,000
Process piping	143,295,000
Instrumentation	163,766,000
Insulation	12,282,000
Electrical	40,942,000
Buildings	409,416,000
Yard improvement	61,412,000
Auxiliary facilities	163,766,000
TPDC	1,464,562,000
Total plant indirect cost (TPIC)	
Engineering	366,141,000
Construction	512,597,000
TPIC	878,737,000
Total plant cost (TPC = TPDC + TPIC)	
TPC	2,343,299,000
Contractor's fee & contingency (CFC)	
Contractor's fee	117,165,000
Contingency	234,330,000
CFC	351,495,000
Direct fixed capital cost (DFC = TPC + CFC)	
DFC	2,694,794,000

12.5 Conclusion

Techno–economic analysis was accomplished for a creation of biotechnological route to produce alkaline protease from glucose through batch fermentation. The discrete methods used in techno–economic analysis include process design, equipment size, also economic estimation. The use of this analysis will surely make the cost-effective manufacturing of enzymes for biofuel applications.

References

1. A.K. Yadav, S. Pandey, A.D. Tripathi, V. Paul, Role of enzymes in biofuel production. bioenergy research: evaluating strategies for commercialization and sustainability, pp. 1–18 (2021)
2. P.V. Harris, F. Xu, N.E. Kreel, C. Kang, S. Fukuyama, New enzyme insights drive advances in commercial ethanol production. Curr. Opin. Chem. Biol. **19**, 162–170 (2014)

Part III
Thermochemical Conversion

Chapter 13
Biomass Briquetting for Gasification: Waste to Wealth

Anjali Kumari, Kritika Guleria, L. C. Meher, Madhulika Pathak, and Madhu Bala

Abstract World is progressing toward renewable and clean energy. The main driving forces are dependency on fossil fuel, limited sources, increasing demand, and environmental concerns. There is a requirement of shifting from petroleum-based fuel to renewable energy sources, and biomass is a potential source of renewable energy. Gasification provides a solution for renewable and clean energy as well as utilizes waste and/or abundant biomass in a better way, where partial combustion in limited supply of oxygen takes place to produce combustible gases called producer gas. The producer gas can directly be burnt to produce heat or can be purified to syn gas for production of electricity or liquid fuel. Biomass are mostly industrial waste, forest residue, farm produce, etc., having low bulk density and are largely varying in physical and chemical properties. Bulk density of biomass ranges from 40 kg/m^3 for loose straw to 250 kg/m^3 for wood residue which is unmanageable for handling, storage, or transportation. Densification increases bulk density by 2–10 folds and can be done with high, low, or medium compaction. Low compaction of biomass always requires binder. Screw press, piston press, hydraulic piston press technologies are in practice for compaction of biomass. Selection of any one of these technologies depends upon the feedstock availability, physical and chemical properties as well as characteristics of the product. Densification necessitates prior evaluation of the physicochemical properties of biomass like moisture, ash, lignin, calorific value, volatile matter, etc. Briquettes properties like low moisture content, good water resistance, high calorific value, good compressive strength, low ash content, etc., are the desirable properties for their acceptance by the users.

Keywords Lignocellulosic biomass · Energy density · Densification · Briquetting technologies · Gasification

---

A. Kumari (✉) · K. Guleria · L. C. Meher · M. Bala
Defence Institute of Bio-Energy Research, DRDO, Haldwani, Uttarakhand, India
e-mail: anjali.diber@gov.in

M. Pathak
Government Degree College, Bagheshwar, Uttarakhand, India

N. Gakkhar et al. (eds.), *Recent Advances in Bio-Energy Research*, Springer Proceedings in Energy, https://doi.org/10.1007/978-981-99-5758-3_13

13.1 Introduction

13.1.1 Renewable Energy

Crude oil is the world's main source of energy generating gasoline, diesel, jet fuel, asphalt, and lubricant oil, used mainly for power production, transportation, and heat generation. The oil reserves were 1732 billion barrels at the end of year 2020, two billion down to the figure of 2019. Demand of world's crude oil in first quarter of 2021 was 93.30 million barrel per day and is expected to increase as per IEA World Oil Supply and Demand Forecast. Road transportation took a major share of the oil in OECD (Organization for Economic Co-Operation and Development) member states. Motor vehicle usage was 35.23 percent of total oil consumption in 2019. Global purchase of imported crude oil in 2020 was US$ 683.1 billion reflecting demand from 115 countries, territories, and islands. Top four countries namely China, United States, India, and South Korea accounted 53.7% of the overall value of imported crude oil in 2020 [1]. India imported crude oil costing US$ 64.6 billion during 2020. Iraq, Saudi Arabia, and UAE were the top 3 suppliers of crude oil generating one-half (54.4%) of total Indian import of crude petroleum [1].

Petroleum products from crude oil are neither sustainable nor renewable and posing environmental threat because of their obnoxious exhaust products [2–8]. Earth climate is rapidly changing and one of the contributors is excessive greenhouse gases in the earth atmosphere. Atmosphere's share of carbon dioxide, one of the greenhouse gases, has increased by 46% since pre-industrial times. Most of the greenhouse gases caused by human activities come from burning of fossil fuel for electricity, transportation, and heating. World total carbon dioxide emissions in 2018 were 35.92 billion metric tons. India ranked third in carbon dioxide emissions with a figure of 2.65 billion metric tons (7%) after China (28%) and United States (15%) [9]. As per the World Resources Institute Climate Analysis Indicators Tool (WRI CAIT), India's 2014 GHG profile was dominated by emissions from the energy sector and accounted for 68.7% of total emissions. Further, extraction and transportation of oil poses environmental and safety risk. Deepwater Horizon disaster of 2010 released three million barrels of oil into the Gulf of Mexico leaving impact on ecosystem for decades.

Besides the concern of environmental impact, fossil fuels are having geopolitical impact too. These are typically found in specific parts of the world, hence making them plentiful for some nations only. World top ten oil producers share 72% of total oil production in the year 2020 as per International Energy Statistics 2021. This imparts monopoly of fossil fuel-rich countries which controls the oil prices as well as having concern of national security which import the oil from fossil fuel-rich countries. Fluctuating prices of crude oil, rapidly changing world political scenario, and adverse environmental impact are always being a big concern for India. Increasing energy demand, depleting resources, and notorious exhaust products of fossil fuel justify renewable, economic, and ecofriendly energy source [10–12].

Renewable energies are those which come from natural sources or processes that are constantly replenished. These are also termed as clean energy. Renewable energy sources are Solar energy, Wind energy, Hydro energy, Geothermal energy, and Biomass energy. Global renewable energy capacity in 2020 was 2.8 million MW [13]. India had 96.96 GW of renewable energy capacity as on July 2021, which is 25.2% of overall installed power capacity. India ranked fourth in wind power, fifth in solar, and fourth in renewable power installed capacity in 2020. Government of India is encouraging Industries/Academia/Organizations to work in the field of renewable energy via policies, subsidies, and other incentives. According to the analytics firm British Business Energy, India ranked 3rd globally in terms of its renewable energy investments and plans in 2020. The Ministry of New and Renewable Energy (MNRE) has set an ambitious target to set up renewable energy capacities to the tune of 227 GW by 2022, of which about 114 GW is planned for solar, 67 GW for wind, and other for hydro and bio among others.

Among all the renewable energy sources, the largest contribution, especially in the short to medium terms is expected from lignocellulosic biomass. According to IREDA "Biomass is capable of supplementing the coal to the tune of about 260 million tons," "saving of about Rs. 250 billion, every year." Biomass energy potential in India is 16,000 MW from biomass energy and 3,500 MW from bagasse cogeneration. In India approx. 19% of the electricity was produced through renewable energy sources and out of which the contribution of biomass energy was 0.62% for the year 2018–19 [14]. Biogas production in 2014–15 was 5% of the total LPG consumption in the country. As per the latest report, the biogas energy capacity in India was approximately 14 MW in 2021 [15]. The current ethanol production capacity in India is 4260 million liters from sugarcane-based distilleries and 2580 million liters from grain-based distilleries [16].

India implemented biomass power and cogeneration program with the main objective of promoting technologies for optimum use of country's biomass resources for grid power generation. Biomass materials used for power generation include bagasse, rice husk, straw, cotton stalk, coconut shells, soya husk, de-oiled cakes, coffee waste, jute wastes, groundnut shells, saw dust, etc. The Study indicated estimated surplus biomass availability at about 230 million metric tons per annum covering agricultural residues corresponding to a potential of about 28 GW. Over 800 biomass power and bagasse/non-bagasse cogeneration projects aggregating to 10,170 MW capacity have been installed in the country for feeding power to the grid.

13.1.2 Lignocellulosic Biomass and Conversion Technologies

The term biomass (Greek bio meaning life + maza meaning mass) refers to any material of biological origin excluding those that have been embedded in geological formations undergoing a process of mineralization. The term lignocellulosic biomass is used for biomass, particular of plant origin like agricultural waste (straw and husk from paddy and other, vegetable residue, fruit waste, etc.), forest residue (palm leaves, pine needles, oak leaves, etc.), industrial waste (corn cobs, oilseed cake, coconut shell, groundnut shell, etc.) [17, 18]. It is mainly composed of cellulose, hemicellulose, and lignin, including a few organic components like lipid and extractives [18–21]. Cellulose consists of long chains of β-glucose monomers which are arranged in highly ordered manner, whereas hemicelluloses consist of pentoses and hexoses sugar monomer units and are amorphous in nature. The third major constituent of lignocellulose biomass, i.e., lignins are phenolic compounds which are formed by polymerization of three types of monomers, i.e., p-coumaryl, coniferyl and synapyl alcohols [18]. The lignocellulosic nature makes biomass rich in energy content [22–24].

The ISO 17225-1: classified biomass feedstock resources into four groups, namely woody, herbaceous, fruit, and aquatic biomass [25]. In developing countries large amount of biomass residues are generated annually as by product of the commercial forestry, agriculture, and industrial sector [26, 27]. The supply of biomass from various sources around the globe is approximately 220 billion ton/year [28]. In USA, more than a billion tons of biomass are available which can replace about 30% of its petroleum consumption [29]. Annual potential of biomass in world is expected in between 200 and 500 EJ/year excluding aquatic biomass by 2050 [30]. In India, 750 million metric tons of biomass is produced every year as per the recent study report sponsored by MNRE. Agricultural residues annually amount to a total of 98 million tons of which 41 million tons can be used as energy source [31].

Biomass can be converted to fuel via biological or thermo-chemical processes [32–36]. Biological conversion includes fermentation to ethanol, biomethanation to methane, etc., whereas thermo-chemical conversion includes combustion to generate heat, pyrolysis to solid and liquid fuel, gasification to combustible gas or liquid fuel via Fischer–Tropsch reaction, etc.

Combustion of biomass for energy is being practiced since discovery of fire, however, this poses environmental threat like global warming because of their exhaust products like CO_2, CO, NOx, etc. [37–39]. Further, these biomass neither burn completely nor harness the total energy of the biomass. Gasification is the emerging and promising technology for harnessing biomass energy in the form of gaseous fuel [40].

13.1.3 Gasification

Gasification is a thermo-chemical process where carbonaceous materials are converted to combustible gases in presence of reagent, typically air, steam, hydrogen, oxygen, or a combination of these at high temperature (>700 °C). Unlike combustion, which takes place in presence of oxygen to give CO_2 and H_2O, gasification takes place in limited supply of oxygen to give combustible gases along with char and tar [10, 12, 41]. The crude combustible gaseous product is called Producer gas which consists of CO_2, CO, H_2O, CH_4, NOx, N_2, etc., and can be burned as a fuel such as in a boiler for heat or in an internal combustion gas engine for electricity generation or combined heat and power (CHP). This gas can further be purified to Syn gas which is mainly consists of CO and H_2. The Syn gas can be used as a fuel for electricity generation or heating and warming appliances or can be converted to liquid fuel via Fischer–Tropsch process, biomethanol, or biohydrogen.

Gasifiers are being used for biomass gasification to combustible gases. These are broadly classified into two types, i.e., fixed bed and fluidized bed gasifiers. Fixed types of gasifiers are updraft and downdraft types. In both type of gasifier material is loaded from the top of the gasifier and reactive agent is passed from the bottom of the gasifier. However, in the first type of gasifier, the combustible gases exit from the top and in the second type it exits from the bottom side. Each gasifier is having its own advantages and disadvantages, however, for small-scale gasification downdraft gasifiers (Fig. 13.1.) are mostly being used because of less complexity, less tar formation, and better performance [42].

Like any technology, gasification is having its own challenges. One of such challenge is feedstock material which is mainly low bulk density lignocellulosic biomass with varying shapes, sizes, and characteristics. Low bulk density of biomass poses difficulty in transportation, storage, and handling as well as affects feeding of biomass to gasifier especially in fixed bed gasifier. Because of low density, less material is

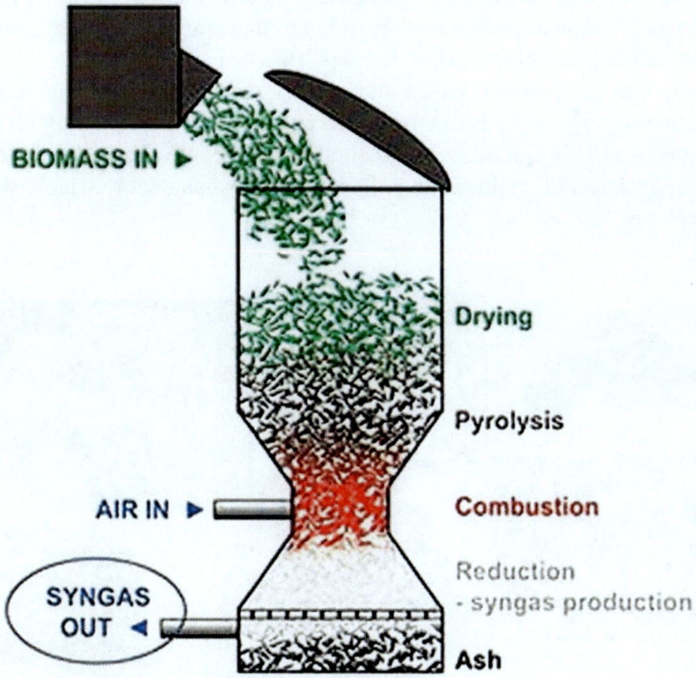

Fig. 13.1 Downdraft gasifier

loaded in the limited space of gasifier. For improving feeding characteristics, densification of biomass is the prime requirement which improves the energy density of biomass, subsequently enhancing the performance of gasifier.

13.2 Briquetting

Briquetting is densification of low-density biomass to get intact solid fuel with improved energy density. Biomass bulk density ranges from 40 kg/m^3 for loose straw to 250 kg/m^3 for wood residue [43]. Voluminous biomass possesses many problems in handling, transportation, and storage [44–53] and decreases transportation and storage efficiency. Densification in form of briquettes minimizes the handling, transportation, and storage problems [51, 52]. Besides this, briquetting also improves physical and mechanical properties of loose biomass [50, 51]. Based on physical and mechanical characteristics, briquettes have been presented to significantly reduce costs of storage and transportation [54–56]. Further, to improve feeding characteristics of these biomass to combustor/gasifier, densification is the prime requirement. Densification improves bulk density 2–10 times [57, 58] with enhanced energy density of the biomass, ultimately enhancing the performance of gasifier [59–63]. In addition, densification improves storage life of biomass as well as more convertibility into liquid fuels as compared to raw biomass [64–67].

Briquette can be classified based on biomass used, shape of briquette as well as size of briquette (Fig. 13.2). Based on the biomass used for briquetting there are rice husk briquette, coal briquette, coconut shell briquette, etc. These briquettes can be prepared in oval, round, cylindrical, hollow, or brick shape with variable sizes as per the requirement.

Fig. 13.2 Briquettes of different shapes and sizes

13.2.1 Potential Biomass and Their Characteristics

Availability. The biomass must be abundant and easy to collect for further processing like rice husk, ground nut shell, corn cobs, coconut shell, oilseed cake, etc. Sufficient suitable biomass must be available in the proximity of the biomass processing unit as well as to the biomass to bio-energy conversion facility to make the whole process sustainable and economic.

Moisture content. Moisture content for densification should not be >10% as high moisture content results in poor quality briquettes [68, 69]. However, the presence of moisture in optimal quantity helps in lowering the Tg of binder (lignin) and promotes solid bridge formation and increases the inter-particle contact via Van der Walls forces [60, 69].

Volatile Matter. It is a mixture of combustible and non-combustible gases consisting of short and long chain hydrocarbons which strongly affect the combustion behavior of briquettes [70]. High amount of volatile matter has positive influence on the sustainability of combustion, and it makes biomass a highly reactive fuel [70].

Ash content. Low Ash content is desirable in view of fuel properties of biomass as it is an indicator of slagging behavior of biomass [69]. In general, biomasses are having low ash content (<5%) but there are some exceptions like tobacco dust (19%) rice husk (22%), etc.

Particle size. Particle size and shape of biomass effects densification. Small particle results in smooth finishing without clogging and jamming of the machine. Very fine particles (<1 mm) are not suitable for screw extruder because of their cohesiveness and non-free flowing properties. The presence of different size particles improves the packing dynamics and contributes to high static strength [43–54].

Lignin content. Lignin shows glassy and brittle behavior below its glass transition temperature (Tg) (65–70 °C), becomes viscous, and spreads above this temperature and upon cooling it re-solidified. Lignin imparts binding properties and helps in densification [29, 51, 60].

Calorific value. High calorific value of biomass is desirable in briquetting for end application. Calorific value of biomass ranges from 3000 to 4700 cal/g for agricultural and forest residue from 5000 to 6000 cal/g for oilseed cake.

13.2.2 Biomass Processing to Briquette and Challenges

Suitability of biomass for briquetting depends on their physical and chemical properties [71–73]. These lignocellulosic biomasses are heterogeneous in nature with varying shapes and sizes. Moisture content and density also varies widely. Hence, prior to densification these biomass must undergo several stages like collection, storage, cleaning, drying, and size reduction. Biomass may be small, medium, or large. Medium and large size particles like pine needle, wheat straw, grasses, etc., are shredded to smaller size of 20–30 mm prior to pulverization. However, smaller particles like rice husk, camelina husk, leaves, etc., are directly pulverized to powder. Pulverizer selection depends upon the shape, size, and density of biomass. As moisture affect the densification process, hence, it must be reduced to acceptable value prior to compaction. Air drying does not add any cost but takes more time and is environment dependent. Low-temperature oven drier or solar drier can be used for reduction of biomass moisture in reduced time frame.

13.2.3 Fundamental Aspect of Briquetting

There are three fundamental aspects of briquetting [69].

Pressure compaction: There are high pressure compaction (100 MPa), medium pressure compaction (5–100 MPa), and low pressure compaction with binder. Under high pressure compaction, some of the components like lignin get activated as binder.

Binding mechanism: Binding mechanism under high pressure are (a) Adhesion and cohesion forces, (b) Attractive forces between solid particles, and (c) Interlocking bonds (Fig. 13.3).

Adhesion forces at the solid–liquid interface and cohesion forces within the solid are used for binding. Viscous binder such as tar form bonds very similar to solid bridges. Finally divided particles attract free atoms or molecules from the surrounding

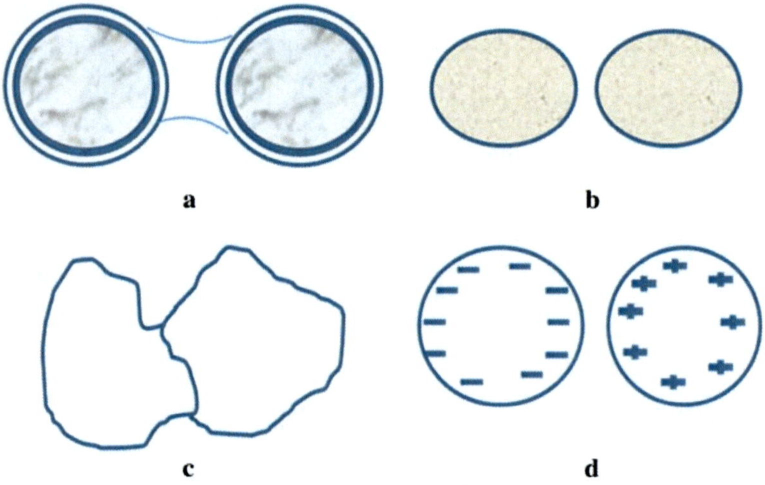

Fig. 13.3 Various binding mechanism: **a** Hardening binder, **b** Van der walls forces, **c** interlocking, **d** electrostatic forces

atmosphere. The thin layer formed is not freely movable and to some extent penetrates each other.

Upon applying external pressure, it further increases the contact area. This increases the molecular forces to transmit high enough and increases the bond strength between the adhering partners. Another important binding mechanism is Van der Walls forces which are very prominent between the adhesion partners separated by short distances. This type of adhesion possibility is much higher for powders. Interlocking or closed bonds are prominent in case of Fibrous and bulky particles.

Mechanism of Compaction. In case of screw extruder compaction occurs by the following mechanism:

(a) reaching the compression zone the biomass gets partially compressed. This leads to closer packing and increased density.
(b) At the compression zone, the biomass becomes relatively soft because of high temperature (200–250 °C). Loss of elasticity resulting in the biomass pressed into voids and the area of inter-particle contact increases. When the particles come together, they form local bridges which selectively support and dissipate the applied pressure. Interlocking of the particles may also take place depending upon the particle shape and size.
(c) The biomass gets further compressed in the tapering die (~280 °C) to form the briquette. The removal of steam and compaction takes place simultaneously during this step. Here, the pressure exerted transmits throughout the material giving uniform pressure and thus uniform density.

Different Briquetting technologies. Various types of briquetting technologies are in use [69, 74, 75]. Three widely used briquetting technology are discussed below:

Screw Press. Biomass in powder form or in recommended size is fed continuously into a screw which forces the material into a cylindrical die. Pressure is applied with augur continuously by passing the material through a screw with diminishing volume with or without heating system. Internal temperature often reaches 250–300 °C. At this temperature lignin of biomass liquifies and acts as binder excluding the external binder requirement. Briquettes form are of high density and good quality but the power requirement per ton is also high when compared with other technologies.

Piston Press. In this technology, material in powder form is fed into a cylinder which is then compressed by a piston mechanically into a slightly tapering die. It uses rotary power to generate a pressing force to shape the material in the form of briquette. Unlike Screw Press, pressure is applied discontinuously. Briquettes form are of good quality but with less density as compared to those produced from Screw Press. Binder may or may not require depending on the type of biomass.

Hydraulic Piston Press. This is similar to Piston Press with only difference that energy to the piston is transmitted from an electric motor via a high-pressure hydraulic oil system. Pressure is applied discontinuously by the action of a piston on material packed into a cylinder.

The briquette processing technology is being selected depending upon the feedstock availability, physical and chemical properties as well as characteristics of the product.

13.2.4 Physical and Chemical Parameters of Briquettes

Moisture content (%). Presence of moisture in briquette has adverse effect on shelf life and will lead to chemical and physical degradation. It has been reported that the moisture should be between 6 and 10% by weight [31]. One of the researchers reported that maximum durability rating of biomass pellets was obtained with moisture content 8.6% [58].

Apparent and bulk Density (kg/m³). The apparent density of a briquette from almost all types of biomass ranges from 1200 to 1400 kg/m^3. The apparent density of briquette is always higher than that of the bulk density of the briquette and usual reduction is approximately 600 to 700 kg/m^3 or sometimes less. Unit and bulk densities of briquettes depend upon material moisture, particle size, process temperature, and pressure [76, 77]. One researcher reported that bulk density of ground switchgrass was 165.5 kg/m^3 whereas, density of its pellets at 6.3% to 17% (wet basis) moisture content varies between 536 and 708 kg/m^3 [58]. Further, the density of blended biomass pellet also increases significantly than the individual agricultural biomass pellets [53]. Higher density leads to higher energy per unit volume and longer burning time.

Compressive strength (MPa). It shows the strength of briquette against compression. It should be more than 0.38 MPa for Industrial application [78].

Durability Index (%). Its significance is response of briquette during transportation. It is determined by tumbling test [80]. High durability index is desirable for commercial application of briquette.

Impact Strength Index (%). This signifies the strength and hardness of briquette. This is determined by dropping the sample three times from a height of 2 m into a metal floor. Impact Strength Index was calculated in percentage and it should be at least 90% [79].

Ash content (%). Ash content normally causes an increase in the combustion remnant, thereby lowering the heating effect [80–87]. Briquettes produced from high ash content biomass have high ash content as reported in case of corn cob and groundnut shell briquettes. In contrary mixture of wood residue of lower ash content to the agro wastes considerably reduced their ash content. Lower ash content is an indication of good-quality briquettes.

Volatile matter (%). Lower volatile matter is an indication that the briquettes might not be easily ignited, but once ignited they will burn smoothly, whereas high volatile matter results in high combustibility with low ash content [85]. As per one of the researchers, briquettes with higher volatile matter had higher specific heat of combustion [86].

Water resistance (Min). It shows the resistance of briquette to water and becomes significant during storage or transportation of briquettes. High water resistance is a desirable property to keep the material intact. It is determined by standard methods ASTM (1993), D870-09 [88].

Mass burning rate (g/sec). It is the time required to completely burn the unit mass of material. It signifies that how long a briquette will burn. It gets influenced by density due to reduced porosity which tends to hamper the rate of infiltration of oxidant and outflow of thermo-combustion products [53].

Calorific value (MJ/kg). It is the amount of energy released when unit mass of a substance burns completely in the presence of sufficient amount of oxygen. For commercial application, it should be more than 17.5 MJ/kg [63].

13.3 Conclusion

Gasification of biomass is an emerging and promising technology for renewable and sustainable energy source. Low bulk density of feedstock materials with varying shapes and sizes is a challenge for biomass gasification. It poses difficulty in transportation, storage, and handling as well as affects the feeding of biomass to gasifier. For improving feeding characteristics, densification of biomass is the prime requirement. Loose biomass bulk density ranges from 40 kg/m^3 for loose straw to 250 kg/m^3 for wood residue. Two to ten times gain in bulk density can be achieved through briquetting. Briquetting not only eases handling, transportation, and storage but also improves energy density. Briquettes can be prepared with high, low, or medium compaction in various shapes and sizes. Low pressure compaction of biomass always requires binder for densification in the form of briquette and is suitable for production of briquettes at smaller scale with lower investment. On the other hand, medium and high-pressure compaction can be achieved with or without binder, however, energy and infrastructure requirement make these processes more cost intensive. Such processes are suitable for large-scale industrial production. Different types of briquetting technologies like Screw press, piston press, hydraulic piston press, etc., are available in the market. Selection of the briquette processing technology depends upon the feedstock availability, physical and chemical properties as well as characteristics of the product. Prior to densification it is necessary to evaluate the physicochemical properties of biomass like moisture content, ash content, calorific value, volatile matter, density, lignin content, and other as these not only affect the densification process but also affect the properties of finished product. Further to this densification of biomass requires shredding to smaller particles or pulverizing to powder depending upon the type of briquetting technologies. Physical and chemical properties of briquettes like less moisture content, good water resistance, high calorific value, good compressive strength, low ash content, etc., are the desirable properties for their acceptance in the market. Regular supply chain of biomass for running large industrial briquetting plant is a critical factor, however, this can be addressed by using wide spectrum of available biomass as well as plantation of short duration woody crops. Cost of briquette production is another critical factor for sustained growth of the technology and need attention. Despite these challenges, biomass briquetting is the only solution to handle the voluminous material in a better way in terms of storage, transportation, and application.

References

1. W. Daniel, Crude oil imports by country. https://www.worldstopexports.com/crude-oil-imports-by-country. Last accessed 1 Nov 2021
2. Fossil Fuels, Environmental and Energy Study Institute (EESI). https://www.eesi.org/topics/fossil-fuels/description. Last accessed 10 Sept 2021
3. N. Christina, https://www.nationalgeographic.com/environment/article/fossil-fuels. Last accessed Sept 2021
4. D. Melissa, Fossil Fuels: The Dirty Facts. Natural Resource Defence Council. https://www.nrdc.org/stories/fossil-fuels-dirty-facts. last accessed 10 Sept 2021
5. Global energy review 2021, Assessing the effects of economic recoveries on global energy demand and CO_2 Emissions in 2021. *International Energy Agency*. Last accessed 10 Sept 2021
6. United Nation Environment Programme Emissions Gap Report 2021. https://www.unep.org/resources/emissions-gap-report-2021. Last accessed 23 Nov 2021
7. J.J. James, K. Peter, Solid-fuel household cook stoves: characterization of performance and emissions. Biomass Bioenergy **33**(2), 294–305 (2009)
8. N. MacCarty, D. Still, D. Ogle, Fuel use and emissions performance of fifty cooking stoves in the laboratory and related benchmarks of performance. Energy Sustain. Develop. **14**(3), 161–171 (2010)
9. https://www.investopedia.com/articles/investing/092915/5-countries-produce-most-carbon-dioxide-co2.asp. Last accessed 10 Sept 2021
10. K.R. Anil, Biomass gasification, in *Alternative energy in Agriculture*, vol. II, ed. by D. G. Yogi (CRC Press, 2014), pp. 83–102
11. G. Foley, G. Barnard, *Biomass Gasification in Developing Countries*, Technical Report No. 1 (Earthscan, London, 1983), p. 174
12. S. Shreya, V. Savita, Study of biomass briquettes, factors affecting its performance and technologies based on briquettes. IOSR J. Environ. Sci. Toxicol. Food Technol. (IOSR-JESTFT) **9**(11), 37–44 (2015)
13. IRENA, Renewable Energy Statistics 2021. The International Renewable Energy Agency, Abu Dhabi (2021). www.irena.org. Last accessed 10 Sept 2021
14. Summary of All India Provisional Renewable Energy Generation, http://www.cea.nic.in/reports/monthly/renewable/2019/renewable-03.pdf. Last accessed 10 Sept 2021
15. M. Jagmohan, Biogas Energy Capacity in India. https://www.statista.com/statistics/1044652/india-biogas-energy-capacity. Last accessed 15 Nov 2022
16. V.S. Kumar, Central Tuber Crop Research Institute bets on cassava as feedstock for bioethanol production Kochi (2021)

17. Z. Aya, P. Gabriel, Lignocellulosic biomass: understanding recalcitrance and predicting hydrolysis. Front. Chem. **7**, 874 (2019)
18. D. Ayhan, *Green Energy and Technology, Fuels from Biomass*, chapter 2 (Springer, 2009), pp. 43–59
19. T. Antonio, A review on biomass: Importance, chemistry, classification, and conversion. Biofuel Res. J. **6**(2), 962–979 (2019)
20. K.R. Navnit, T.R. Tr., S. Renganathan, Production of bioethanol by an innovative biological pre-treatment of a novel mixture of surgical waste cotton and waste card board. Energy Sources Part A: Recovery, Utilization Environ. Effects **42**(8), 942–953 (2020)
21. K.M. Ranjeet, M. Kaustubha, Characterization of non-edible lignocellulosic biomass in terms of their candidacy towards alternative renewable fuels. Biomass Convers. Biorefinery **8**, 799–812 (2018)
22. D. Ayhan, Combustion characteristics of different biomass fuels. Progress Energy Combust. Sci. **30**(2), 219–230 (2004)
23. D. Idalina, A. Umit, F. José, C.-L. Luisa, S. Ali, S. Sirri, E. Bruno, Calorific power improvement of wood by heat treatment and its relation to chemical composition. Energies **13**, 5322 (2020)
24. C. Telmo, J. Lousada, The explained variation by lignin and extractive contents on higher heating value of wood. Biomass Bioenerg. **35**, 1663–1667 (2011)
25. ISO 17225-1, *Solid Biofuels—Fuel Specifications and Classes—Part. 1: General Requirements* (ISO: Geneva, Switzerland, 2014)
26. N. Mary, K. Nancy, P. Gordon, M. John, M. Patrick, G. Kuria, M. Beatrice, *Community-Based Energy Briquette Production from Urban Organic Waste at Kahawa Soweto Informal Settlement, Nairobi, Urban Harvest Working Paper Series No. 5* (International Potato Center: Lima, Peru, 2009)
27. P. Sugumaran, S. Seshadri, *Biomass Charcoal Briquetting, Technology for Alternative Energy Based Income Generation in Rural Areas* (Shri AMM Murugappa Chettiar Research Centre: Taramani, Chennai, 2010), pp 1–20
28. K. Anil, K.N. Nitin, B. Prashant, S. Ashish, A review on biomass energy resources, potential, conversion, and policy in India. Renew. Sustain. Energy Rev. **45**, 530–539 (2015)
29. S.T. Jaya, Effect of process variables on the density and durability of the pellets made from high moisture corn stover. Biosys. Eng. **119**, 44–57 (2014)
30. Strategic insight, *World Energy Resources: Bioenergy, World Energy Council, Bioenergy*, chapter 7, (2013)
31. M.S. Khardiwar, K.D. Anil, D.M. Mahalle, S. Kumar, Study on Physical and chemical properties of crop Residues briquettes for gasification. Int. J. Renew. Energy Technol. Res. **2**(11), 237–248 (2013). ISSN: 2325–3924
32. C.C. Antonio, M. Palumbo, P.M. Pelagagge, F. Scacchia, Economics of biomass energy utilization in combustion and gasification plants: effects of logistics variables. Biomass Bioenerg. **28**(1), 35–51 (2005)
33. Y. Lin, S. Tanaka, Ethanol fermentation from biomass resources: current state and prospects. Appl. Microbiol. Biotechnol. **69**(6), 627–642 (2006)
34. T. Yoshioka, S. Hirata, Y. Matsumura, K. Sakanishi, Woody biomass resources and conversion in Japan: the current situation and projections to 2010 and 2050. Biomass Bioenergy, **29**(5), 336–346 (2005)
35. V. Dipti, S. Ashish, L. Banwari, M.S. Priyangshu, Conversion of biomass-generated syngas into next-generation liquid transport fuels through microbial intervention: potential and current status. Curr. Sci. **110**(3), 329–336 (2016)
36. K. Qian, A. Lise, T. Tianwei, D. Raf, Bioethanol from lignocellulosic biomass: current findings determine research priorities. Sci. World J **298153**, 13 (2014)
37. S.J. Gerssen-Gondelach, D. Saygin, B. Wicke, M.K. Patel, A.P.C. Faaij, Competing uses of biomass: assessment and comparison of the performance of bio-based heat, power, fuels and materials. Renew. Sustain. Energy Rev. **40**, 964–998 (2014)
38. K. Sunde, A. Brekke, B. Solberg, Environmental impacts and costs of woody biomass-to-liquid fuel (btl) production and use: a review. Forest Policy Econ. **13**(8), 591–602 (2011)

39. R.S. Dhillon, V.W. George, Mitigation of global warming through renewable biomass. Biomass Bioenerg. **48**, 75–89 (2013)
40. W. Lijun, L.W. Curtis, D.J. David, A.H. Milford, Contemporary issues in thermal gasification of biomass and its application to electricity and fuel production. Biomass Bioenerg. **32**(7), 573–581 (2008)
41. S. Sammy, Gasification, producer gas and syn gas, agriculture and natural resources. FSA 1051 (2009)
42. D. Ayhan, Green energy and technology. Biohydrogen **6**, 163–219 (2009)
43. M. Sudhagar, G.T. Lope, S. Shahab, Effect of compressive force, particle size, and moisture content on mechanical properties of biomass pellets from grasses. Biomass Bioenerg. **30**, 648–654 (2006)
44. S. Shahab, M. Sudhagar, B. Xiaotao, Z. Parisa, T. Lope, *Binderless Pelletization of Biomass, An ASAE Meeting Presentation*, Paper Number: 056061 (2005)
45. S. Shahab, K. Amit, F.T. Antony, Development and implementation of integrated biomass supply analysis and logistics model (IBSAL). Biomass Bioenergy **30**(10), 838–847 (2006)
46. M. Pettersson, T. Nordfjell, Fuel quality changes during seasonal storage of compacted logging residues and young trees. Biomass Bioenerg. **31**, 782–792 (2007)
47. T.A. Muhammad, H.B. Alemayehu, S. Shahab, M. Waseem, Storage of comminuted and uncomminuted forest biomass and its effect on fuel quality. BioResources **5**(1), 55–69 (2010)
48. E.M. Jan, D.K. Pieter, Influence of particle size and moisture content on tendency to bridge in biofuels made from willow shoots. Biomass Bioenerg. **24**, 429–435 (2003)
49. H.V. Pallavi, S. Srikantaswamy, B.M. Kiran, D.R. Vyshnavi, C.A. Ashwin, Briquetting agricultural waste as an energy source. J. Environ. Sci. Comput. Sci. Eng. Technol. **2**(1), 160–172 (2013)
50. S.T. Jaya, T.W. Christopher, J. H. Richard, L.K. Kevin, A review of biomass densification systems to develop uniform feedstock commodities for bioenergy application. Biofuels, Bioprod. Biorefining **5**, 683–707 (2011)
51. N. Kaliyan, R.V. Morey, Factors affecting strength and durability of densified biomass products. Biomass Bioenerg. **33**(3), 337–359 (2009)
52. Y.K. Sunday, F.Z. Mohamad, A.M. Latifah, M.R. Ahmad, A review of technical and economic aspects of biomass briquetting. Sustainability **12**, 4609 (2020)
53. C. Joel, *Combustion Characteristics of Biomass Briquettes*. Ph.D. Thesis Submitted to University Nottingham (2010)
54. Y.H. Noorfidza, T.A. Mohammad, Effect of particle size on mechanical properties made from biomass blends, in *4th International Conference on Process Engineering and Advanced materials, Procedia Engineering*, vol 148 (2016), pp. 93–99
55. K.T. Arun, P.V.R. Iyer, C.K. Tara, A techno-economic evaluation of biomass briquetting in India. Biomass Bioenergy **14**(5/6), 479–488 (1988)
56. T. Chesta, Producing fuel briquettes from sugarcane waste, in *EWB-UK National Research and Education Conference* (2011), pp. 39–45
57. P.S. Lam, S. Sokhansanj, X. Lim, C. Bi, L.J. Naimi, M. Hoque, S. Mani, A.R. Womac, X.P. Ye, S. Narayan, Bulk density of wet and dry wheat straw and dry particles. Appl. Eng. Agric. **24**(3), 351–358 (2008)
58. Z. Colley, O.O. Fasina, D. Bransby, Y.Y. Lee, Moisture effect on the physical characteristics of switchgrass pellets. Am. Soc. Agric. Biol. Eng. **49**(6), 1845–1851 (2006)
59. S. Reuben, M.M. Daniel, Production of loose biomass briquettes from agricultural and forestry residues. Sci. Direct Procedia Manuf. **7**, 98–105 (2017)
60. S.T. Jaya, Effect of pellet die diameter on density and durability of pellets made from high moisture woody and herbaceous biomass. Carbon Resourc. Conv. **1**, 44–54 (2018)
61. S. Poddar, M. Kamruzzaman, S.M.A. Sujan, M. Hossain, M.S. Jamal, M.A. Gafur, M. Khanam, Effect of compression pressure on lignocellulosic biomass pellet to improve fuel properties: higher heating value. Fuel **131**, 42–48 (2014)
62. C. Ndiema, P. Manga, R. Ruttoh, Densification and characteristics of rice straw briquettes. J. Inst. Energy **75**(502), 11–13 (2002)

63. K. Ajit, R. Madhuka, K. Krishnendu, Densification of biomass by briquetting: a review. Int. J. Recent Sci. Res. **8**(10), 20561–20568 (2017)
64. R. Binod, C. Lgathinathane, K. Bishnu, Y. Manlu, W.P. Scott, Combined effect of pelleting and pretreatment on enzymatic hydrolysis of switchgrass. Bioresource Technol. **116**, 36–41 (2012)
65. E.R. Allison, N.H. Amber, N. Nick, C. Xiaowen, L.G. Garold, Effect of pelleting on the recalcitrance and bioconversion of dilute-acid pretreated corn stover under low- and high-solids conditions. Biofuels **4**(3), 271–284 (2013)
66. J.W. Edward, J.N. Nicholas, M.N. Ryan, J.P. Darren, E.R. Allison, M.S. Daniel, The effect of biomass densification on structural sugar release and yield in biofuel feedstock blends. Bioenergy Resource **10**(2), 478–487 (2017)
67. K. Linoj, T. Zahara, S. Shahab, N.S. Jack, Does densification influence the steam pretreatment and enzymatic hydrolysis of softwoods to sugars? Bioresource Technol. **121**, 190–198 (2012)
68. N. Arzola, A. Gomez, S. Rincon, The effect of moisture content, particle size and binding agent content on oil Palm shell pellet quality parameters. Ingenieria E Investigacion **32**(1), 24–29 (2012)
69. P.D. Grover, S.K. Mishra, Biomass Briquetting: Technology and Practices, Regional Wood Energy Development Programme in Asia, GCP/RAS/154/NET, Field Document No. 46, Food and Agriculture Organization of The United Nation, Bangkok (1996)
70. S.V. Loo, J. Koppejan, *The Handbook of Biomass Combustion and Cofiring* (Earthscan, London, 2008)
71. S.T. Jaya, T.W. Christopher, L.K. Kevin, J.H. Richard, *A technical review on biomass processing: densification, preprocessing, modelling and optimization*. ASABE meeting paper, Paper No. 1009401 (2010)
72. J. Jindaporn, L. Charoenporn, Influences of mixing ratios and binder types on properties of biomass pellets. Sci. Direct, Energy Procedia **138**, 1147–1152 (2017)
73. S.H. Sengar, A.G. Mohod, Y.P. Khandetod, S.S. Patil, A.D. Chendake, Performance of briquetting machine for briquette fuel. Int. J. Energy Eng. **2**(1), 28–34 (2012)
74. R.N. Singh, P.R. Bhoi, S.R. Patel, Modification of commercial briquetting machine to produce 35 mm diameter briquettes suitable for gasification and combustion. Renew. Energy **32**(3), 474–479 (2007)
75. S. Eriksson, M. Prior, The briquetting of agricultural wastes for fuel. FAO Environ. Energy **11**, 141 (1990)
76. S. Mani, L.G. Tabil, S. Sokhansanj, An overview of compaction of biomass grinds. Power Handling Process. **15**(2), 160–168 (2003)
77. M. Sudhagar, G.T. Lope, S. Shahab, Specific energy requirement for compacting corn stover. Biores. Technol. **97**(12), 1420–1426 (2006)
78. P. Jitthep, M. Akarawit, Properties of solid fuel briquettes produced from rejected material of municipal waste composting, in *The 3rd International Conference on Sustainable Future for Human Security SUSTAIN 2012, Procedia Environmental Sciences*, vol. 17 (2013), pp. 603–610
79. B. Gabriel, S. Witold, W. Katarzyna, Effect of starch binder on charcoal briquette properties. Int. Agrophys. **31**, 571–574 (2017)
80. O.A. Sotannde, A.O. Oluyege, G.B. Abah, Physical and combustion properties of charcoal briquettes from neem wood residues. Int. Agrophys. **24**, 189–194 (2010)
81. A.C. Adetogun, K.M. Ogunjobi, D.B. Are, Combustion properties of briquettes produced from maize cob of different particle sizes. J. Res. For. Wildlife Environ. **6**(1), 28–38 (2014)
82. O.J. Akowuah, F. Kermausuor, J.S. Mitchual, Physicochemical characteristics and market potential of sawdust charcoal briquette. Int. J. Energy Environ. Eng. **3**, 18–26 (2012)
83. E.A. Emerhi, Physical and combustion properties of briquettes produced from sawdust of three hardwood species and different organic binders. Adv. Appl. Sci. Res. **2**(6), 236–246 (2011)
84. I.I. Ikelle, C. Anyigor, Comparative thermal analysis of the properties of coal and corn cob briquettes. IOSR J. Appl. Chem. **7**(6), 93–97 (2014)
85. A.R. Ige, C.M. Elinge, L.G. Hassan, I.A. Adegoke, H. Ogala, Effect of binder on physicochemical properties of fuel briquettes produced from watermelon peels. AASCIT J. Energy **5**(2), 23–27 (2018)

86. C.F. Babjide, I.J. Victoria, L.A. Oluwaseyi, D.N. Ravi, Performance evaluation of the physical and combustion properties of briquettes produced from agro-wastes and wood residues. Recycling **3**, 37 (2018)
87. K. Abia, K. Nicolas, K. Simon, B.K. Hussein, T. Peter, Potential of densification of mango waste and effect of binders on produced briquettes. Agric. Eng. Int: CIGR J. **16**(4), 146–155 (2014)
88. ASTM D 870, ASTM D 2247. https://www.trl.com/accelerated_water. Last accessed 1 Nov 2021

Chapter 14
Status and Future Thrusts of Sugarcane Processing Waste to Energy Conversion in Sugar Mills of India: Comprehensive Review

Omkar Karpe and P. Subramanian

Abstract Sugarcane and sugar play an important part in India's economy since sugar is the country's second-largest agriculture-based industrial sector, next to cotton. India is the largest sugar consumer and second-largest producer in the world. Sugar derived from sugarcane has a recovery rate of 12% throughout the sugar-making process. The remaining (other than crushed) sugar is found in molasses. The molasses is processed at the distillery unit to generate different grades of alcohol, including extra neutral alcohol, rectified spirits, impure alcohol and ethanol. During the crushing season, electricity is generated by burning bagasse as a fuel, and during the off-season, electricity is generated by cogeneration using conserved bagasse or other biomass. Bio earth is made by composting sugar industrial solid waste, and it is being used in farming operations as a bio-fertiliser and soil improver. The utilization of sugarcane processing waste to extract energy will benefit the industry and farmers economically, also it will lower the dependence on conventional and fossil fuels, and reduce environmental stress by reducing trash and reuse of waste.

Keywords Sugar mill · Cogeneration · Bagasse · Agro-industrial waste · Distillery · Press mud · Molasses

14.1 Introduction

Sugarcane (*Saccharum officinarum*) is a commonly produced crop in India, belonging to the Gramineae (Poaceae) family. It contributes to employing over a million people while also contributing significantly to the national budget [1]. Sugar is the country's

O. Karpe · P. Subramanian (✉)
Department of Renewable Energy Engineering, TNAU, Coimbatore, India
e-mail: manianpasu@yahoo.com; manianpasu@tnau.ac.in

O. Karpe
e-mail: omkarkarpe09@gmail.com

N. Gakkhar et al. (eds.), *Recent Advances in Bio-Energy Research*, Springer Proceedings in Energy, https://doi.org/10.1007/978-981-99-5758-3_14

second-largest agro-based sector. Over 5 crore farmers and their dependents work to grow sugarcane on nearly 50 lakh hectares of land, and together, the sugar industry and sugar cane have an influence on their quality of life [2]. India is the world's largest consumer and second-largest producer of sugar [3]. Sugarcane production averages roughly 35.5 million metric tonnes a year, with about 3 million tonnes of sugar produced [4]. Over 5000 LMT (Lakh Metric Tonnes) of sugarcane were harvested in India during the sugar crushing season October–September 2021–22, and 3574 LMT were crushed to produce 394 LMT of sugar. 35 LMT of this sugar was directly utilized for ethanol production, leaving 359 LMT for sugar industries to manufacture, with this India has surpassed all sugar processing countries to become the world's largest producer, user and exporter of sugar. These sugar exports brought in roughly Rs. 40,000 crores in foreign money for the nation [5].

14.2 Current Status of Sugar Mills Wastes to Energy Conversion

India contributes significantly to the global sugar market by producing about 15% and 25% of the world's sugar and sugarcane, respectively [6]. Each year, India produces 370–400 million tonnes (MT) of cane, 27–30 MT of white sugar, 6–8 MT of jaggery and khandsari. 3.2 billion litres of alcohol, 4700 MW of electricity, and other chemicals are also produced. Around 3500 MW of power export capacity by the industry to the national grid [7].

The sugar business consumes a lot of energy. With sugarcane as a raw material, steam and power is required to run the mill [8]. Largely the sugar industries have co-generation facility for the generation of steam and electricity [9]. As the low-pressure steam is used for heating, the cogeneration plant's efficiency is increased to 75 and 90%, compared with the traditional plant's efficiency of 35% [10]. During the sugar-making process, cane typically recovers 12% of its sugar content. The remaining sugar is present in the process's by-product, i.e. molasses [11].

Different types of spirit which include rectified, extra neutral, impure alcohol and ethanol can be produced in distillery [12]. The distillery having made up of several stages of fermentation chambers that are distilled in stagged manner with separation columns for distillation of different grades of alcohol [13]. Spent wash which is by-product of distillation process and has high fructose can be used for the biogas plants as feedstock [11]. Compost fertiliser can be produced by using press mud and spent wash by addition of microbial culture. Mixing and aeration machine is used to reduce the composting time [14]. A flow diagram of sugar mill process is shown in Fig. 14.1.

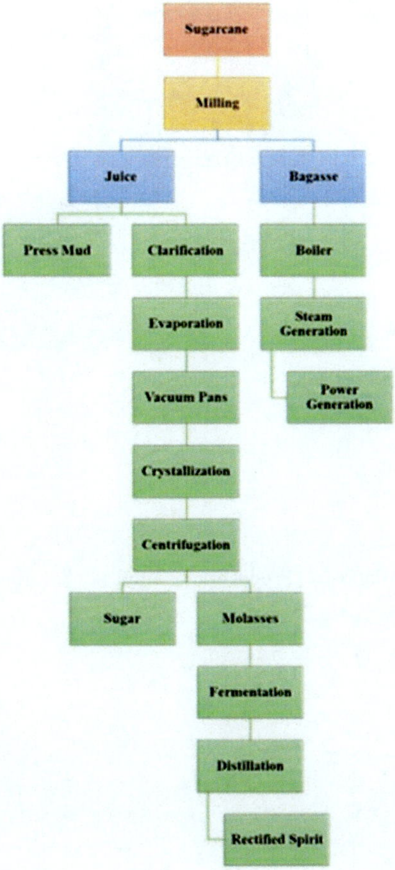

Fig. 14.1 Flow diagram of sugar mill process [15]

14.2.1 Alcohol Form Molasses

Molasses is a precious by-product of sugarcane, about 1 tonne of sugarcane produces 4% by weight of molasses by the sugar processing industry in every single run. Molasses is used for making alcohol. Brewer's yeast (Saccharomyces cerevisiae) is used to ferment the molasses, which has sugar and minerals as constituents, and can be used to produce beer that contains 6–10% ethanol [16]. Alcohol is used for making Extra Neutral Alcohol and Ethyl alcohol [17]. From 1 MT of molasses, 270 L of alcohol is produced [18]. A flow diagram of production of ethyl alcohol from cane molasses is shown in Fig. 14.2.

India presently has over 70 lakh tonnes of extra sugar, so there is plenty of room to shift extra sugarcane toward ethanol production without compromising the quantity of sugar needed to satisfy domestic demand [20]. One tonne of sugarcane produces 70 lit. of ethanol, as is well known [21]. The Ethanol Blended Petrol (EBP) initiative was started by the Indian government in 2003 with the goal of promoting environmentally

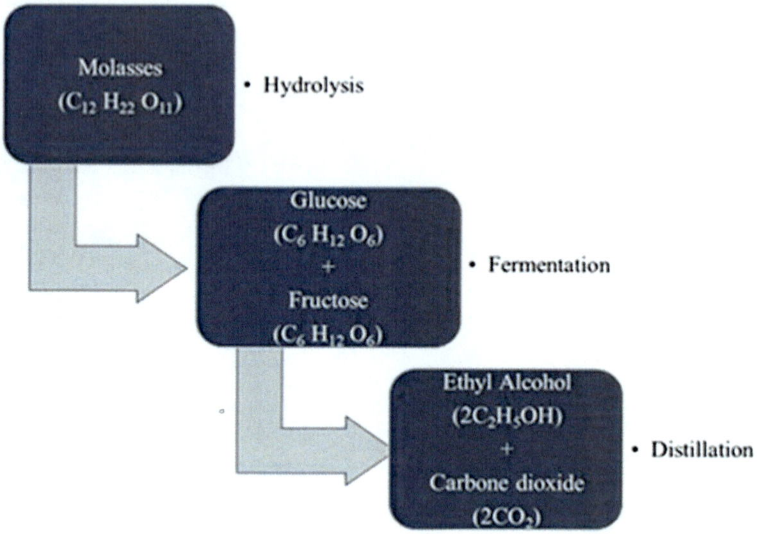

Fig. 14.2 Production of ethyl alcohol from cane molasses [19]

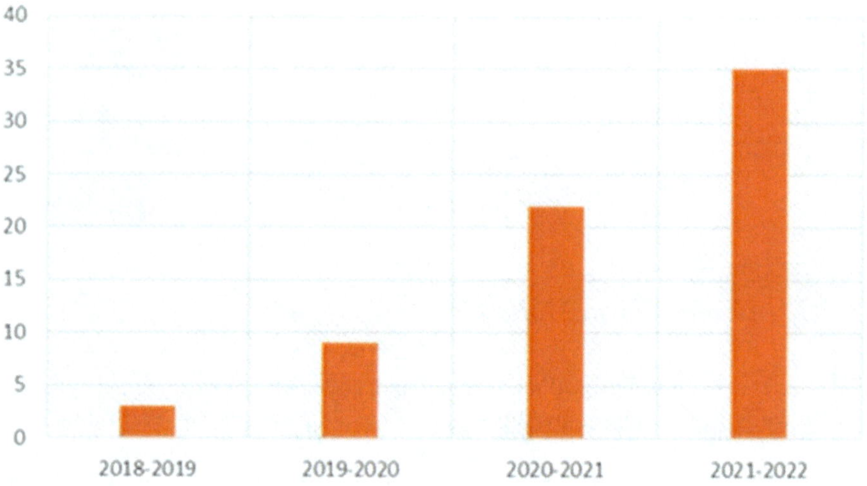

Fig. 14.3 Diversion of sugar for ethanol production (LMT)

friendly fuels by increasing the use of ethanol and lowering energy imports [22]. The EBP programme provides consistent ethanol demand, which injects money into the sugarcane industry [23]. This facilitates prompt payment to cane growers and reduces accumulated arrears for them. The National Policy on Biofuels 2018 expands the parameters for obtaining the raw materials needed to produce ethanol. The strategy planned to achieve a 20% mixing proportion by 2029–2030, [24]. The EBP initiative, by the Ministry of Petroleum and Natural Gas is now implementing, aims to attain a 10% ethanol mix ratio in petrol by 2021–2022 [25]. The target of blending of 10% ethanol by 2021–22 and the 20% by 2029–30, a mid-term target of 15% blending could be explored for 2024–25 [23]. In order to facilitate sugar mills to pay the cane dues of farmers on schedule and to be in a better financial position to continue their operations, the government is consistently pushing mills to divert sugar to ethanol production and to export excess sugar. The sugar industry has gained profit by production of ethanol as a biofuel over the past five years since sugar's usage in ethanol has improved sugar mills financial standing through quicker payments, lower working capital needs and less blockage of cash due to less surplus sugar with mills. The sale of ethanol generated revenue for sugar mills and distilleries in 2021–2022 of around 18,000 crore, which also contributed to the early payment of farmers' cane debts. In accordance with the Ethanol Blending with Petrol (EBP) Programme, the annual ethanol production capacity of molasses/sugar-based distilleries has expanded to 605 crore litres, and progress is still being made toward the 2025 objective of 20% blending. From 35 to 50 LMT of sugar is anticipated to be diverted to ethanol in the next season, bringing in roughly 25,000 crores in revenue for sugar mills [5].

14.2.2 Cogeneration from Bagasse

Bagasse is the fibres of sugarcane obtained after the extraction of juice from sugarcane crushing [16, 21]. Three tonnes of wet bagasse is generated for every ten tonnes of sugarcane crushed [26]. Wet bagasse contains moisture in the range of 48–52% [27]. In terms of net calorific value, dried bagasse is more valuable to the economy and industry. The task force on sugarcane suggests offering incentives to sugarcane mills so they can recycle bagasse [23]. Bagasse serves a variety of different purposes besides being a biofuel. Currently, it is used in the production of construction materials, pulp and paper products, and as a biofuel [28].

Initially bagasse often contains 40–50% moisture, which makes it impossible to utilize as fuel [29]. Bagasse is often kept before being processed further. It is maintained damp, where the modest exothermic process gently breaks down any remaining sugars and dries the bagasse pile, which is then used to produce power [30]. It is often stored moist for paper and pulp manufacture for the removal of the small pith fibres, which obstruct the paper making process, and to remove any leftover sugar. Bagasse is typically used as a fuel source in sugar mills. It can be burnt in big enough amounts to provide all the thermal energy required by a typical sugar mill [31].

Bagasse cogeneration uses the fibrous sugarcane waste, bagasse to cogenerate heat and electricity at higher efficiency in sugar mills. Electricity is produced by burning bagasse as fuel in the crushing season and from surplus bagasse or another biomass during off-season [32] (Fig. 14.4).

An average sugar and ethanol plant consumes 30 kWh of electricity per tonne of crushed sugarcane. Several electric motors and the machines both utilize this energy.

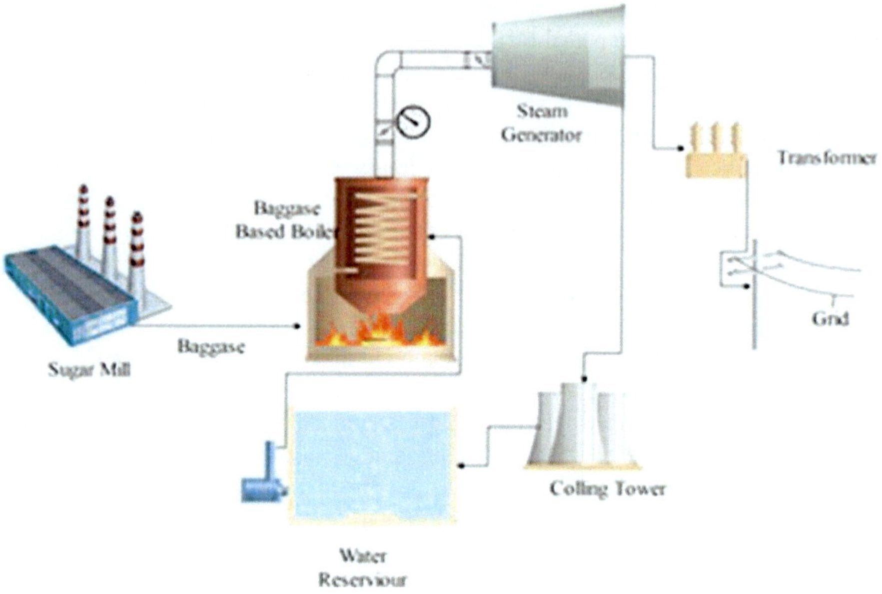

Fig. 14.4 Sugarcane bagasse cogeneration plant [33]

About 300 kWh/tonne of crushed cane (300 kWh/tc) of thermal energy is consumed during manufacture, which is 10 times more than the amount of electrical energy used. Sugarcane mills mostly employ cogeneration because it produces more thermal energy (TE) in the form of steam than electric power (EP) [34].

14.2.3 Press Mud

Press mud is a type of industrial waste produced by sugar mills. About 12 million tonnes of press mud (filter cake) are produced by sugar mills in India as a by-product of the twofold sulphitation process [35]. The proximate analysis of crude press mud showed the presence of moisture (67.95–76.53), nitrogen (1.63–2.29), ash (19.28–30.76), sugar (12.10–13.29) and crude wax (6.70–11.01) [36]. As a by-product, around 3 tonnes of press mud cake are left behind for every 100 tonnes of sugarcane crushed [37]. Bio earth is produced by composting press mud, and it is a biological oxidation process in which sugar industrial solid waste is mixed with microbial culture in warm, humid and aerobic conditions for decomposition [38]. Degradable organic substrate is transformed physically and chemically during the process to create a stable humidified product. The substance is useful in agriculture as a soil enhancer and an organic fertiliser. Recycling organic waste is becoming increasingly important. By adding early nitrogen, improving nutrient availability, boosting water

Table 14.1 Composition of bio earth [39]

Nutrients	Bio earth (%)
Nitrogen	2.0–2.5
Phosphorus	1.8–2.2
Potassium	3.0–4.0

retention, and adding colloidal nitrogen, phosphorous, potassium, calcium, sulphur, and micronutrients, Bio Earth boosts the soil's microorganism population [39]. The application of sugarcane press mud treatments boosted soil fertility, according to the study [40]. The composition of Bio Earth is shown in Table 14.1.

14.3 Future Thrusts of Sugarcane Waste to Energy Conversion

Biofuels are gaining more scientific and public attention as a way to meet future energy demands, and they may be the only viable option to ensure and strengthen the energy security by reducing the world's reliance on finite fossil fuels [41]. With the rise in the standard of living in recent years, per capita energy utilization has risen drastically. By 2025, global energy consumption is predicted to increase by a factor of ten, majorly occurring in quickly growing economies [42]. Bagasse pith has recently been used for bioethanol synthesis, which involves pre-treatment, hydrolysis, fermentation and dehydration [43]. Microbial Electrolysis Cell (MEC) is encouraging green technology which can be implemented for production of energy from sugar industrial wastewater [44]. Anaerobic co-digestion (AcoD) is technically feasible for sugarcane waste [45]. The Government of India has advanced the target for 20% ethanol blending in petrol (also called E20) to 2025 from 2030. E20 will be rolled out from April 2023 [46]. Study shows that the addition of bagasse to the co-digestion of press mud and distillery wastewater can boost biogas generation [47] (Fig. 14.5).

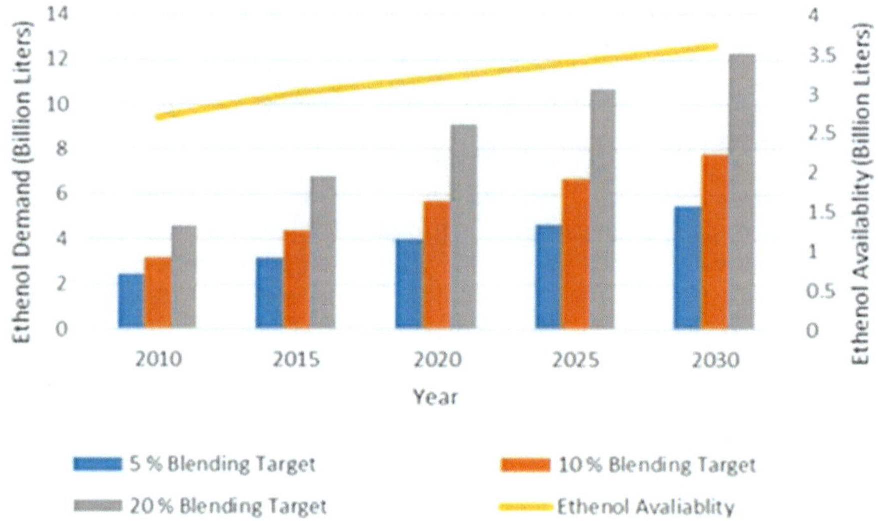

Fig. 14.5 Ethanol blending and availability graph of India [48]

14.4 Conclusion

Utilization of sugarcane processing waste to extract the energy is beneficial in terms of economic benefit to industry and farmers. It also reduces the burden on conventional and fossil fuel, reduces stress on environment by reducing waste and reusing the waste. It is also beneficial in agriculture as good quality manure in the form of Bio Earth. Thus, the sugarcane waste to energy generation and its technological advancement can be the breakthrough solution for the economic as well as environmental concerns.

References

1. N. Gowda, Sugarcane scenario in India: a view. Int. J. Soc. Sci. Econ. Res. (2019). Accessed: 04 Feb 2022. [Online]. Available: www.ijsser.org
2. NITI Aayog, Report of the task force on sugarcane and sugar industry
3. Sugar | WWF India. https://www.wwfindia.org/about_wwf/making_businesses_sustainable/sugar/. Accessed 04 Feb 2022
4. Niti Aayog panel recommends linking sugarcane prices to sugar rates: the economic times. https://economictimes.indiatimes.com/news/economy/agriculture/niti-aayog-panel-recommends-linking-sugarcane-prices-to-sugar-rates/articleshow/77658185.cms
5. India emerges as the world's largest producer and consumer of sugar and world's 2nd largest exporter of sugar. https://pib.gov.in/PressReleasePage.aspx?PRID=1865320. Accessed 04 Nov 2022
6. S. Solomon, Sugarcane agriculture and sugar industry in India: at a glance. Sugar Tech. **16**(2), 113–124 (2014). https://doi.org/10.1007/S12355-014-0303-8

7. S. Solomon, M. Swapna, Indian sugar industry: towards self-reliance for sustainability. Sugar Tech **24**(3), 630–650 (2022). https://doi.org/10.1007/S12355-022-01123-5/TABLES/3
8. A.N. Pathak, Energy conservation in sugar industries. J. Sc. Indus. Res. **58**, 76–82 (1999)
9. S.D. Mane, Cogeneration in Indian sugar industry: a review. Int. J. Sci. Eng. Appl. Sci. (IJSEAS) **2** (2016). [Online]. Available: www.ijseas.com
10. Co-generation. http://edugreen.teri.res.in/explore/renew/cogen.htm. Accessed 16 Feb 2022
11. Resource Recovery from Waste: Business Models for Energy, Nutrient and Water ...—Google Books. https://books.google.co.in/. Accessed 16 Feb 2022
12. Distillery Products. https://vsil.co.in/distillery-products/. Accessed 16 Feb 2022
13. R. Katzen, P.W. Madson, G.D. Moon, Ethanol distillation: the fundamentals of a distilling system
14. B. Ketan et al., Dynamic soil, dynamic plant compost from sugarmill pressmud and distillery spentwash for sustainable agriculture. Accessed: 16 Feb 2022. [Online]. Available: www.cpcb.nic.in
15. V. Saravanan, R.B. Kumar, Mathematical modelling and controller design using electromagnetic techniques for sugar industry process. **62**(2), 155–162 (2019). https://doi.org/10.1080/00051144.2019.1653667
16. T.R. Brown, M.M. Wright, Y. Román-Leshkov, R.C. Brown, Techno-economic assessment (TEA) of advanced biochemical and thermochemical biorefineries. Adv. Biorefineries: Biomass Waste Supply Chain Exploitation 34–66 (2014). https://doi.org/10.1533/9780857097385.1.34
17. G. Karthiga Devi, K. Vignesh, S. Chozhavendhan, Effective utilization of sugarcane trash for energy production. Refining Biomass Residues for Sustain. Energy Bioprod. Technol. Adv. Life Cycle Assessment Econ. 259–273 (2020). https://doi.org/10.1016/B978-0-12-818996-2.00012-0
18. Siddapur distilleries limited project report for capacity enhancement from 60 klpd to 70 klpd by keeping the effluent generation constant (2016)
19. P. Singh, Sugar industry: a hub of useful bio-based chemicals, in *Sugar and Sugar Derivatives: Changing Consumer Preferences* (2020) pp. 171–194. https://doi.org/10.1007/978-981-15-6663-9_11
20. Sugar mills: Mills exports 26.5 lakh tonnes of sugar till Jan in 2021–22 market year—the economic times. https://economictimes.indiatimes.com/news/economy/agriculture/mills-exports-26-5-lakh-tonnes-of-sugar-till-jan-in-2021-22-mkt-year/articleshow/89303750.cms?from=mdr. Accessed 16 Feb 2022
21. P. Murali, B. Ram, P. Prathap, K. Hari, V. Venkatasubramanian, Sugarcane based ethanol production for fuel ethanol blending program in India (2021)
22. Implementation of the Ethanol Blended Petrol Programme in India and its Policy Outcomes—India Foundation. https://indiafoundation.in/articles-and-commentaries/implementation-of-the-ethanol-blended-petrol-programme-in-india-and-its-policy-outcomes/. Accessed 16 Feb 2022
23. G. of I. NITI Aayog, Repot on the task force on sugarcane and sugar industry
24. Cabinet approves National Policy on Biofuels—2018. https://pib.gov.in/Pressreleaseshare.aspx?PRID=1532265. Accessed 16 Feb 2022
25. Refining. https://mopng.gov.in/en/refining/ethanol-blended-petrol. Accessed 16 Feb 2022
26. A.S. Amarasekara, Handbook of Cellulosic Ethanol. Google Books. https://books.google.co.in. Accessed 16 Feb 2022
27. S.I. Anwar, Determination of moisture content of bagasse of jaggery unit using microwave oven. J. Eng. Sci. Technol. **5**(4), 472–478 (2010)
28. T.J. Rainey, G. Covey, Pulp and paper production from sugarcane bagasse. Sugarcane-Based Biofuels Bioprod. 259–280. https://doi.org/10.1002/9781118719862.CH10
29. Bagasse Moisture. https://www.moisttech.com/applications/bioenergy-moisture/bagasse-moisture/. Accessed 16 Feb 2022
30. M.C. Detroja, Bagasse ash brick (2018). Accessed: 16 Feb 2022. [Online]. Available: www.ijariie.com

31. F.A. Salehi, Bagasse as a fuel for combined heat and power (CHP): an assessment of options for implementation in Iran. Accessed: 16 Feb 2022. [Online]. Available: https://bradscholars.brad.ac.uk/handle/10454/5303
32. C. Dinakaran, S. Purushotham, S.M. Harikrishna, Study of a cogeneration plant in sugar mill by using bagasse as a fuel (2016). [Online]. Available: http://www.ieejournal.com/
33. R.G.D. Molin Filho et al., Characterization of different sugarcane bagasse ashes generated for preparation and application as green products in civil construction. Clean Technol. Environ. Policy **21**(8), 1687–1698 (2019). https://doi.org/10.1007/S10098-019-01740-X/FIGURES/1
34. M. do Fernanda Scaranto Amaral, J. Roberto Ribas, S. Carlos, Economic viability study of an energy cogeneration project from sugarcane bagasse challenges and maturity of production engineering: competitiveness of enterprises, working conditions, environment
35. D. Munde, S. Sutar, Novateur publications international journal of innovations in engineering research and technology [IJIERT] vermicomposting of pressmud from sugar industry
36. M.M. Saleh-e-In, S. Yeasmin, B.K. Paul, M. Ahsan, M.Z. Rahman, S.K. Roy, Chemical studies on press mud: a sugar industries waste in Bangladesh. Sugar Tech **14**(2), 109–118 (2012). https://doi.org/10.1007/S12355-012-0139-Z
37. N. Gupta, S. Tripathi, C. Balomajumder, Characterization of pressmud: a sugar industry waste. Fuel **90**(1), 389–394 (2011). https://doi.org/10.1016/J.FUEL.2010.08.021
38. M. Ayilara, O. Olanrewaju, O. Babalola, O. Odeyemi, Waste management through composting: challenges and potentials. Sustainability **12**(11), 4456 (2020). https://doi.org/10.3390/su12114456
39. Cogeneration. https://sugar.maharashtra.gov.in/1040/1165/COGENERATION. Accessed 16 Feb 2022
40. V. Kumar, A.K. Chopra, Effects of sugarcane pressmud on agronomical characteristics of hybrid cultivar of eggplant (*Solanum melongena* L.) under field conditions. Int. J. Recyc. Organic Waste Agric. **5**(2), 149–162 (2016). https://doi.org/10.1007/S40093-016-0125-7/FIGURES/15
41. M.S. Khan, G. Mustafa, F.A. Joyia, S.A. Mirza, Sugarcane as future bioenergy crop: potential genetic and genomic approaches. Sugarcane Biotechnol. Biofuels (2021). https://doi.org/10.5772/INTECHOPEN.97581
42. Sugarcane as Energy Crop—Sugarcane Breeding Institute, Coimbatore, India. https://sugarcane.icar.gov.in/index.php/en/2014-04-28-11-31-50/sugarcane-as-energy-crop. Accessed 22 Feb 2022
43. E. Martinez-Hernandez, M.A. Amezcua-Allieri, J. Sadhukhan, J.A. Anell, Sugarcane bagasse valorization strategies for bioethanol and energy production. Sugarcane Technol. Res. (2017). https://doi.org/10.5772/INTECHOPEN.72237
44. K. Keruthiga, S.N. Mohamed, N.N. Gandhi, K. Muthukumar, Sugar industry waste-derived anode for enhanced biohydrogen production from rice mill wastewater using artificial photo-assisted microbial electrolysis cell. Int. J. Hydrogen Energy **46**(39), 20425–20434 (2021). https://doi.org/10.1016/J.IJHYDENE.2021.03.181
45. O. Mendieta, L. Castro, J. Rodríguez, H. Escalante, Management and valorization of waste from a non-centrifugal cane sugar mill via anaerobic co-digestion: technical and economic potential. Bioresour. Technol. **316**, 123962 (2020). https://doi.org/10.1016/J.BIORTECH.2020.123962
46. India's ethanol roadmap: the targets and challenges. https://www.downtoearth.org.in/blog/energy/india-s-ethanol-roadmap-the-targets-and-challenges-77360. Accessed 22 Feb 2022
47. M.C. Almendrala et al., Codigestion of press mud and distillery waste water with sugarcane bagasse for enhanced biogas production. https://doi.org/10.5220/0008692000460051
48. P. Purohit, S. Dhar, P. Purohit, S. Dhar, Lignocellulosic biofuels in India: current perspectives, potential issues and future prospects. AIMS Energy **6**(3), 453–486 (2018). https://doi.org/10.3934/ENERGY.2018.3.453

Chapter 15
Assessment of Thermal Behavior and Pyrolytic Kinetics of Selected Agro-residues through Thermogravimetric Analysis

Bhautik Gajera, Anil Kumar Sarma, and Mithilesh Kumar Jha

Abstract The physicochemical characteristics and kinetics of sugarcane bagasse (SCB) and black gram straw (BGS) have been studied using a thermogravimetric analyzer under nitrogen atmosphere at 20 °C/min heating rates. The analysis showed that SCB has higher volatile matter (81.32 wt. %), lower ash (3.50 wt. %), and moisture content (5.36 wt. %) while significantly higher HHV (18.10 MJ/kg). The TG and DTG curves displayed four significant zones mainly attributed to moisture removal, decomposition of hemicellulose and cellulose, and lignin, respectively. The Coats-Redfern model fitting method was used to determine the kinetic and thermodynamic parameters of the active pyrolysis zone. The activation energies of SCB and BGS were 88.74 and 96.74 kJ/mol, respectively. The following thermodynamic parameters were found: change in enthalpy is 83.65 and 92.14 kJ/mol, Gibbs free energy is 126.64 and 144.69 kJ/mol, and change in entropy is −73.29 and −94.92 J/mol.K, for SCB and BGS, respectively. These physicochemical properties and kinetic parameters confirm that SCB and BGS are promising feedstocks for pyrolysis.

Keywords Biomass residues · Physicochemical characterization · TG/DTG analysis · Reaction mechanism · Kinetic analysis

B. Gajera (✉) · A. K. Sarma
Chemical Conversion Division, Sardar Swaran Singh National Institute of Bio-Energy, Kapurthala, Punjab, India
e-mail: bhautik.ccd@nibe.res.in

B. Gajera · M. K. Jha
Centre for Energy and Environment, Dr. B R Ambedkar National Institute of Technology, Jalandhar, Punjab, India

N. Gakkhar et al. (eds.), *Recent Advances in Bio-Energy Research*, Springer Proceedings in Energy, https://doi.org/10.1007/978-981-99-5758-3_15

15.1 Introduction

Currently, renewable energy sources are gaining attention because fossil fuels are environmentally unfriendly and are set to deplete soon. Due to its availability, easy processing capabilities, and environmental friendliness, biomass is considered the most attractive renewable energy source. Despite being widely dispersed, it's already the world's fourth-biggest energy source. As opposed to coal, biomass can be processed easily to produce fuels and chemicals because of its high reactivity and volatility [1, 2]. In biomass materials, sunlight is converted into carbon dioxide and organic compounds via photosynthesis. This means that they are significantly less responsible for increasing CO_2 levels in the environment [3]. When biomass materials (and associated fuels) are burned, they cause considerably less environmental damage and health issues than conventional fuels. The use of biomass as alternative energy source also helps resolve issues associated with waste disposal.

As opposed to other renewable energy sources, biomass can be used as a raw material for liquids, gases, and solids. In general, there are two ways to accomplish the conversion: (i) biochemically, which occurs through the action of enzymes; (ii) thermochemical, which involves heating or oxidizing biomass [4]. The three most common thermochemical processes are pyrolysis, gasification, and liquefaction. There are several energy-producing methods, but pyrolysis is at the forefront as it allows for high fuel-to-feed ratios [2]. Biomass pyrolysis is the thermal breakdown of biomass in the absence of oxygen at moderate temperatures (400–600 °C) to get liquid products (bio-oil). Additionally, pyrolysis is used as an initial step in the combustion, gasification, and liquefaction processes of biomass [5]. For this reason, it is essential to understand the thermal behavior and kinetics of the pyrolysis process to determine biomass behavior in order to design efficient and effective pyrolysis and gasification systems.

SCB is among the most widely available biomass resources in India. The sugar mill industry's waste material has fulfilled the energy needs of developing nations such as Brazil, India, China, and Mexico. Currently, SCB generated from the sugar mill industry is used as a fuel for combined heat and power systems [6]. India is the world's second-largest sugarcane producer after Brazil, with 303.83 MMT. In India, sugarcane farming covers approximately 5.11 Mha areas with average productivity of 70 t/ha [7]. SCB is one of the biggest cellulosic agro-industrial-based byproducts. Sugar mills generate up to 30 wt. % residual lignocellulosic SCB from sugarcane, which is significant [8]. SCB comprises around 40–50% cellulose, 19–25.7% hemicellulose, 17–25% lignin, and 2–4% ash [7]. Most commonly, sugar mills use it as a primary energy source. SCB is utilized as a feedstock for various processing and manufacturing applications. The SCB can be used to produce biogas, biofuel, and bio fertilizer. Additionally, SCB can also be utilized as a feedstock for co-generating steam and electricity [9]. At the same time, black gram is a popular pulse crop in India, where it is commonly called Urad. Approximately 70% of India's global black gram is harvested [10]. In addition to being utilized as food for humans and livestock, it is also used to fix atmospheric nitrogen and maintain soil quality. Therefore, a large amount of black gram residue is generated after harvesting, like stalk and straw, which can be used for biofuels generation.

The purpose of this study is to achieve a greater understanding of the pyrolytic properties and kinetics of biomass residue, principally sugarcane bagasse and black gram straw. Thermogravimetric analyses were carried out for this purpose, and the kinetics triplets and thermodynamic parameters were assessed using the model fitting Coats-Redfern method. Moreover, the analyzed biomass residues were compared with the recent literature for their suitability as pyrolysis feedstock.

15.2 Materials and Methodology

SCB and BGS samples were acquired from the local market of the Kapurthala region (Punjab), India. The collected SCB and BGS were ground to particle size <212 and pre-dried in an open environment for two days before being stored in polyethylene bags.

15.2.1 Physicochemical Properties

The proximate, ultimate and compositional analysis and calorific values were carried out as per the protocol reported by Varma ad Mondal [11]. Volatile matter, moisture, and ash contents were estimated according to ISO 18134–3, ISO 18122, and ASTM E872-82 standard methods, respectively. A value of FC was estimated by subtraction of VM, MC, and Ash contents from 100 wt. %. Using an elemental analyzer (Vario MICRO Cube (Elementar, Germany)), elemental carbon, hydrogen, nitrogen, and sulfur were determined. Elemental oxygen was calculated by subtracting carbon, hydrogen, nitrogen, and sulfur percentages from 100 wt. %. According to BS 1016–5:1967, the HHV of SCB and BGS samples were determined using an oxygen bomb calorimeter model CC01/M2A (Toshniwal Technologies, India). To facilitate accurate results, experiments were performed in a repetition of three for the same sample and the mean value was accepted as the final value.

15.2.2 TG and DTG Analysis

PerkinElmer (STA 6000) TG/DTA instrument was employed to analyze and record the thermal behavior of selected biomass samples during pyrolysis. Experiments were run non-isothermally at a single heating rate of 20 °C/min. As a carrier, nitrogen gas with a purity of 99.99% was used at a 20 mL/minute flow rate. During each experiment, 10 $\pm$ 3 mg of sample mass was analyzed. Testing was conducted at a temperature range between 30 and 700 °C. The heating rate of 20 °C/min was used to estimate the kinetic triplets such as activation energy (E_α), pre-exponential factor (A), and reaction order (n). Each experiment was conducted twice, and average outcomes were recorded.

15.2.3 Kinetics Modeling

The kinetic modeling was performed using Coats-Redfern. Under iso-conversion methods, the mass of M_0 can be transformed to the mass of M_t at time t. The DTG curve gives a conversion ratio a, which is as follows:

$$i = \frac{M_0 - M_t}{M_0 - M_f} \tag{15.1}$$

The degree of transformation is i, the mass at the initial stage is M_0, the mass at time t is M_t, and the final mass is M_f.

Equation 15.3 can be used to determine the pyrolysis ratio with integrated Eq. 15.2

$$f(i) = (1 - i)^n \tag{15.2}$$

$$\frac{di}{dt} = Kf(i) = K(1 - i)^n \tag{15.3}$$

K is the Arrhenius Equation that expresses the rate constant as Eqs. 15.4 and 15.5 can be derived

$$K = Ae^{-\frac{E}{RT}} \text{ and } dT = \beta dt \tag{15.4}$$

$$\frac{di}{dT} = \frac{A}{\beta}e^{-\frac{E}{RT}}(1 - i)^n \tag{15.5}$$

where "n" is the order of reaction, "A" is the pre-exponential factor, "E" is the energy of activation, "R" is the gas constant, "T" is the temperature, and "t" is the time.

Panwar et al. [12] describe pyrolysis as a one-step global reaction, hence n = 1, so model the pyrolysis kinetic equation as Eq. 15.6 from Eq. 15.5,

$$\ln\left[-\frac{(1 - i)}{T^2}\right] = \ln\left[\frac{A_i R}{\beta E_i}\right] - \frac{E_i}{RT}\,(for\; n = 1) \tag{15.6}$$

Most reactions have minimum values of $\frac{RT}{E_i}$ (i.e., $\frac{RT}{E_i} \ll 1$), and therefore the following two formulas can be determined by

$$\ln\left[\frac{1 - (1 - i)^{1-n}}{T^2(1 - n)}\right] = \ln\left[\frac{A_i R}{\beta E_i}\right] - \frac{E_i}{RT}\,(for\; n \neq 1) \tag{15.7}$$

For each case, the following Y and X notations were used:

$$Y = \ln\left[\frac{1-(1-i)^{1-n}}{T^2(1-n)}\right] and\ X = \frac{1}{T}\ (for\ n \neq 1) \tag{15.8}$$

$$Y = \ln\left[-\frac{(1-i)}{T^2}\right] and\ X = \frac{1}{T}\ (for\ n = 1) \tag{15.9}$$

Therefore, both approaches fulfill the following equation:

$$Y = -\frac{E_i}{R}X + \ln\left[\frac{A_i R}{\beta E_i}\right] \tag{15.10}$$

The value of the order of reaction, activation energy, and pre-exponential factor has been calculated using the above expression.

15.2.4 Thermodynamic Parameters

Based on the Açıkalın et al. [13] equation, the change of enthalpy (ΔH), entropy (ΔS), and Gibbs free energy (ΔG) have been calculated, which is described as follows:

$$\Delta H = E_i - RT \tag{15.11}$$

$$\Delta G = E_i + RT_{\max} \ln\left(\frac{K_B T_{\max}}{hA}\right) \tag{15.12}$$

$$\Delta S = \left(\frac{\Delta H - \Delta G}{T_{\max}}\right) \tag{15.13}$$

Here, T_{max} is DTG peak temperature (K); K_B is the Boltzmann constant (1.381 $\times$ 10^{-23} J/K); h is the Plank constant (6.626 $\times$ 10^{-34} J.s). The E_i and A values were calculated using CR methods to calculate the above-mentioned thermodynamic parameters.

15.3 Results and Discussion

15.3.1 Physicochemical Properties of SCB and BGS

Proximate Analysis

Biomass fuel can be classified by its proximate and ultimate analysis. In Table 15.1, SCB, BGS, and wheat straw are compared in terms of proximate analysis, ultimate analysis, and heating value. The proximate analysis provides a brief analysis of biomass's heating values and burning properties [14]. It gives the mass percentages of moisture, volatile, ash, and fixed carbon. A significant variation exists in biomass moisture content, which is influenced by biomass's type and storage method. The low-moisture biomass is suitable for thermal conversion, while high-moisture biomass is more suitable for enzymatic or hydrothermal processes, as suggested by Chutia et al. [15]. For the pyrolysis process, the moisture content of the analyzed SCB and BGS samples is about 5.36 and 5.50%, respectively, which is within the desired range (<10%). In addition, moisture in the biomass sample makes pyrolysis more difficult, as it requires extra heat to extract the moisture from the biomass [14]. In addition, SCB and BGS biomass samples contain higher volatile matter (81.32 and 75.49%, respectively) and less ash content (3.50 and 5.65%, respectively). This indicates that biomass is well suited to be pyrolyzed due to its high reactivity and easy devolatilization and combustion. Due to high ash content, incomplete combustion can occur, leading to high operating costs, waste disposal issues, and lower conversion rates [6]. The SCB and BGS biomass samples contain a bit more fixed carbon (9.82 and 13.45%, respectively) than wheat straw (see Table 15.1).

It was found that the HHV of SCB and BGS biomass samples were about 18.10 and 16.90 MJ/kg, which were comparable or even more significant than those of the other reported crop residues such as Wheat straw (16.10 MJ/kg), walnut shell (17.85 MJ/kg), Sugarcane bagasse (18.10 MJ/kg), and Pistachio shell (17.39 MJ/kg) [16–19].

Ultimate Analysis

As a result of the ultimate analysis of the SCB biomass sample, its carbon, hydrogen, nitrogen, sulfur, and oxygen contents were 45.52, 5.67, 0.21, 0.05, and 48.55%, respectively, and 43.22, 6.30, 0.10, 0.02, and 50.36%, respectively, for the BGS biomass sample. The analysis results are compared with the wheat straw sample, as shown in Table 15.1. Compared to fossil fuel, SCB and BGS biomass samples contain higher amounts of hydrogen and oxygen than carbon because carbon–oxygen and carbon-hydrogen bonds have lower energy content than carbon–carbon bonds [20]. As compared to wheat straw, SCB and BGS contain a small amount of sulfur and nitrogen, so they produce significantly fewer toxic emissions such as SO_X and NO_X during the thermochemical conversion process [21]. SCB sample exhibited molar ratios of H/C and O/C around 1.48 and 0.48, respectively, whereas BGS showed 1.73 and 0.48, respectively. Also, the C/N ratio of SCB (252.78) and BGS

Table 15.1 Physicochemical properties of selected biomass residues

Sample	SCB	BGS
Proximate analysis (wt. %)		
M.C	5.36	5.50
V.M	81.32	75.49
Ash content	3.50	5.65
F.C[a]	9.82	13.45
HHV(MJ/kg)	18.10	16.90
Ultimate analysis (wt. %)		
Carbon	45.52	43.22
Hydrogen	5.67	6.30
Nitrogen	0.21	0.10
Sulfur	0.05	0.02
Oxygen[a]	48.55	50.36
Molar ratio		
H/C ratio	1.48	1.73
O/C ratio	0.80	0.87
C/N ratio	252.78	504.02
Fiber analysis (%)		
Hemicellulose	25.32	22.71
Cellulose	41.85	27.54
Lignin	22.17	38.95
Extractives	10.66	10.80

(504.02%) biomass samples as compared to wheat straw (99.93%) shows that clean fuel combustion is possible, as reported by Rathore et al. [16].

Lignocellulosic Composition

The composition analysis of biomass samples shows that SCB has the highest cellulose content, which is 41.85%, while BGS has 27.54%. The hemicellulose content varied from 22.71 to 25.32%, lignin content around 39% in the BGS sample, which was higher than the SCB sample. The BGS biomass had a higher lignin content than SCB (22.17%), leading to a higher level of char formation, as shown in Fig. 15.1. As shown in Table 15.1, SCB and BGS extractive contents were around 10%, which was very close to the other reported biomass values [12]. The lignocellulosic composition influences biomass thermal pyrolysis behavior. Biomass with high hemicellulose and cellulose and relatively low lignin content has a faster rate of thermal degradation during pyrolysis because these compounds are degraded at a lower temperature range than the lignin [21].

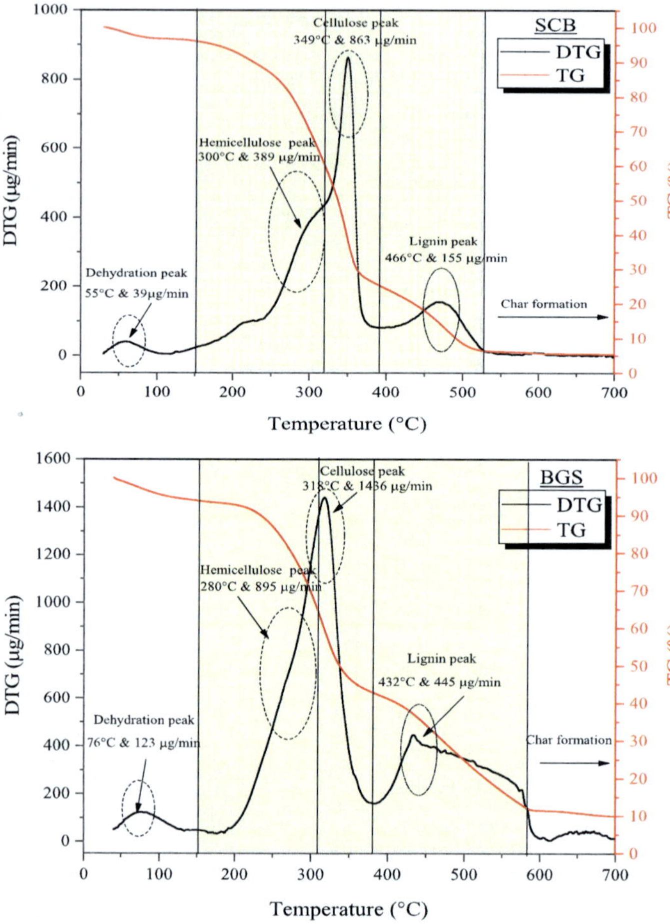

Fig. 15.1 TG and DTG thermograms for the SCB and BGS

15.3.2 TG and DTG Analysis

Figure 15.1 shows the thermograms of mass loss (TG) and derivative mass loss (DTG) for SCB and BGS at 20 °C/min. Generally, the pyrolysis process involves four stages: free moisture removal, the primary devolatilization process in which hemicellulose and cellulose degradation take place, and continuous slight devolatilization [22]. A slight weight loss was observed during the first stage (<150 C). A total weight loss of about 83.17% for SCB and 76.10% for BGS was observed within the range of active zones (see Fig. 15.1 and Table 15.2). Shah et al. [23] reported that around 72 to 80% of total mass loss occurred within this temperature zone. The SCB sample showed a more significant weight loss than BGS, which correlates with higher volatile and lower ash content (Table 15.1). The position of the peaks in the DTG curves indicates

Table 15.2 Thermal properties of biomass samples by using TG analysis at 20 °C/min heating rate

Stage	Dehydration	Hemicellulose degradation	Cellulose degradation	Lignin degradation	Char formation
SCB					
T (°C)	30–150	150–320	320–390	150–525	>525
TG (%)	4.05	32.45	34.92	15.80	–
DTG_{max} (μg/min)	39	389	863	155	–
T_{max} (°C)	55	300	349	466	–
BGS					
T (°C)	30–150	150–310	310–380	150–580	>570
TG (%)	7.2	22.18	27.64	26.15	–
DTG (μg/min)	123	895	1436	445	–
T_{max} (°C)	76	280	318	432	–

the temperatures at which the maximum weight loss rate occurs. The SCB and BGS samples exhibit different TG and DTG values, indicating different reactivity. As shown in Table 15.2, the SCB and BGS samples have significantly different decomposition behaviors (Table 15.2). As shown in Fig. 15.1 and Table 15.2, the sample of SCB displayed three clearly visible peaks with a very light shoulder at about 300 °C and two others at 349 and 466 °C. In contrast, the BGS sample exhibited two peaks at approximately 318 and 432 °C in the DTG curves. Biomass composed of lignocellulosic materials consists primarily of hemicellulose, cellulose, and lignin, all of which affect pyrolysis behavior [8]. Most of the lignocellulosic materials decompose between 200 and 500 °C. Decomposition begins with lignin at a lower temperature and a slower rate and continues until approximately 900 °C. DTG curves exhibit four major peaks that can be qualitatively explained as follows: The first peak could result from the removal of free moisture, the second (light shoulder) from the decomposition of hemicellulose and some lignin, and the third one from the decomposition of cellulose and the remaining lignin. The final decomposition of the lignin fraction occurs at high temperatures (>550 °C), resulting in minimal weight loss [24].

15.3.3 Kinetics and Thermodynamic Parameters

Designing effective and efficient gasifiers/pyrolizers for producing biofuels from biomass requires understanding the pyrolysis kinetics of SCB and BGS biomass samples. To analyze the pyrolysis kinetics in this study, only the active pyrolysis zone (light red color) was considered. Table 15.3 displays the values of the pre-exponential factor (A), apparent activation energy (Eα), reaction order (n), and correlation coefficient (R-Square) for each stage of decomposition for SCB and BGS samples. The reaction order (n) was determined through trial calculations for a wide range of n values from 0.1 to 2. Use the maximum R-square values from the linearization process to calculate the values of Eα and A. The plot in Fig. 15.2 illustrates the linear data for both the SCB and the BGS samples in accordance with Eq. (15.13). The results show that the R-square values are > 0.90, which fits well with the experimental data.

Table 15.3 Kinetics triplets of selected agro-residues biomass through CR method

Sample	Stage	n	R-Square	E (kJ/mol)	A (s^{-1})
SCB	1st	0.8	0.952	46.00	5.93×10^{03}
	2nd	1.2	0.947	131.06	3.63×10^{09}
	Average	1	–	88.53	1.81×10^{09}
BGS	1st	1.5	0.973	87.89	2.40×10^{08}
	2nd	1.4	0.935	105.70	1.39×10^{07}
	Average	1.4	–	96.74	1.27×10^{08}

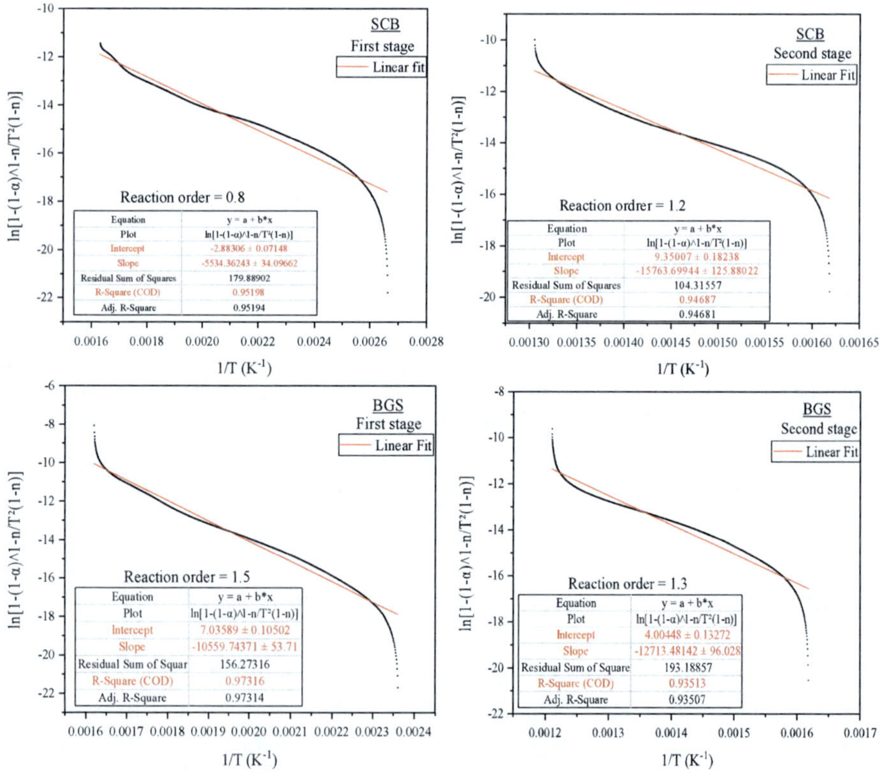

Fig. 15.2 Linear plots for SCB and BGS sample by using CR model fitting method

During the first decomposition stage, temperatures ranged from 150 to 300 °C, mostly hemicellulose and partly cellulose decompose within this range, with activation energy and pre-exponential factor 46.00 kJ/mol and 5.93 × 10^{03} s^{-1} respectively for SCB and 87.89 kJ/mol and 2.40 × 10^{08} s^{-1} respectively for BGS. Kristanto et al., [25] found that this temperature range was very close to that for the breakdown of hemicellulose and cellulose. It was also noted by Wang et al. [21] that in the temperature range of 200–280 °C, pure hemicellulose was broken down, and pure cellulose decomposed in a range of 250–360 °C. The second stage of biomass decomposition occurred in the temperature range of 300 to 550 °C. The apparent activation energy (Eα) and pre-exponential factor (A) were relatively high in the second stage. The activation energy for this stage was 131.06 kJ/mol and 105.70 kJ/mol, and the pre-exponential factor was and 3.63 × 10^{09} s^{-1} and 1.39 × 10^{07} s^{-1} for SCB and BGS samples, respectively. As de Palma et al. [18] stated, pyrolysis reactions with a high average Eα require a higher reaction temperature or a longer reaction time. The order of reaction (n) values for SCB and BGS samples are summarized in Table 15.3. It can be seen that the values of n are close to first-order reactions, within the range of 0.8 to 1.5. Although these values can differ with the method and approach used, the results obtained are in accordance with previously reported studies by other authors (Table 15.5).

Table 15.4 summarizes the calculated thermodynamic parameters (ΔH, ΔG, and ΔS). The concept of enthalpy refers to the difference in energy levels between reactants and intermediates, along with the nature of the chemical reaction [10]. Moreover, throughout the degradation stage, a positive value of enthalpy was found for both samples, verifying that reaction is endothermic in which molecules utilize the heat energy to decompose and form new chemical bonds. The difference between enthalpy and activation energy values is minimal for both the samples, which confirms that pyrolysis of selected biomass for energy generation is feasible [15]. Gibb's free energy indicates the total available energy of a system. The Gibbs free energy found in this study was 126.64 kJ/mol and 144.69 kJ/mol in SCB and BGS, respectively. According to the results obtained, endothermic or nonspontaneous reactions occur throughout the process. Rathore et al. [16] studied the kinetics and thermodynamics characteristics of pigeon pea stalk and found a Gibbs free energy of 165–172 kJ/mol. At the heating rate of 20 C/min, the calculated average entropy (ΔS) for SCB and BGS was −73.29 and −94.92 J/mol.K, respectively. The negative sign of ΔS reflects less disorder in products formed through bond dissociation as compared to primary reactants. Gajera and Pawar [10] reported comparable results for pyrolysis kinetics of black gram straw.

Table 15.4 Thermodynamic parameters of selected agro-residues biomass

Sample	ΔH (kJ/mol)	ΔG (kJ/mol)	ΔS (J/mol. K)
SCB	83.65	126.64	−73.29
BGS	92.14	144.69	−94.92

Table 15.5 Comparison of kinetics parameters with previously reported agro-residues

Sample	Method	E (kJ/mol)	A (s^{-1})	References
SCB	CR	88.53	1.81×10^{09}	This study
BGS	CR	96.74	1.27×10^{08}	This study
Cotton stalk	CR	128.70	1.78×10^{10}	Panwar et al. [12]
Walnut shell	CR	45.6–78.4	10^{03}	Açıkalın [17]
Pistachio shell	CR	156–370	10^{15}–10^{20}	Açıkalın [19]
Pine needle	CR, FWO, KAS	70–79	10^{10}–10^{11}	Kristanto et al. [25]
Bagasse	DAEM	175	10^{10}–10^{20}	Palma et al. [18]
Bagasse	FWO, KAS	66–125	10^{10}–10^{12}	Varma and Mondal [11]
Black gram straw	FWO, KAS, Starink	135–210	10^{10}–10^{16}	Gajera and Panwar [10]
Mesua ferrea L deoiled cake	CR, FWO, KAS	64–250	10^{10}–10^{20}	Chutia et al. [15]

15.4 Conclusion

This work analyzes the physicochemical properties and kinetics of SCB and BGS using a non-isothermal TG analyzer. As a result of physicochemical properties, SCB has higher volatile matter, higher HHV value, and lower ash content than BGS. The TG analysis observed four main zones: moisture removal, hemicellulose, cellulose decomposition, and lignin decomposition. More than 70% of the mass loss occurred in active pyrolysis zones. Both SCB and BGS samples had similar behaviors but different reactivities because of their different crystalline structures and elemental compositions. The activation energies associated with the active pyrolysis zone were 88.53 kJ/mol for SCB and 96.74 kJ/mol for BGS, respectively. Based on the result, SCB was highly suitable for pyrolysis processes.

Acknowledgements The authors are thankful to the Sardar Swaran Singh National Institute of Bio-Energy, Punjab, India (An Autonomous Institute of Ministry of New and Renewable Energy, Government of India) and National Institute of Technology, Jalandhar, India.

References

1. L. Dai, Y. Wang, Y Liu, R. Ruan, C. He, Z. Yu, L. Jiang, Z. Zeng, X. Tian, Integrated process of lignocellulosic biomass torrefaction and pyrolysis for upgrading bio-oil production: a state-of-the-art review. Renew. Sustain. Energy Rev. **107**, 20–36 (2019)
2. J.A. Okolie, S. Nanda, A.K. Dalai, J.A. Kozinski, Chemistry and specialty industrial applications of lignocellulosic biomass. Waste Biomass Valorization. **12**(5), 2145–2169 (2021)

3. A. Abdullah, A. Ahmed, P. Akhter, A. Razzaq, M. Hussain, N. Hossain, M.S.A. Bakare, S. Khurram, K. Majeed, Y.K. Park, Potential for sustainable utilisation of agricultural residues for bioenergy production in Pakistan: an overview. J. Clean. Prod. **287**, 125047 (2021)
4. J. Yan, O. Oyedeji, J.H. Leal, B.S. Donohoe, T.A. Semelsberger, C. Li, A.N. Hoover, E. Webb, E.A. Bose, Y. Zeng, C.L. Williams, K.D. Schaller, N. Sun, A.E. Ray, D. Tanjore, Characterizing variability in lignocellulosic biomass: a review. ACS Sustain. Chem. Eng. **8**(22), 8059–8085 (2020)
5. V. Dhyani, T. Bhaskar, A comprehensive review on the pyrolysis of lignocellulosic biomass. Renew. Energy. **129**, 695–716 (2018)
6. M. Hiloidhari, K. Araújo, S. Kumari, D.C. Baruah, T.V. Ramachandra, R. Kataki, I.S. Thakur, Bioelectricity from sugarcane bagasse co-generation in India-an assessment of resource potential, policies and market mobilization opportunities for the case of Uttar Pradesh. J. Clean. Prod. **182**, 1012–1023 (2018)
7. K.S. Konde, S. Nagarajan, V. Kumar, S.V. Patil, V.V. Ranade, Sugarcane bagasse based biorefineries in India: potential and challenges. Sustain. Energy Fuels. **5**(1), 52–78 (2021)
8. S. Quereshi, T.K. Naiya, A. Mandal, S. Dutta, Residual sugarcane bagasse conversion in India: current status, technologies, and policies. Biomass Convers. Biorefin. 1–23 (2020)
9. G. Athira, A. Bahurudeen, S. Appari, Thermochemical conversion of sugarcane bagasse: composition, reaction kinetics, and characterisation of byproducts. Sugar Tech. **23**(2), 433–452 (2021)
10. B. Gajera, N.L. Panwar, Pyrolysis and kinetic behaviour of black gram straw using thermogravimetric analysis. Recovery Util. Environ. Eff. 1–14 (2019)
11. A.K. Varma, P. Mondal, Physicochemical characterization and pyrolysis kinetic study of sugarcane bagasse using thermogravimetric analysis. J. Energy Resour. Technol. **138**(5) (2016)
12. N.L. Panwar, B. Gajera, S. Jain, B.L. Salvi, Thermogravimetric studies on co-pyrolysis of raw/torrefied biomass and coal blends. Waste Manag. Res. **38**(11), 1259–1268 (2020)
13. K. Açıkalın, G. Gözke, Thermogravimetric pyrolysis of onion skins: determination of kinetic and thermodynamic parameters for devolatilization stages using the combinations of isoconversional and master plot methods. Bioresour. Technol. **342**, 125936 (2021)
14. J. Cai, Y. He, X. Yu, S.W. Banks, Y. Yang, X. Zhang, Y. Yu, R. Liu, A.V. Bridgwater, Review of physicochemical properties and analytical characterization of lignocellulosic biomass. Renew. Sustain. Energy Rev. **76**, 309–322 (2017)
15. R.S. Chutia, R. Kataki, T. Bhaskar, Thermogravimetric and decomposition kinetic studies of Mesua ferrea L. deoiled cake. Bioresour. Technol. **139**, 66–72 (2013)
16. N.S. Rathore, A. Pawar, N.L. Panwar, Kinetic analysis and thermal degradation study on wheat straw and its biochar from vacuum pyrolysis under non-isothermal condition. Biomass Convers. Biorefin. 1–13 (2021)
17. K. Açıkalın, Thermogravimetric analysis of walnut shell as pyrolysis feedstock. J. Therm. Anal. Calorim. **105**(1), 145–150 (2011)
18. K.R. de Palma, N. Garcia-Hernando, M.A. Silva, E. Tomaz, A. Soria-Verdugo, Pyrolysis and combustion kinetic study and complementary study of ash fusibility behavior of sugarcane bagasse, sugarcane straw, and their pellets—case study of agro-industrial residues. Energy Fuels. **33**(4), 3227–3238 (2019)
19. K. Açıkalın, Pyrolytic characteristics and kinetics of pistachio shell by thermogravimetric analysis. J. Therm. Anal. Calorim. **109**(1), 227–235 (2012)
20. T. Kan, V. Strezov, T.J. Evans, Lignocellulosic biomass pyrolysis: a review of product properties and effects of pyrolysis parameters. Renew. Sustain. Energy Rev. **57**, 1126–1140 (2016)
21. S. Wang, G. Dai, H. Yang, Z. Luo, Lignocellulosic biomass pyrolysis mechanism: a state-of-the-art review. Prog. Energy Combust. Sci. **62**, 33–86 (2017)
22. J. Chen, X. Fan, B. Jiang, L. Mu, P. Yao, H. Yin, X. Song, Pyrolysis of oil-plant wastes in a TGA and a fixed-bed reactor: thermochemical behaviors, kinetics, and products characterization. Bioresour. Technol. **192**, 592–602 (2015)
23. M.A. Shah, M.N.S. Khan, V. Kumar, Biomass residue characterization for their potential application as biofuels. J. Therm. Anal. Calorim. **134**(3), 2137–2145 (2018)

24. A.K. Varma, P. Mondal, Physicochemical characterization and kinetic study of pine needle for pyrolysis process. J. Therm. Anal. Calorim. **124**(1), 487–497 (2016)
25. J. Kristanto, M.M. Azis, S. Purwono, Multi-distribution activation energy model on slow pyrolysis of cellulose and lignin in TGA/DSC. Heliyon **7**(7), e07669 (2021)

Chapter 16
Slow Pyrolysis of Rice Husk in a Lab-Scale Batch Reactor: Influence of Temperature on the Products Yield and Bio-oil Composition

Hari Kiran Tirumaladasu, Piyush Pratap Singh, Anurag Jaswal, and Tarak Mondal

Abstract Pyrolysis of agricultural biomass to produce bio-oil and biochar is a sustainable route for energy and fuel production. The rice husk (RH) is a major agricultural residue produced in larger amounts in India. In this study, lab-scale pyrolysis of RH was performed to evaluate the type of products and their composition at different temperatures (200–400 °C) in a batch reactor. Furthermore, RH was characterized to understand its physical as well as chemical properties thoroughly via various methods, for instance, proximate analysis, ultimate analysis, calorific value analysis, and thermogravimetric analysis (TGA). Along with this, the effect of pyrolysis temperature on the yields of products and the composition of bio-oil was also investigated. The experimental results revealed that the maximum bio-oil yield of 33.4 wt.% was obtained at 400 °C. Acid pre-treatment using hydrochloric acid (HCl) was also done to remove the impurities like Alkali and Alkaline Earth Metals (AAEMs) present in the RH. The pre-treatment was found to increase the yield of bio-oil to 37.3 wt.% and reduce the yield of biochar and gases by 2.1 wt.% and 1.8 wt.%, respectively. Gas Chromatography−Mass Spectroscopy (GC−MS) and Fourier Transform−Infrared (FT−IR) analysis were used to identify the chemical composition and functional groups present in the bio-oil obtained at various temperatures. A complex combination of acids, alcohols, furans, ketones, phenols, sugars, and other compounds was discovered in the bio-oil.

Keywords Pyrolysis · Rice husk · Bio-oil · Batch reactor · Biomass conversion

---

H. K. Tirumaladasu · P. P. Singh · A. Jaswal · T. Mondal (✉)
Department of Chemical Engineering, Indian Institute of Technology Ropar, Rupnagar, Punjab−140001, India
e-mail: tarakmondal@iitrpr.ac.in

N. Gakkhar et al. (eds.), *Recent Advances in Bio-Energy Research*, Springer Proceedings in Energy, https://doi.org/10.1007/978-981-99-5758-3_16

16.1 Introduction

Fossil fuels have been the most important source of chemicals, fuels, and energy for human civilization for over two centuries, accounting for most of our energy and chemical demands. The rapid urbanization and industrialization have resulted in a population expansion in recent years, fueling a tremendous increase in worldwide energy consumption [1]. Presently, fossil fuels—crude oil, natural gas, and coal—supply nearly 83% of the global energy demand, with renewables supplying just 10% of the energy demand [2]. This over-reliance on fossil fuels has prompted fears of supply chain disruptions, price volatility, and/or market manipulation in the not-too-distant future. Their usage also leads to extraordinarily large-scale emissions of greenhouse gases which contribute to global climate change, as well as extensive environmental damage in the form of land, air, and water pollution. As a result, a move to alternate green energy sources is a critical step for continuing sustainable growth [3]. Renewable energy sources are environmentally friendly and will not run out soon due to their steady long-term availability, these sources are also more price-stable than fossil fuels. So, biomass is a promising alternative to fossil fuels as a new energy source since it is renewable and carbon neutral.

Biomass has historically been regarded as a viable and sustainable energy source capable of creating a diverse range of chemicals and materials. Biomass is a generic term for any organic material—plant and animal based—that is of recent origin. As a major consequence of such a broad definition is that biomass feedstocks have a tremendous diversity which makes it difficult to categorize them as a whole. Broadly, biomass feedstocks can be categorized as forestry based, agri-industrial based, energy crops, food crops, etc. [3]. Biomass has several advantages over conventional fossil fuels, including minimal sulfur and nitrogen content, as well as no net CO_2 emissions into the atmosphere [4]. Biomass resources cover many materials such as forest waste, organic residues, and agricultural waste. Agricultural biomass, a widely available residue, is produced on a global scale every year but is often unused [5]. While it is hard to get reliable numbers, reports suggest that India, China, the United States of America, and Brazil together produced greater than one billion tons of biomass every year, a large fraction of which is agricultural waste such as rice husk and straw, sugarcane, etc. [6].

India is a global agricultural powerhouse and produces a huge amount of waste residues. Thus, there is a vast scope for generating renewable energy, producing valuable chemicals (like biofuels, platform chemicals, transportation fuels, etc.) from biomass in India. RH was chosen as the feedstock in this study because it is a significant agricultural biomass source in India that is readily available in huge amounts. While some of the produced rice husk is utilized by the farmers for various purposes, a significant amount is left behind as it is often surplus to their need. Owing to the narrow timeline between the harvesting of one crop and the sowing of another crop as well as the high cost of collecting the residues and the unavailability of labor, the surplus residue is commonly burned off in the open for quick disposal [7]. However, the open burning of residues in the fields is discouraged because it emits significant volumes of particulate matter, and gases (NO_x, CO, CO_2, and SO_2) into the environment [8]. As a result, the focus of research in this domain in recent years has been on identifying potential strategies and/or technologies that could allow the utilization of this carbon resource for energy production. Among the various possible strategies, thermochemical conversion is one that has exhibited great potential. The thermochemical conversion of biomass can be carried out in a number of ways—pyrolysis, hydrothermal liquefaction, gasification, and torrefaction [9]. Among these, pyrolysis is by far the most promising one as it employs comparatively milder conditions for converting biomass. Pyrolysis is carried out under an inert atmosphere at temperatures ranging from 200 to 600 °C, resulting in the formation of three products—gases, biochar, and bio-oil—which can themselves be used for energy production as well as the production of chemicals and fuels [9]. Its operational simplicity, use of relatively milder conditions, and the wide range of products that can be obtained make pyrolysis a particularly appealing strategy and has resulted in its application at an industrial scale [9].

Pyrolysis conversion technology can be approached in two ways. The first, also known as classical or classic pyrolysis, aims to improve biochar formation at low temperatures and low heating rates, or to boost fuel gas yield under high temperature, medium heating rate, and extended vapor residence time. Flash or fast pyrolysis aims to maximize the liquid product yield under the following conditions: (1) extremely high ramp rate (about 100 – 200 °C/min), (2) grinded biomass which has a very small size (1−2 mm), (3) temperature around 550 °C, and (4) rapid quenching of the produced vapors (very little vapor residence time) to produce bio-oil [10].

The majority of the previous studies on pyrolysis of RH focused on fixed/fluidized bed reactors as shown in Table 16.1 and are usually conducted at very high temperatures (>500 °C) and high heating rates (>100 °C /min) which consume a lot of energy and are associated with high costs. There is no extensive investigation on the slow pyrolysis of RH in a batch reactor setup at low temperatures with an emphasis on the examination of bio-oil generated, as far as we know. The main motive of this study is to see how different pyrolysis reaction temperatures (200–400 °C) affect the slow pyrolysis of RH so that comparative studies can be made under the same conditions. Experiments were conducted out at temperatures of 200, 250, 300, and 350 degrees Celsius in a nitrogen atmosphere at a heating rate of 10 °C/min. Analytical techniques such as GC–MS and FT–IR were used to characterize the obtained bio-oil.

Table 16.1 Previous studies on thermal pyrolysis of RH at various reaction conditions

Process variants	Reactor	Operating conditions			Product yields (wt.%)			References
		Temp (°C)	RT (min)	HR (°C / min)	Bio-oil	Syngas	Biochar	
Fast pyrolysis	Fixed bed	500	2	200	36	16	48	Tsai et al. [10]
Fast pyrolysis	Fluidized bed	465	–	–	56	15	29	Ji-lu [11]
Slow pyrolysis	Fixed bed	500	–	60	32	33	35	Natarajan [12]
Fast pyrolysis	Conical spouted bed	450	–	–	70	4	26	Alvarez et al. [13]
Fast pyrolysis under vacuum	Tubular quartz	400	20	–	47	19	34	Téllez et al. [14]
Fast pyrolysis	Auger reactor	500	–	–	51	34	15	Yu et al. [15]

Temp: Optimum temperature; RT: Reaction time; HR: Heating rate.

16.2 Methodologies

16.2.1 Materials

The RH biomass was purchased from a local rice mill in Rupnagar, Punjab. The acquired feed was cleaned with water to remove any impurities present and then sun-dried. The dried feed was grinded and sieved using an appropriately sized mesh to obtain particle sizes ranging from 0.8 to 2 mm.

16.2.2 Experimental Procedure

Prior to beginning the experiments, the feed was dried for 12 h in an air oven at 105 °C to eliminate any pre-absorbed moisture. 50 g of dried feed is taken and heated to 200 °C in a three-neck round-bottomed flask using a heating mantle for 90 min as shown in Fig. 16.1. The entire pyrolysis process has been done in nitrogen with flowrate kept fixed at 80 ml min^{-1}, and the water temperature in the chiller for condensation was kept at 4 °C [16]. The vapors generated from the heating of the feed were passed into the condenser where the liquid product is collected in the conical flask. The non-condensable gases are escaped through an outlet pipe into the atmosphere and the biochar is collected from the round-bottomed flask after the

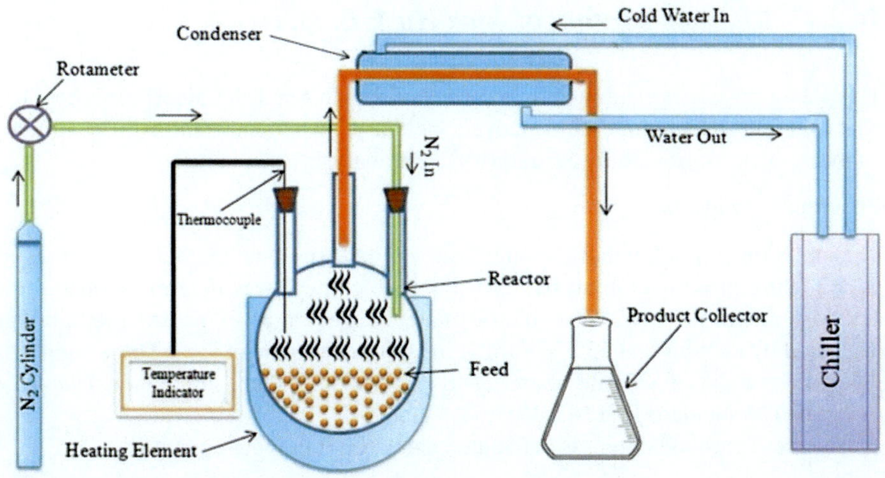

Fig. 16.1 Schematic diagram of the batch pyrolysis experimental setup

experiment is completed. The same procedure was repeated for other temperatures (250–400 °C). The experiment for pre-treated RH was performed only at 400 °C as the bio-oil yield was highest at this temperature. The acid pre-treatment was carried out to investigate the influence of acid on product yield and quality.

The yields of products [21] have been calculated using the below approach:

$$\text{Bio-oil yield (wt.\%)} = \frac{\text{Amount of bio-oil formed}}{\text{Amount of feed taken}} \times 100 \tag{16.1}$$

$$\text{Biochar yield (wt.\%)} = \frac{\text{Amount of biochar formed}}{\text{Amount of feed taken}} \times 100 \tag{16.2}$$

$$\begin{aligned}&\text{Non-condensable gases yield (wt.\%)}\\&= 100\text{wt.\%} - (\text{bio-oil yield (wt.\%)} + \text{biochar yield (wt.\%)})\end{aligned} \tag{16.3}$$

16.2.3 Pre-treatment Method

The feed was washed with 0.5 M HCl solution as HCl has good removal efficiency of these alkali and alkaline earth metals which are predominantly present in RH [17]. 100 g of RH sample was dissolved in 1000 ml 0.5 M HCl solution and stirred at 25 °C for 6 h. The feed was then rinsed with deionized water until the pH reached neutral. Next, the feed was dried in the sun for 48 h to remove the moisture.

16.2.4 Characterization of Rice Husk Biomass

The RH biomass was examined for a variety of physical and chemical properties. The characterization techniques and analysis include (i) proximate analysis, (ii) ultimate analysis, (iii) calorific analysis, and (iv) TGA analysis.

Proximate Analysis

The moisture content, volatile matter, ash, and fixed carbon content were all determined using proximate analysis. The moisture content was determined using the ASTM standard E871–82. The volatile matter concentration of the dried sample was evaluated by heating it at 550 °C for 2 h in an N_2 environment [18]. The weight loss gives the content of volatile matter. The ash content of the sample was evaluated using ASTM standard D–1102.

The fixed carbon content is calculated using the following equation:

$$\%FC = 100\% - (\%VM + \%A) \tag{16.4}$$

where FC = Fixed Carbon, VM = Volatile Matter, and A = Ash.

Ultimate Analysis

The carbon, hydrogen, and oxygen contents of RH biomass were found using empirical correlations developed by Channiwala et al. [19] as shown below:

$$\%C = (0.637 \times FC) + (0.455 \times VM) \tag{16.5}$$

$$\%H = (0.052 \times FC) + (0.062 \times VM) \tag{16.6}$$

$$\%O = (0.304 \times FC) + (0.476 \times VM) \tag{16.7}$$

where, FC = Fixed Carbon (%) and VM = Volatile Matter (%)

Calorific Value Analysis

Proximate analysis data was used to calculate the HHV of RH in this study. The following correlation developed by Channiwala et al. [20] was used for HHV determination.

$$\text{HHV, MJ/kg} = (0.3536 \times \text{FC}) + (0.1559 \times \text{VM}) - (0.0078 \times \text{A}) \tag{16.8}$$

where FC = Fixed Carbon (%), VM = Volatile Matter (%), and A = Ash (%)

TGA Analysis

The thermogravimetric analysis was carried out using TA SDT 650 instrument. The tests were carried out using 10–20 mg of RH biomass at various ramp rates ranging from 10 to 40 °C min^{-1} in an N2 environment at a flow rate of 100 ml min^{-1}, with a temperature range of 25–900 °C.

16.2.5 Product Analysis Description

GC–MS analysis

GC/MS, Agilent 7890 B instrument was used to examine the compounds found in the bio-oil. The carrier gas used was Helium (He), and the separation was performed on an HP-5 column. An oven program was set at 35 °C for 5 min followed by heating at a rate of 10 °C min^{-1} and a hold time of 2 min, and finally heated at a rate of 15 °C min^{-1} till 280 °C.

FT–IR analysis

The FT–IR analysis of the bio-oil samples was done by using ATR assembly in Nicolet iS50 FT–IR Spectrometer, provided by Thermo Fischer Scientific at a scan rate of 32 with a resolution of 4 cm^{-1} over a wavelength range of 400–4000 cm^{-1}.

16.3 Results and Discussions

16.3.1 Biomass Analysis and Characterization

Proximate Analysis, Ultimate Analysis, Higher Heating Value

The proximate analysis, ultimate analysis, and higher heating value (HHV) of RH are shown in Table 16.2. As shown in Table 16.2, the values calculated are comparable to the literature values. The major component was found to be volatile matter which is 71.3%. This high volatile matter indicates that the RH contains a good amount of pyrolyzable chemicals which can produce highly valuable bio-oil using pyrolysis.

Table 16.2 Proximate analysis, ultimate analysis, and HHV of RH

Parameter	Value	Parameter	Value	Parameter	Value
Proximate analysis (wt.%)		Ultimate analysis (wt.%)		Calorific value (MJ/kg)	
Moisture	9.5	C	41.2	HHV	15.9
Volatile matter	71.3	H	5.1		
Ash	14.9	O	38.1		

TGA Analysis

In an N_2 environment, the RH thermogravimetric behavior at different rap rates is illustrated in Fig. 16.2. According to the reports available in the literature, the weight loss during biomass pyrolysis can be considered to take place in three distinct temperature ranges, as a result of which the pyrolysis process of pine needles could be roughly divided into three stages. Between room temperature and roughly 100 °C, a relatively small amount of weight loss was observed which could be attributed to the loss of free and bound water. Following that, a modest weight loss up was noted between 250 and 300 °C, based on the ramp rate utilized, and then a significant loss of weight loss was observed between 370 and 400 °C. It has been noted in the available literature that hemicellulose, cellulose, and lignin undergo degradation at temperatures ranging from 210–325, 310–400, and 160–900 °C, respectively. Consequently, the weight loss between 150 and 500 °C, which constitutes the active pyrolysis zone, can be attributed to the degradation of hemicellulose and cellulose fractions, resulting in the evolution of volatiles [21, 22]. Beyond roughly 500 °C lies the region called the passive pyrolysis zone, which corresponds to the decomposition of lignin [21, 22]. Owing to its high thermal stability, lignin is known to decompose slowly and over a wide temperature range, which can be seen from the formation of trailing tail up to 800 °C.

On increasing the heating rate, the decomposition temperatures were observed to have shifted to higher temperatures (Fig. 16.2). This could be attributed to the fact that at higher heating rates, the decomposition of the feedstock occurs in a smaller span of time. Additionally, a small reduction was observed in the amount of mass loss in the active pyrolysis zone which could also be attributed to the fact that the use of higher heating rate shifts the region of maximum weight loss to higher temperatures owing to the lower amount of time available [23]. As the ramp rate increases, the maximum peak degradation increases, and the peaks shift toward the right. This is because of the limitation of the transfer of heat. During TGA, a lower ramp rate provides more immediate heat energy to the system, and the carrier gas may take longer to attain equilibrium with the sample temperature. Higher heating rates have

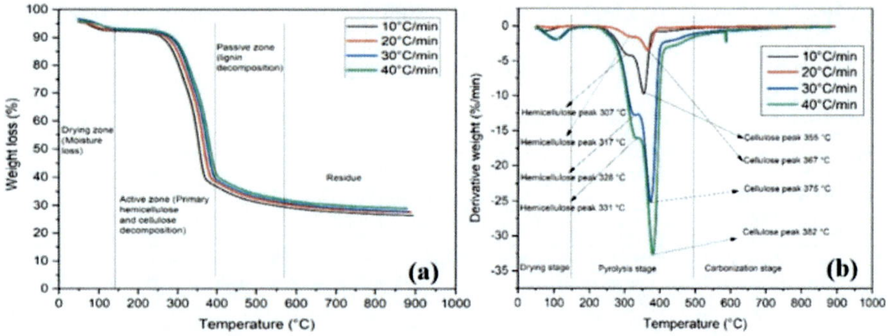

Fig. 16.2 Weight loss (%) **a** and derivative weight change **b** of RH biomass at different heating rates

a faster response time and require a higher temperature for the sample to degrade at the same time and in the same temperature range [23]. The DTG peaks shift toward the right as a result which matches well with the reported work [23].

16.3.2 Product Yields

Table 16.3 shows the influence of temperature and acid pre-treatment on pyrolysis product yield in a batch reactor. As seen from Fig. 16.3, as the temperature rises from 200 to 400 °C, the yield of bio-oil and syngas increases while the biochar yield is reduced. Temperature has been observed to have a major effect on the yields of the products. This is because as the temperature rises, more volatiles will be generated due to the decomposition of biomass constituents (i.e., hemicellulose, cellulose, and lignin). This can also be inferred from the TGA analysis of RH biomass which confirms the temperature ranges for the decomposition of hemicellulose, cellulose, and lignin. The highest bio-oil yield of 33.4 wt.% was obtained at 400 °C indicating the optimum temperature to produce bio-oil.

It can be seen that the bio-oil yield slightly increased after the acid pre-treatment of the feed. This might be because HCl removed a considerable quantity of undesirable minerals, mainly AAEMs, from RH, increasing bio-oil yield by 3.9 wt.%, while decreasing biochar yields by 2.1 wt.%. Furthermore, the removal of a considerable proportion of biomass components (particularly hemicellulose and lignin) might be another cause for the rise in bio-oil yield following HCl pre-treatment, resulting in increased cellulose content in the HCl pre-treated RH. This increased cellulose content could be the reason for the rise in the bio-oil yield because a higher content of cellulose leads to a more bio-oil product, whereas a higher hemicellulose and lignin contents lead to more gases and biochar, respectively [24]. Thus, if the goal is to increase the bio-oil yield, the HCl pre-treatment of RH is advantageous. Similar trends in the variation of the yield after acid pre-treatment with HCl were also observed by other researchers [25].

Table 16.3 Yield of pyrolysis products

Temperature (°C)	Bio-oil (wt.%)	Syngas (wt.%)	Biochar (wt.%)
200	7.4	14.3	77.6
250	22.1	22.6	53.3
300	30.2	25.7	44.1
350	31.3	26.5	42 0.2
400	33.4	27.9	38.7
400 (Pre-treated)	37.3	26.1	36.6

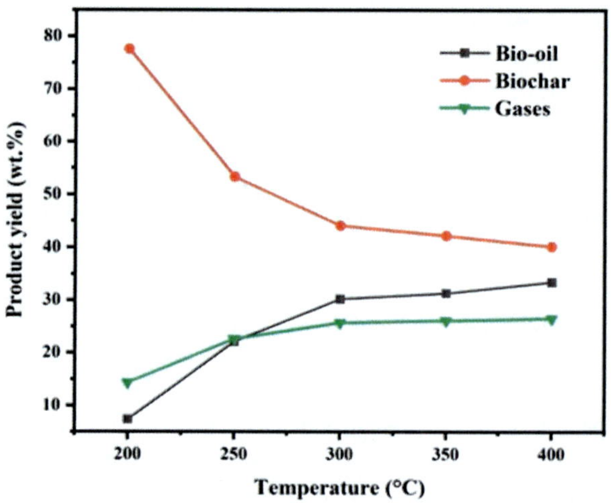

Fig. 16.3 Product yields in a batch reactor at various temperatures

16.3.3 Bio-oil Characterization

GC–MS analysis of bio-oil at different temperatures

Table 16.4 depicts the various components identified in bio-oil collected at various temperatures and in the pre-treated case. Acetic acid was found to be the main component in bio-oil in all cases, as shown in the table. The other major components include Furfural which was obtained at 200 and 250 °C with an area of 7.05% and 6.68%, respectively, and 2-Furanmethanol in which the peak area decreased from 7.10 to 3.53% as the temperatures were increased. Phenol area increased from 1.62 to 2.91% as the temperature increased, whereas for Phenol, 2-methoxy-, the area decreased from 3.12 to 2.38% with the increase in temperature. Catechol, which is a class of Phenol, was obtained at 200 °C and its area increased from 1.61 to 3.79%. This increase in the area might be because of the thermal degradation of lignin which generally occurs at high temperature. Next major class of compounds, namely, were sugars mainly found to be 1,4:3,6-Dianhydro-. alpha. -d-glucopyranose, beta. -d-glucopyranose, 1,6-anhydro-. As the temperature was raised from 200 to 400 °C, the peak area of the sugars increased.

Peak Areas of Major Class of Components in the Untreated RH Bio-oil Obtained by GC−MS Analysis

The following is a discussion on the major class of components found in the untreated RH bio-oil: at the optimum temperature, i.e., at 400 °C. The major class of components can be divided into these functional groups: (1) acids, (2) alcohols, (3) furans, (4) ketones, (5) nitrogen compounds (6) phenols (7) sugars, and (8) others as shown in Table 16.5. The peak areas of acids, alcohols, furans, ketones, nitrogen compounds, phenols, sugars, and others were found to be 29.6%, 10.2%, 6.9%, 19.7%, 11.3%,

Table 16.4 Components found in bio-oil with a peak area of ≥ 1% as determined by GC–MS analysis

Component	Untreated RH					Pre-treated RH
	Peak area (%)					
	200 °C	250 °C	300 °C	350 °C	400 °C	400 °C
Acetic acid	22.22	21.54	14.65	12.02	22.07	20.28
2-Propanone, 1-hydroxy	6.51	6.18	6.99	7.14	4.81	4.61
Propanoic acid	–	1.85	–	1	0.55	0.78
1-Hydroxy-2-butanone	2.21	3.14	2.54	3.15	1.66	1.80
Hydrazine, ethyl-	10.4	11.71	–	–	–	–
Furfural	7.05	6.68	–	–	–	–
2-Furanmethanol	7.10	6.29	5.38	4.7	3.53	2.46
2-Propanone, 1-(acetyloxy)-	1.98	1.68	1.43	1.38	1.49	1.56
Cyclopentanone	2.40	–	–	3.27	–	–
2-Furancarboxal-dehyde, 5-methyl-	2.02	1.92	1.95	–	–	–
Phenol	1.62	1.77	1.96	2.76	2.61	1.30
1,2-Cyclopentane-dione, 3-methyl-	1.24	–	–	–	2.32	2.17
Phenol, 2-methoxy-	3.12	3.07	2.46	2.41	2.38	1.05
Pentanal	3	2.76	2.16	2.03	1.92	–
1,4:3,6-Dianhydro-.alpha.-d-glucopyranose	2.38	2.17	2.70	1.67	2.93	1.01
Catechol	–	1.61	2.46	3.17	3.79	2.59
Beta.-D-Glucopyranose, 1,6-anhydro-	–	2.10	5.96	5.99	6.82	8.35
1H-Pyrazole, 1,5-dimethyl-	–	–	5.70	6.70	–	–
Hydroquinone	–	–	–	1.05	1.24	–
Pyrazole,1,4-dimethyl	–	–	–	–	5.20	4.76
1,2-Cyclopentaedione	–	1.48	1.61	2	2.37	2.34

5.3%, 9.4%, and 7.6%, respectively. Other compounds include a small amount of aldehydes, esters, and hydrocarbons. As shown in Table 16.6, the major component was found to be acetic acid which has a peak area of 22.07%.

The other major components were found to be 2-Propanone, 1-hydroxy, Pyrazole,1,4-dimethyl, 2-Furanmethanol, 3-Aminoisoxazole, 1,2-Cyclopentaedione, phenol, 1,2-Cyclopentaedione, 3-methyl, Phenol, 2-methoxy, Cyclopropyl carbinol, Catechol, and.beta.-D-Glucopyranose,1,6,-anhydro-. Acetic acid was generated due to the degradation of hemicellulose. Other compounds such as anhydro sugars, particularly levoglucosan (.beta.-D-Glucopyranose,1,6,-anhydro-) formed due to the decomposition of cellulose, hemicellulose decomposition leads to light oxygenates like acetic and formic acids, and lignin decomposition leads to phenolic compounds such as guaiacyl, and p-hydroxyphenyl derivates during pyrolysis [26].

Table 16.5 Major functional groups in bio-oil at 400 °C

Category	Peak area (%)	
	Untreated RH bio-oil	Pre-treated bio-oil
Acids	29.6	22.1
Alcohols	10.2	14.1
Furans	6.9	10.4
Ketones	19.7	17.6
Nitrogen compounds	11.3	7.6
Phenols	5.3	4.6
Sugars	9.4	11.2
Others	7.6	12.4

Effect of Acid Pre-Treatment on the Peak Areas of Components in the Bio-Oil Obtained by GC – MS Analysis

The following is a discussion on the influence of HCl pre-treatment on the components found in the HCL pre-treated bio-oil: at the optimum temperature, i.e., at 400 °C. As similar to untreated RH bio-oil, the major class of components can be divided into these functional groups: (1) acids, (2) alcohols, (3) furans, (4) ketones, (5) nitrogen compounds (6) phenols (7) sugars, and (8) others. As shown in Table 16.5, the peak areas of acid, alcohols, furans, ketones, nitrogen compounds, phenols, sugars, and others were found to be 22.1%, 14.1%, 10.4%, 17.6%, 4.6%, 11.2%, and 12.4%, respectively, as shown in Fig. 16.4. The major component was found to be acetic acid which has a peak area of 20.28%. Acetic acid was generated due to the degradation of hemicellulose, although pre-treatment with HCl hindered the generation of acetic acid from some hemicellulose acetyls. The esterification interaction between acid and ethanol might be the source of the acid concentration reduction [27]. The area of sugars slightly increased, indicating that the acid washing with HCl affected the chemical composition of RH biomass. Thus, it was discovered that pre-treatment with HCl improved the quality of bio-oil by reducing the quantity of light oxygenates such as acids and alcohols while boosting sugar yields.

FT–IR Spectra of Bio-Oil

Figure 16.5 illustrates the FT–IR spectra of bio-oil at various temperatures and for the pre-treated case. The peaks are almost similar for all temperatures indicating that the bio-oil contains similar functional groups but vary in the peak intensity. The presence of alcohol groups and phenols is indicated by O–H stretch vibration peak at 3360–3300 cm^{-1}. C–H stretch vibrations at wavenumbers of 3000–2850 cm^{-1} suggest the existence of alkane groups. Aldehydes, ketones, and carboxylic acids are detected by a significant C = O stretching absorbance between 1720–1640 cm^{-1}. Aromatic groups are represented by C = C stretch vibrations at wavenumbers of 1640–1400 cm^{-1} [28], whereas carboxylic acids, esters, and ether are detected by C = O stretch vibrations with wavenumbers of 1370–1020 cm^{-1}. These functional

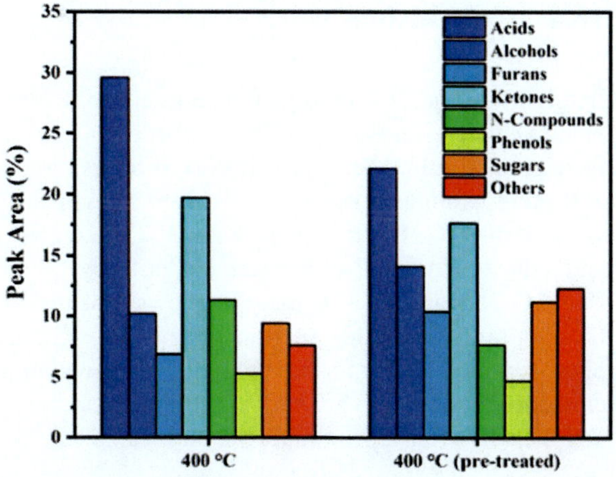

Fig. 16.4 Functional groups in bio-oil at 400 °C with their peak areas (%)

groups have shown increasing trends with temperature which matches well with the studies performed in this direction [29]. Thus, the FT–IR analysis concludes that the obtained bio-oil is a highly oxygenated complex mixture of alcohols, acids, aromatics, alkanes, esters, ethers, ketones, and phenols which was also confirmed from GC–MS analysis in the previous section[11, 13, 15].

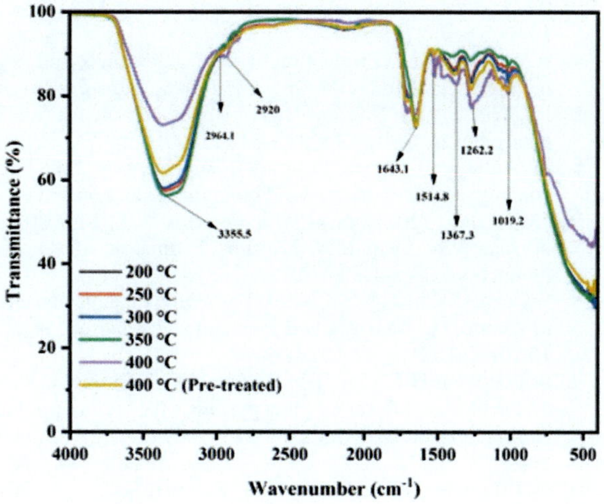

Fig. 16.5 FT–IR spectra of bio-oil

16.4 Conclusion

In this study, the slow pyrolysis experiments of RH at low temperatures were carried out in a batch reactor system. The yield of pyrolysis products and the quality of bio-oil were greatly influenced by the pyrolysis temperature. The yield of bio-oil increased as the temperatures goes from 200 to 400 °C, with the highest bio-oil yield being 33.4 wt.% at 400 °C. Pre-treatment with HCL increased the bio-oil yield to 37.3 wt.% and reduced the yields of biochar and gases by 2.1 wt.% and 1.8 wt.%, respectively. The GC–MS and FT–IR analysis shows that the bio-oil contains a good number of complex compounds such as aromatics, alcohols, hydrocarbons, ketones, and phenols which can be further upgraded using techniques like distillation, extraction, cracking, hydro deoxygenation, hydrogenation, etc. to produce valuable chemicals and fuels like transportation fuels, biofuels, alternative fuels, etc.

References

1. A.V. Bridgwater, Review of fast pyrolysis of biomass and product upgrading. Biomass Bioenerg. **38**, 68–94 (2012). https://doi.org/10.1016/j.biombioe.2011.01.048
2. BP, Statistical review of world energy globally consistent data on world energy markets and authoritative publications in the field of energy. BP Energy Outlook. **70**, 8–20 (2021)
3. A. Jaswal, P.P. Singh, T. Mondal, Furfural-a versatile, biomass-derived platform chemical for the production of renewable chemicals. Green Chem. **24**, 510–551 (2022). https://doi.org/10.1039/d1gc03278j
4. A.V. Bridgwater, Production of high grade fuels and chemicals from catalytic pyrolysis of biomass. Catal Today **29**, 285–295 (1996). https://doi.org/10.1016/0920-5861(95)00294-4
5. P.T. Williams, N. Nugranad, Comparison of products from the pyrolysis and catalytic pyrolysis of rice husks. Energy **25**, 493–513 (2000). https://doi.org/10.1016/S0360-5442(00)00009-8
6. M. Antar, D. Lyu, M. Nazari, A. Shah, X. Zhou, D.L. Smith, Biomass for a sustainable bioeconomy: an overview of world biomass production and utilization. Renew Sustain. Energy Rev **139**, 110691 (2021). https://doi.org/10.1016/J.RSER.2020.110691
7. N. Jain, V.K. Sehgal, N. Kaushik, Estimation of surplus crop residue in India for biofuel production. (New Delhi, 2018)
8. F. Duan, C. Chyang, Y. Chin, J. Tso, Pollutant emission characteristics of rice husk combustion in a vortexing fluidized bed incinerator. J. Environ. Sci. **25**, 335–339 (2013). https://doi.org/10.1016/S1001-0742(12)60054-0
9. A.T. Hoang, H.C. Ong, I.M.R. Fattah, C.T. Chong, C.K. Cheng, R. Sakthivel et al., Progress on the lignocellulosic biomass pyrolysis for biofuel production toward environmental sustainability. Fuel Process. Technol. **223**, 106997 (2021). https://doi.org/10.1016/J.FUPROC.2021.106997
10. W.T. Tsai, M.K. Lee, Y.M. Chang, Fast pyrolysis of rice husk: product yields and compositions. Bioresour. Technol. **98**, 22–28 (2007). https://doi.org/10.1016/J.BIORTECH.2005.12.005
11. Z. Ji-lu, Bio-oil from fast pyrolysis of rice husk: Yields and related properties and improvement of the pyrolysis system. J. Anal. Appl. Pyrolysis. **80**, 30–35 (2007). https://doi.org/10.1016/J.JAAP.2006.12.030
12. E. Natarajan, S.E. Ganapathy, Pyrolysis of rice husk in a fixed bed reactor. World. Acad. Sci. Eng. Technol. **56**, 504–508 (2009)
13. J. Alvarez, G. Lopez, M. Amutio, J. Bilbao, M. Olazar, Bio-oil production from rice husk fast pyrolysis in a conical spouted bed reactor. Fuel **128**, 162–169 (2014). https://doi.org/10.1016/J.FUEL.2014.02.074

14. J.F. Téllez, M.P. Silva, R. Simister, L.D. Gomez, V.C. Fuertes, J.M. De Paoli et al., Fast pyrolysis of rice husk under vacuum conditions to produce levoglucosan. J. Anal. Appl. Pyrolysis. **156**, 105105 (2021). https://doi.org/10.1016/J.JAAP.2021.105105
15. Y. Yu, Y. Yang, Z. Cheng, P.H. Blanco, R. Liu, A.V. Bridgwater et al., Pyrolysis of rice husk and corn stalk in auger reactor. 1. Characterization of char and gas at various temperatures. Energy Fuels. **30**, 10568–74. https://doi.org/10.1021/acs.energyfuels.6b02276
16. B.B. Krishna, B. Biswas, J. Kumar, R. Singh, T. Bhaskar, Role of reaction temperature on pyrolysis of cotton residue. Waste. Biomass. Valorization. **7**, 71–78 (2016). https://doi.org/10.1007/s12649-015-9440-x
17. K. Wang, J. Zhang, B.H. Shanks, R.C. Brown, The deleterious effect of inorganic salts on hydrocarbon yields from catalytic pyrolysis of lignocellulosic biomass and its mitigation. Appl. Energy. **148**, 115–120 (2015). https://doi.org/10.1016/J.APENERGY.2015.03.034
18. A.R. Sahito, R. Mahar, Z. Siddiqui, K. Muhammad Brohi, Estimating calorific values of lignocellulosic biomassfrom volatile and fixed solids. Int. J. Biomass. Renewables. 1–6 (2013)
19. J. Parikh, S.A. Channiwala, G.K. Ghosal, A correlation for calculating elemental composition from proximate analysis of biomass materials. Fuel **86**, 1710–1719 (2007). https://doi.org/10.1016/j.fuel.2006.12.029
20. J. Parikh, S.A. Channiwala, G.K. Ghosal, A correlation for calculating HHV from proximate analysis of solid fuels. Fuel **84**, 487–494 (2005). https://doi.org/10.1016/J.FUEL.2004.10.010
21. C.A. Koufopanos, A. Lucchesi, H.G. Maschio, Kinetic modelling of the pyrolysis of biomass and biomass components. Can. J. Chem. Eng. **67**, 75–84 (1989). https://doi.org/10.1002/cjce.5450670111
22. F. Shafizadeh, Pyrolysis and combustion of cellulosic materials. Adv. Carbohydr. Chem. **23**, 419–474 (1968). https://doi.org/10.1016/S0096-5332(08)60173-3
23. K. Slopiecka, P. Bartocci, F. Fantozzi, Thermogravimetric analysis and kinetic study of poplar wood pyrolysis. Appl. Energy. **97**, 491–497 (2012). https://doi.org/10.1016/j.apenergy.2011.12.056
24. R. Kumar, V. Strezov, H. Weldekidan, J. He, S. Singh, T. Kan et al., Lignocellulose biomass pyrolysis for bio-oil production: a review of biomass pre-treatment methods for production of drop-in fuels. Renew. Sustain. Energy. Rev. **123** (2020). https://doi.org/10.1016/j.rser.2020.109763
25. H. Tan, S.R. Wang, Experimental study of the effect of acid-washing pretreatment on biomass pyrolysis. J. Fuel. Chem. Technol. **37**, 668–672 (2009). https://doi.org/10.1016/S1872-5813(10)60014-X
26. F. Lin, C.L. Waters, R.G. Mallinson, L.L. Lobban, L.E. Bartley, Relationships between biomass composition and liquid products formed via pyrolysis. Front. Energy Res. **3** (2015). https://doi.org/10.3389/fenrg.2015.00045
27. J.H. Lee, I.G. Lee, J.Y. Park, K.Y. Lee, Efficient upgrading of pyrolysis bio-oil over Ni-based catalysts in supercritical ethanol. Fuel **241**, 207–217 (2019). https://doi.org/10.1016/J.FUEL.2018.12.025
28. H. Yang, R. Yan, H. Chen, D.H. Lee, C. Zheng, Characteristics of hemicellulose, cellulose and lignin pyrolysis. Fuel **86**, 1781–1788 (2007). https://doi.org/10.1016/j.fuel.2006.12.013
29. L.D. Kasmiarno, S. Steven, J. Rizkiana, E. Restiawaty, Y. Bindar, Kinetic studies and performance analysis of Indonesian rice husk pyrolysis. IOP Conf. Ser.: Mater. Sci. Eng. **1143**, 012067 (2021). https://doi.org/10.1088/1757-899x/1143/1/012067

Part IV
Biomass and Energy Management

Chapter 17
Characterization of Biomass Materials and Identification of their Energy Potential

Rajesh Kumar, Rickwinder Singh, and Ashish Kumar Srivastava

Abstract Being a renewable energy resource, biomass materials are a commendable candidate for GHGE reduction and have the capacity to reduce the reliance on fossil fuels. Unexplored biomass materials (wasteland residues, organic matter, and agricultural waste, etc.) in India have a huge potential for clean energy production by employing advanced technologies that, despite receiving little attention, can contribute to the nation's economic growth. This work explored the eight biomass materials available in surplus quantity in India for energy generation. The first part of the work highlights the potential of biomass materials, whereas in the second part selected biomass materials were characterized through laboratory tests. The characterizations of fuels have prodigious importance as they help to predict combustion behavior and process outcome. The results obtained indicated that the selected biomass materials are valuable sources of heat and power.

Keywords Biomass · Bioenergy · Thermogravimetric analysis (TGA) · Derivative thermogravimetric (DTG) · Combustion

17.1 Introduction

Clean and sustainable energy is one of the key components needed to meet future energy and environmental goals, as pollution in India is at an all-time high. The majority of India's energy needs are met by conventional energy sources including hydropower, coal, gas, oil, etc. (about 82.2% of energy is produced from traditional sources as indicated in Fig. 17.1. [1]). To provide power independence and a healthy

R. Kumar (✉)
Chitkara University Institute of Engineering and Technology, Chitkara University, Punjab, India
e-mail: rajesh.k@chitkara.edu.in

R. Singh
Centre for Energy and Environment, Malaviya National Institute of Technology, Jaipur, India

A. K. Srivastava
Mechanical Engineering Department, Manipal University Jaipur, Jaipur, Rajasthan, India

N. Gakkhar et al. (eds.), *Recent Advances in Bio-Energy Research*, Springer Proceedings in Energy, https://doi.org/10.1007/978-981-99-5758-3_17

environment to each individual in the country, it is essential to explore renewable energy resources like biomass materials, wind, and solar. Presently India is driving the world's largest renewable energy expansion program and is focused to achieve the 227 GW target of renewable energy by 2022 and to generate 40% of electric power installed capacity from non-fossil fuel by 2030. After wind energy, biomass as a renewable energy source has the maximum installed capacity (approximately 12% of total installed capacity or approximately 8GW biomass power) [2]. The maximum Indian population depends upon biomass for their energy requirement and approximately 32% of total energy use in country is generated from biomass. As per the Ministry of New and Renewable Energy (MNRE) of India, approximately 120 to 150 metric tons per annum of biomass is available as surplus quantity has 18 GW energy potential. To meet the energy generation target and to get benefitted from surplus biomass power, it is necessary to identify the available energy potential biomass material and their effective energy conversion process (combustion, torrefaction, pyrolysis, and gasification). The proper and effective utilization of biomass material (agricultural residues and domestic waste, organic matter, wasteland residue, and forest waste, etc.) can participate in energy freedom, pollution control, and economic growth of the country. Kumar et al. [3] reviewed the biomass energy resource, energy conversion, potential, and policy for promotion implemented by the Government of India. Authors reported that in 2015, 2665 MW of power being generated from biomass out of which 1666 MW is generated by cogeneration or co-firing.

In the present study, the potential of different biomass materials is analyzed by critical review or laboratory tests for clean and sustainable energy generation. The biomass was selected on the basis of surplus availability (throughout the year), the possibility of biomass to use as a co-fuel, energy potential of waste materials.

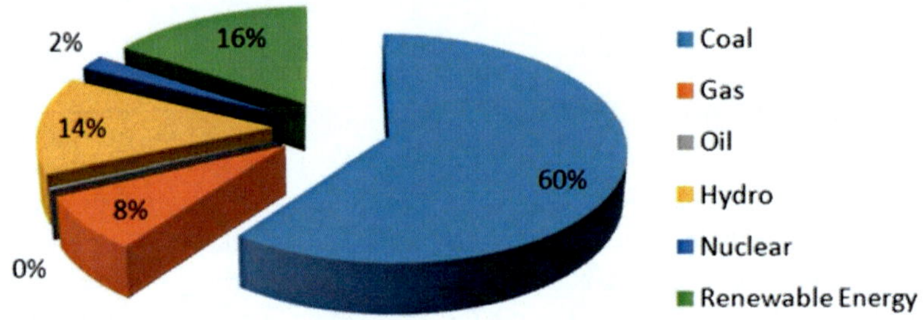

Fig. 17.1 Power generation sources in India [1]

17.2 Prominent Biomass Materials for Energy Generation in India

India, being a diversified country has an enormous reserve of biomass materials which includes agricultural waste, wasteland residue, organic matter, forest waste, etc. As per Bisht and Thakur [4] in India, approximately 145,026.6 kT/year of agriculture biomass and 59,678 kT/year of forest biomass are available in surplus quantity. The biomass materials are selected from a wide range of different categories: forest biomass (Pine needles), wasteland biomass (Prosopis Juliflora), agriculture residues (mustard husk, guar stalks, Castor seeds shell, Cotton Stalk), agro-industrial residue (rice husk), and plant litter.

17.2.1 Pine Needles (PN)

PNs are forest biomass material available abundantly in the Himalayan region of North India. In Himachal Pradesh (Indian State), pine forest spread over approximately 1500 square km whereas in Uttarakhand (Indian State) it is spread over 3400 square km. As per the MNRE report [5] published in 2012, 1 m^2 of pine forest will yield 1.19 kg of PN every year and a 100 kW gasifier running for 24 h would require 4500 kg of PN. In the above-said regions, the atmospheric temperature increased in summers, due to which PN catches fire. These PN are generally non-biodegradable and highly volatile in nature and have an adverse effect on biodiversity, economy, and environment when catches fire. One can be benefited by converting this forest waste to energy by employing some suitable methods. Few of the authors predicted the thermal energy potential of pine needles for different states of India. Kala and Subarao [6, 7] predicted that Uttarakhand has the potential to generate 18–32 petajoules of thermal energy annually from pine needles whereas Sharma and Sharma [8] estimated 1.16–1.34 petajoules thermal energy production annually from pine forest of Himachal Pradesh. Kumar and Singh [9–11] successfully co-fired the pine needle with coal and other biomass in a bubbling fluidized bed (BFB). A blend of 75%coal/25%PN achieved 97.05% combustion efficiency under oxy-fired conditions.

17.2.2 Prosopis Juliflora (PJ)

Prosopis Juliflora is wasteland trees introduced in India in mid of eighteenth-century to stop desertification, presently these are available abundantly in rural India. PJ are very hard plants that required very less water to grow and the growth of these plants is very fast. As per the report of MNRE, India published in 2015 [12] PJ yield/year is 29.3 megatons (approximately) out of which 3.5 megatons/year (approximately) is in surplus quantity and the rest is used for domestic purposes. These wasteland residues are available throughout the year. In India every year 12.5 tons/hectare of PJ wood is produced from three years old trees [13]. PJ woods has favorable emission and combustion characteristics. It has the capability of fixing large amounts of CO_2 from the atmosphere and has a higher heating value which makes it an attractive biomass source. Kumar and Singh [10] co-fired the PJ with coal and other biomass materials in BFB combustor. Combustion in co-fired conditions was observed smooth in BFB and 97.09% combustion efficiency was observed for 75%coal/25%PJ under oxygen-enriched condition. Kumar and Singh [14] investigate the co-combustion PJ with municipal sewage sludge (MSS) (50%/50% and 75%/25%) in an oxy-fired bubbling fluidized bed. A blend of 50% MSS/50% PJ achieved a maximum of 97.1% combustion efficiency for BFB combustor under oxygen-enriched condition whereas 25%MSS/75%PJ burned more smoothly without any trouble inside BFB combustor.

17.2.3 Mustard Husk (MH)

Mustard Husk is an agricultural residue that is the byproduct of the Mustard biomass (seasonal biomass) which is available in the month of Feb-May in India. A. et al. [3] estimated the biomass power generation potential in India. Authors predicted that Mustered husk and Mustard stalk have the potential to generate 8657 kt/year power. MH per year production in Rajasthan was predicted to be 6.9 megatons (approximately) by "Rajasthan Renewable Energy Corporation Limited India," [12] in 2015. Out of which 1.1 MT/year is the surplus quantity and leftover MH is used for domestic purposes. Overuse of MH in biomass-based power plants (mainly in MH-based power plant) create a shortage of these fuels in 2015. One of the solutions to tackle such a problem is to co-fire this biomass with another biomass, coal, or with any waste. Kumar and Singh [14] investigate the co-combustion MH with municipal sewage sludge (MSS) (50%/50% and 75%/25%) in an oxy-fired BFB. MH during co-firing shows favorable combustion and emission characteristics under both air-fired and oxy-fired conditions. A blend of 50% MSS/50% MH achieved 89% combustion efficiency under oxygen-enriched conditions. Kumar et al. [15] studied a 30 MW BFB combustor based on co-firing biomass (MH, RH, etc.) and coal. Authors found that combustion during co-firing is smooth and emission is under permissible limit.

17.2.4 Guar Stalk (GS)

Guar Stalk is an agricultural residue that is available in the month of Oct-Dec in India. India is a world leader in the production of guar by producing almost 80% of total world guar production. Guar is mainly grown in five states of India like Punjab, Haryana, Gujrat, Madhya Pradesh, and parts of Utter Pradesh. Approximately 233.3 kT/year of guar is generated out of which 163.3kT/year is available as a surplus quantity [4]. This surplus quantity of guar stalk has a great potential for power generation. Kumar and Singh [14] investigate the co-combustion GS with municipal sewage sludge (MSS) (50%/50% and 75%/25%) in an oxy-fired BFB. GS during co-firing with MSS under both air-fired and oxy-fired conditions gives birth to agglomeration which further defluidized the bed. GS as a co-fuel in fluidized bed requires more investigation with different bed materials.

17.2.5 Castor Seeds Shell (CSS)

CSS is an agriculture residue available after the harvest of caster seed, having good potential for energy generation. Generally, these particles are less than 5 mm and do not require any preparation before combustion. An experimental investigation was carried out to explore the co-firing potential of CSS with municipal sewage sludge (MSS) in an oxygen-enriched BFB combustor. Two ratios of blends (25%MSS/75%CSS and 50%MSS/50%CSS) were examined to observe the combustor and emission performance. CSS shows some promising results as a co-fuel with municipal sewage sludge. The detail of the setup and procedure is given elsewhere [10, 14].

17.2.6 Cotton Stalk (CS)

Cotton Stalk (CS) a byproduct of the cotton crop is one of the major crop residues available in surplus quantity in, India. More than 23 million tons of the cotton stalk are produced annually within the country [16]. Research on the appropriate uses of this surplus amount of CS is going on. However, a small portion is used as domestic fuel, fodder and the rest surplus amount is used widely in biomass-based power plants for energy generation [17, 18]. Few of the authors predicted the potential of CS in fluidized bed as a mono-fuel [19, 20] and co-fuel. Kumar et al. [15] studied a 30 MW co-firing BFB combustor and observed smooth combustion during co-firing of biomass/coal. Kumar and Singh [21] examined the combustion characteristics of CS with MSS in a BFB under air-fired and oxygen-enriched mode. CS as a co-fuel burned positively with MSS under BFB and the burnout of the blend is improved significantly under oxygen-enriched condition. Approximately 94.1% combustion efficiency could be achieved under O-E conditions in BFB combustor by taking CS as a co-fuel.

17.2.7 Rice Husk (RH)

Presently India is holding the second position in rice production after China globally. Rice husk an agro-industrial residue and rice straw an agriculture waste residues are produced due to rice cultivation. The total estimated growth of RH and rice straw (RS) in India is 29.1 and 145.5 million metric tons per year (MMTPA) [23]. Rice husk is a byproduct of the rice seed removal process [22] which has high energy potential in energy industries but its utilization is still limited due to scarcity of the behavior of RH inside the combustor. Singh and Kumar [24, 25] investigate co-firing behavior of RH with coal in a 20 kW oxy-fired BFB. It observed that blend of coal and RH burned successfully in FBC under O_2/N_2/RFG conditions, and the burning of the blend is improved significantly with oxygen enrichment. Kumar and Singh [10] co-fired RH with different fuels and found maximum exergy efficiency of 57.4 with a blend of RH, coal, plant litter, and pine needles.

17.2.8 Plant Litter (PL)

Plant litters are dead plant biomass material (leaves, bark, needles, and twigs) fallen to the ground usually collected by municipalities and disposed of by land-filling, burial, or direct incineration. This biomass constitutes a renewable energy source after gone through washing treatment and mechanical dehydration treatment. Nurmatov et al. [26] reported a 60% reduction in ash contents of leaf litter after washing treatment and a significant improvement in the chemical composition of washed leaf litter by mechanical dehydration treatment. Co-firing of PL with other biomass or coal in a fluidized bed was observed quite smooth [10, 24].

17.3 Materials and Methods

17.3.1 Properties of Biomass Materials

The properties of different fuels presented in this report are shown in Table 17.1. The raw fuels are first dried in direct sunlight and then chopped and crushed by cutter, ball mill, and crushers to prepare for analysis. The moisture content of the fuels is very less as proximate and ultimate analyses are executed on a dry basis. Moisture, volatile matter, fixed carbon, and ash contents present in biomass materials are extracted from proximate analysis which prescribed the combustion behavior of selected biomass materials. It observes that a major portion of biomass fuel is volatile matter. The percentage of ash content is observed lower for all cases. The ultimate analysis of the fuels gives an indication of flue gas emissions but cannot predict the real-time combustion process. The higher heating value (HHV) of the fuels is obtained from ultimate analysis by applying the following equation given by Friedl et al. [27].

$$\mathbf{HHV} = \frac{3.55\mathbf{C}^2 - 232\mathbf{C} - 2230\mathbf{H} + 51.2(\mathbf{C} \times \mathbf{H}) + 131\mathbf{N} + 20600}{1000} \tag{17.1}$$

Table 17.1 Properties of biomass materials

Proximate analysis of fuels (wt.%)					
Fuels	Moisture	Volatile matter	Ash	Fixed carbon	HHV (MJ/kg)
PN	1.11	25	19.12	25	19.12
PJ	0.91	31.99	16.02	31.99	16.02
MH	5.02	30.8	15.96	30.8	15.96
GS	5.34	56.4	10.36	27.9	19.62
CSS	8.1	69.2	8	14.6	16.29
CS	8	65.89	4	22	16.58
RH	1.43–5.22	60.57–66.4	13–15	14–25	15.67
PL	1.8–5.5	58.2–75	8–14	11–26	17.49
Ultimate analysis of fuel (wt.%)					
	Carbon	Hydrogen	Nitrogen	Sulfur	Oxygen
PN	48.01	6.29	0.31	–	33.85
PJ	39.67	5.95	1.63	–	39.17
MH	39.5	5.6	1.2	–	26.3
GS	50	5.3	–	–	29.2
CSS	40.13	4.9	1.1	–	30.6
CS	41	4	0.4	0.1	25
RH	35	3	1.25	0.18	34.45
PL	43.71	5.66	1.59	0.01	31.21

Lower heating value was estimated by applying the following equation

$$\mathbf{LHV} = \mathbf{HHV} - 206\mathbf{H} \tag{17.2}$$

17.3.2 TGA and DTG

TGA and DTG are the most common technique generally employed to investigate the combustion characteristics and performance of fuels [28]. Combustion characteristics and properties of various fuels are monitored on weight loss basis through thermal analysis, where combustion performance was investigated by finding some important parameter like ignition index (Which estimates the combustion reactivity), burnout index (which represents the characteristics of solid fuel combustion), and combustion characteristic factor (measure the fuel combustion performance). A higher ignition index is an indicator of harder reaction proceeds and a greater burnout index indicates the improved combustion reactivity. The following equations are used to calculate the ignition index (D_i), burnout index (D_b), and combustion characteristic factor (CCF) [29, 30].

$$D_i = \frac{DTG_{max}}{t_c.t_i} \tag{17.3}$$

$$D_b = \frac{DTG_{max}}{\Delta t_{1/2}.t_c.t_b} \tag{17.4}$$

$$CCF = \frac{(dm|dt)_{max}.(dm|dt)_{mean}}{T_i^2.T_b} \tag{17.5}$$

17.4 Result and Discussion

17.4.1 Ultimate and Proximate Analysis

Proximate and ultimate analysis results with high heat values are given in Table 17.1. Of all the selected fuels, GS contains the highest value for fixed carbon and high heat value. Low nitrogen is observed (0.3 to 1.63 wt. %) for all the tested samples, which means selected biomass is contributing very little to NO_x emission. Further, very less amount sulfur contents are found in RH, PL, CS, and no sulfur contents are observed in other selected biomass materials. Generally, sulfur content present in biomass materials contributed to slagging and fouling. The high heat value of selected fuels (calculated by using equation-2) is varying from 15 to 19 MJ/kg. Figure 17.2 represents the Van Krevelen diagram which is generally used to analyze the composition characteristics of various fuels [31]. A compositional variation is observed in different fuels used, GS is more carbonaceous than others and has the highest HHV. The elemental distribution of CSS, PL, PN, and MH is very close whereas the RH and PJ are away in different zones, respectively.

17.4.2 Tga/dtg

PerkinElmer TGA-4000 analyzer was used to investigate the thermal decomposition characteristics of selected biomass samples. The fuel samples were heated from 30 to 800 °C, at a heating rate of 40 °C/min under an air-fired environment. Table 17.2 represented the combustion characteristics determined from TGA/DTG analysis of selected biomass. TGA and DTG curves are shown in Fig. 17.3a-h giving the different decomposition profiles of all the selected biomass for the same heating

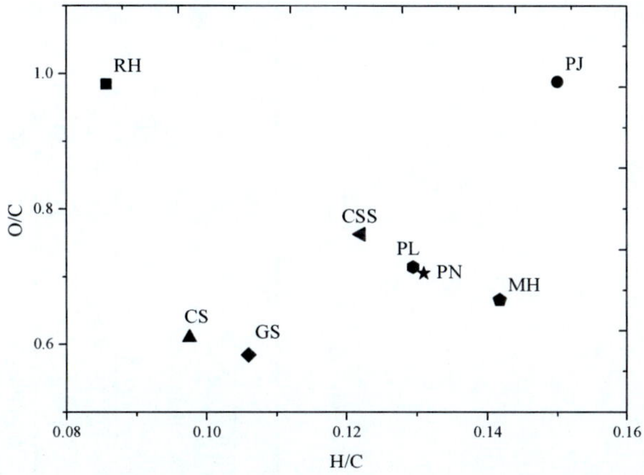

Fig. 17.2 Van krevelen diagram for the fuels

rate. The combustion of the biomass takes place in four stages: stage-1 evaporation of moisture; stage-2 biomass devolatilization/volatile combustion; Stage-3 Char combustion; stage-4 low rate oxidation of char. In stage one, 2 to 4% weight loss was observed for all selected fuels at a temperature up to 100 °C approximately, whereas maximum weight loss was observed in the second stage at a temperature up to 450 °C approximately when volatile matters were released.

DTG curves showed three individual peaks for almost all selected biomass, peak-1 is due to moisture release, peak-2 due to volatile matter release (owe to thermal decomposition of hemicellulose and partial decomposition of lignin), and peak-3 is due to char combustion (due to remaining lignin and fixed carbon), similar peaks were also reported in other works [32]. Whereas the burning profile of the PJ shown in Fig. 17.3b is quite different from others which show only two peaks. Similarly, two peaks were observed by Song et al. [33]. It means the first two stages occur together for PJ and it is not possible here to differentiate between the first two stages.

Combustion parameters (ignition temperature, burnout temperature, combustion time, $DTG_{max,}$ and time corresponding to $DTG_{max.}$) of different biomass used were also predicted by analyzing the DTG curve. Ignition index, burnout index, and combustion characteristics factors were calculated by using the generated data. Combustion characteristic factor demonstrated that the combustion performance of selected biomass is PJ > PN > PL > CSS > RH > GS > MH > CS in that the combustion performance of PJ is highest and CS is lowest. Other real-time effects of this biomass can be predicted by using this biomass in an advanced combustor, under different environments.

Table 17.2 Thermogravimetric characteristics of selected biomass used at a heating rate 40°C/min

Fuel	Temperature (°C)							Time			Ignition index (Di)	CCF	Order
	T_i	P_1	P_2	P_3	T_b	DTG_{max} (%/min)	DTG_{mean} (%/min)	t_c	t_i	t_b			
PN	298	125	350	450	530	2.1	0.415	10.5	6.7	12.5	1.32962E-05	1.8517E-08	2
PJ	280	160	337	455	598	2.35	0.58	10.5	6.25	14.2	1.40349E-05	2.9072E-08	1
MH	275	115	347	452	528	0.98	0.16	8	6.13	12.45	6.74931E-06	3.9269E-09	7
GS	230	105	340	402	550	0.89	0.167	10.5	5	13	7.03557E-06	5.1084E-09	6
CSS	340	150	360	540	725	1.65	0.533	13	7.75	17.1	6.69371E-06	1.0493E-08	4
CS	305	125	255	352	510	0.7	0.131	8	6.8	12	4.50016E-06	1.9329E-09	8
RH	320	110	352	452	540	1.2	0.253	7.95	7.25	12.75	6.94444E-06	5.4905E-09	5
PL	300	105	350	496	555	2.2	0.372	8	6.75	13.13	1.32122E-05	1.6384E-08	3

Ti-ignition temperature; P_1-peak 1; P_2-Peak 2; P_3-Peak 3; T_b-burnout temperature; t_c-time corresponding to DTG max; t_i-ignition time; t_b-bumout time; CCF-combustion characteristics factor.

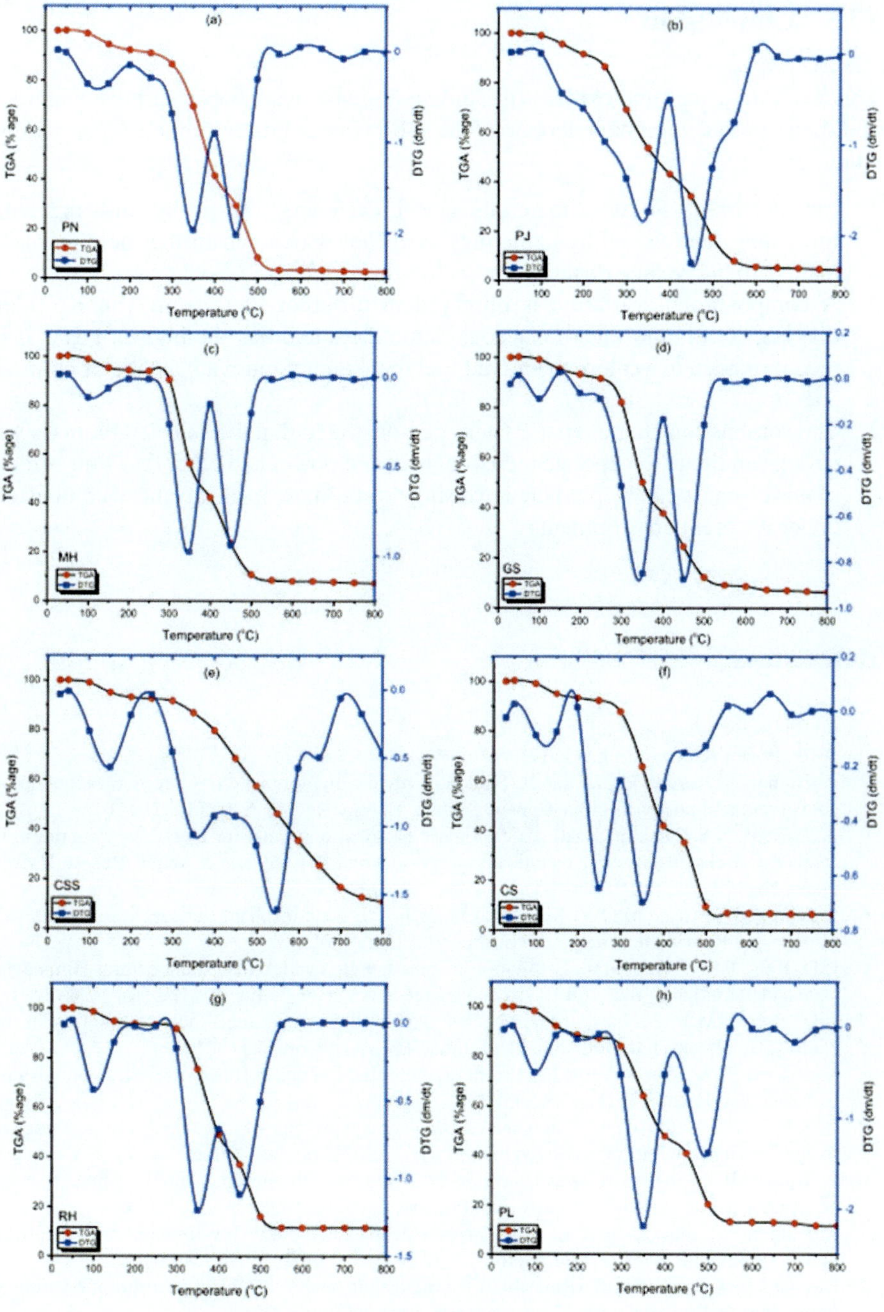

Fig. 17.3 TGA and DTG curve for **a** PN, **b** PJ, **c** MH, **d** GS, **e** CSS, **f** CS, **g** RH, **h** PL

17.5 Conclusion

This work describes and explains the Indian biomass available in surplus quantities and their combustion characteristics. The following points are concluded from this study.

1. Indian surplus biomass materials give great energy scope by co-firing with low-grade fuel or rapidly depleting fossil fuel which can further help in waste management, reduce dependency on fossil fuel, etc.
2. A compositional variation is observed in different fuels from Van Krevelen diagram; GS is more carbonaceous than others and has the highest HHV. It is recommended to explore the mixed fuel to take maximum advantage of biomass energy.
3. The combustion characteristic factor demonstrated that the selected biomass is a strong candidate to generate industrial heat and power and other real-time effects of this biomass can be predicted by using this biomass in an advanced combustor, under different environments.

References

1. Power sector at a glance all India (2017)
2. IBEF, Power sector in India (2017)
3. A. Kumar, N. Kumar, P. Baredar, A. Shukla, A review on biomass energy resources, potential, conversion and policy in India. Renew. Sustain. Energy Rev **45**, 530–539 (2015)
4. A.S. Bisht, N.S. Thakur, Small scale biomass gasification plants for electricity generation in India: resources, installation, technical aspects, sustainability criteria & policy. Renew. Energy Focus **28**, 112–126 (2019)
5. V. Jain, S. Srinivas, emPOWERing rural India the RE way: inspiring success stories. Ministry of New and Renewable Energy, Government of India (2012).
6. L.D. Kala, P.M.V. Subbarao, Estimation of pine needle availability in the central Himalayan state of Uttarakhand, India for use as energy feedstock. Renew Energy **128**, 9–19 (2018)
7. L.D. Kala, P.M.V. Subbarao, Pine needles as potential energy feedstock: availability in the central Himalayan state of Uttarakhand, India. E3S Web Conf. **23** (2017).
8. V. Sharma, R.K. Sharma, Pine needle energy potential in conifer forest of western himalayan. Environ. Nat. Resour. J. **18**, 55–65 (2020)
9. R. Kumar, R.I. Singh, Combustion and emission characteristics of municipal sewage sludge in oxygen-enriched bubbling fluidized bed, in: *GCRE2016*, pp. 6–9 (2016)
10. R. Kumar, R.I. Singh, An investigation in 20 kWth oxygen-enriched bubbling fluidized bed combustor using coal and biomass. Fuel Process. Technol. **148**, 256–268 (2016)
11. R. Kumar, I. Singh, Co-firing of coal with pine needles in a 20kw oxy-fired bubbling fluidized bed: experimental investigation. Mater. Today Proc. **5**, 23007–23013 (2018)
12. Biomass fuel supply study (Rajasthan) for Rajasthan renewable energy corporation limited prepared by ABI energy consultancy services private limited (2015)
13. A. Chandrasekaran, S. Ramachandran, S. Subbiah, Determination of kinetic parameters in the pyrolysis operation and thermal behavior of Prosopis juliflora using thermogravimetric analysis. Bioresour. Technol. **233**, 413–422 (2017)

14. R. Kumar, R.I. Singh, An investigation of co-combustion municipal sewage sludge with biomass in a 20 kW BFB combustor under air-fired and oxygen-enriched condition. Waste Manag. **70**, 114–126 (2017)
15. H. Kumar, R.I. Singh, S.K. Mohapatraa, Study of a 30 MW bubbling fluidized bed combustor based on co-firing biomass and coal. Sadhana **40**, 1283–1299 (2015).
16. P.V.V.P.G. Patil, R.M. Gurjar, A.J. Shaikh, R.H. Balasubramanya, K.M. Paralikar, Cotton plant stalk–an alternate raw material to board industry, in *World Cotton Research Conference* (CIRCOT Mumbai India, 2007)
17. J. Singh, B.S. Panesar, S.K. Sharma, Energy potential through agricultural biomass using geographical information system-A case study of Punjab. Biomass Bioenerg. **32**, 301–307 (2008)
18. A.K. Tripathi, P.V.R. Iyer, T.C. Kandpal, K.K. Singh, Assessment of availability and costs of some agricultural residues used as feedstocks for biomass gasification and briquetting in India. Energy Convers. Manag. **39**, 1611–1618 (1998)
19. Z. Sun, J. Shen, B. Jin, L. Wei, Combustion characteristics of cotton stalk in FBC. Biomass Bioenerg. **34**, 761–770 (2010)
20. Z.Ã. Sun, B. Jin, M. Zhang, R. Liu, Y. Zhang, Experimental studies on cotton stalk combustion in a fluidized bed. Energy **33**, 1224–1232 (2008)
21. R. Kumar, R.I. Singh, Combustion characteristics of municipal sewage sludge with cotton stalk, in *International Conference on Sustainable Energy and Environmental Challenges* (Mohali, 2017)
22. S. Chakma, A. Ranjan, H.A. Choudhury, P. Kumar, V. Moholkar, Bioenergy from rice crop residues: role in developing economies. Clean Technol. Environ. Policy **18**, 373–394 (2016)
23. S.M. Shafie, A review on paddy residue based power generation: energy, environment and economic perspective. Renew. Sustain. Energy Rev **59**, 1089–1100 (2016)
24. R.I. Singh, R. Kumar, Current status and experimental investigation of oxy-fired fluidized bed. Renew. Sustain. Energy Rev **61**, 398–420 (2016)
25. R.I. Singh, R. Kumar, Experimental analysis of splash zone of fluidized bed combustor using rice husk and coal under variable oxygen conditions, in *22nd FBC* (Turku, Finl, 2015), pp. 314–320
26. N. Nurmatov, D. Armando, L. Gomez, F. Hensgen, L. Bühle, High-quality solid fuel production from leaf litter of urban street trees. Sustain. **8** (2016)
27. A. Friedl, E. Padouvas, H. Rotter, K. Varmuza, Prediction of heating values of biomass fuel from elemental composition. Anal. Chim. Acta. **544**, 191–198 (2005)
28. X.G. Li, Y. Lv, B.G. Ma, S.W. Jian, H.B. Tan, Thermogravimetric investigation on co-combustion characteristics of tobacco residue and high-ash anthracite coal. Bioresour. Technol. **102**, 9783–9787 (2011)
29. Z. Yang, S. Zhang, L. Liu, X. Li, H. Chen, H. Yang, X. Wang, Combustion behaviours of tobacco stem in a thermogravimetric analyser and a pilot-scale fluidized bed reactor. Bioresour. Technol. **110**, 595–602 (2012)
30. C. Wang, F. Wang, Q. Yang, R. Liang, Thermogravimetric studies of the behavior of wheat straw with added coal during combustion. Biomass Bioenerg. **33**, 50–56 (2009)
31. Y. Zhang, F. Chen, D. Chen, K. Cen, J. Zhang, X. Cao, Upgrading of biomass pellets by torrefaction and its influence on the hydrophobicity, mechanical property, and fuel quality. Biomass Convers. Biorefinery (2020).
32. W. Qing, X. Hao, L. Hongpeng, J. Chunxia, B. Jingru, Thermogravimetric analysis of the combustion characteristics of oil shale semi-coke/biomass blends. Oil Shale **28**, 284–295 (2011)
33. C.-Z. Song, J.-H. Wen, Y.-Y. Li, H. Dan, X.-Y. Shi, S. Xin, Thermogravimetric assessment of combustion characteristics of blends of lignite coals with coal gangue. **105**, 490—495

Chapter 18
Effect of Covering Basin Area with Float Wick on the Performance of Single Slope Solar Still: An Experimental Study

Himanshu and M. K. Mittal

Abstract In this study, conventional single slope basin type solar still is modified by placing a float wick in the basin of the still. The performance of the modified still is compared with the conventional basin type still of the same dimensions under similar ambient conditions which is referred to as the reference still in this study. The effect of covering the evaporating surface area of the basin in different proportions with float wick is investigated. The variation of cumulative distillate output, temperatures of various components of stills, and productivity gain of the modified still over the reference still are analyzed. The overall daily distillate outputs of both reference and modified still, when 75% of the evaporating surface area of the modified still is lying under float wick are 2.74 kg/m^2 and 3.31 kg/m^2, respectively.

Keywords Solar still · Single slope · Float wick

18.1 Introduction

Water demand for domestic, agricultural, and industrial purposes has increased considerably these days. In coastal regions and arid areas where sea water or slightly salty water is available but there is a scarcity of potable water. There are many ways to treat this impure water using conventional energy. Different types of available distillation technologies are mechanical vapor compression, multi-stage flash distillation, multi-effect distillation, thermal vapor compression, solar still, solar chimney, osmosis, ion-exchange resin, membrane distillation, capacitive deionization, electrodialysis, hydration, and secondary refrigerant freezing. But in remote areas where conventional energy is costly or not available, solar distillation is the only method

Himanshu (✉)
Sardar Swaran Singh National Institute of Bio-Energy, Kapurthala, Punjab 144602, India
e-mail: himanshu.ra@nibe.res.in

M. K. Mittal
Mechanical Engineering Department, Thapar Institute of Engineering and Technology, Patiala, India

N. Gakkhar et al. (eds.), *Recent Advances in Bio-Energy Research*, Springer Proceedings in Energy, https://doi.org/10.1007/978-981-99-5758-3_18

that can be used to get potable water. The main advantages of solar distillation are low maintenance, economical, no moving parts, and use of clean and free energy due to which this technology is eco-friendly. Hanson et al. found solar stills to be promising in the removal of bacteria and non-volatile contaminants from brackish water, and thus making it suitable for drinking [1].

El-Sebaii et al. theoretically and experimentally analyzed the performance of single basin single slope solar still having movable baffle suspended absorber with vents in the basin [2]. The experimental results showed a 20% increment in the daily productivity of the still having baffle suspended absorber over the conventional still. Rababa'h conducted various experiments to compare the performance of basin type solar still having sponge cubes of different sizes in the basin with another conventional still of similar dimensions in the same ambient conditions [3]. It was found that the sponge cubes in the basin caused an appreciable increment in the productivity of the still up to 273%.

Murugavel et al. investigated the performance of double sloped basin type solar still modified by using various materials such as cotton cloth, jute cloth, sponge, and porous materials namely quartzite rocks and washed stones [4]. The maximum value of hourly distillate output was obtained when the cotton cloth was placed in the basin of the still among all materials. Murugavel and Srithar fabricated a single basin double sloped still having various wick materials such as coir mate, sponge sheet, cotton cloth, and waste pieces of cotton [5]. Among different wick materials, black cotton cloth was found to be more effective in improving the distillate output of the still. Srivastava and Agrawal theoretically and experimentally studied the performance of a single slope basin type still in which multiple porous absorbers having low thermal capacity were floating adjacent to each other with the help of thermocol insulation [6].

Hansen et al. analyzed the performance of inclined solar still having wire mesh and various wick materials [7]. It was found that water coral fleece was the most suitable material for higher productivity of the still. Matrawy et al. modified the conventional basin type still with the use of corrugated wick and inclined reflector [8]. The results demonstrated that the glass cover temperature of the simple basin type still was lower than that of the still having corrugated wick because of the higher temperature of the wick surface. Alaian et al. experimentally investigated the performance of solar still having pin finned wick in the basin [9]. An improvement of 23% in productivity was obtained by augmenting pin finned wick in the conventional basin type solar still.

Alawee et al. analyzed the performance of a pyramid solar still using various wick materials along with a number of cracks/cords [10]. An improvement of 122% in productivity was reported over the conventional still, while using 25 cords of jute wick. Jobrane et al. experimentally and numerically analyzed the performance of an improved solar still equipped with wick type design [11]. The results indicated that the efficiency of the modified still was 32% higher as compared to the conventional single slope basin type solar still. Darbari and Rashidi investigated the performance of a single slope solar still using porous float wicks of different geometries [12]. The hourly productivity of modified stills was found to be around 0.6 kg/m^2/h for the majority of their operation time.

In the present study, the conventional single slope basin type solar still is modified by placing multiple float wicks in the basin and referred as proposed still in the study. The performance of the proposed still is compared with the conventional basin type still of the same material and dimensions. The effect of covering the basin area of the still with float wick in different proportions is also investigated.

18.2 Experimental Setup

A schematic diagram of single basin single slope solar still having float wick in the basin is shown in Fig. 18.1. The frame of the solar still was made up of mild steel because of its high tensile strength. The material of construction of the basin of the solar still was stainless steel as it has high resilience and ductility. The evaporating surface area of the basin was 1.5 * 0.68 m. The top of the basin was covered with transparent glass inclined at an angle of 300, which is nearly equal to the latitude of Patiala, India (30,021'N, 76,022'E). The back plate and both triangular sidewalls of the basin were consisted of reflecting surfaces so that radiations can be reflected into the basin.

Distillate outputs from both the stills were collected hourly in the plastic containers, between 9:00 AM and 5:00 PM, and weighed on a digital weighing balance having the least count of 0.1 g. The nocturnal distillates collected from 5:00 PM to 9:00 AM were measured the next day in the morning. The effect of covering

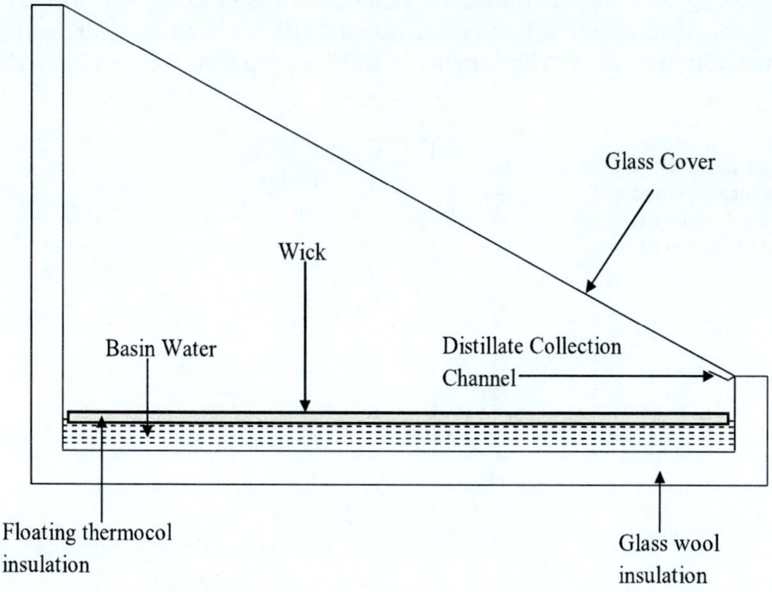

Fig. 18.1 Schematic diagram of modified still

the evaporating surface area with float wick in different proportions at the basin water depth of 5 cm is studied in the present work. Four cases, namely, case 1, case 2, case 3, and case 4 depending on the percentage of the basin area covered with a floating absorber were considered. In case 1 complete (100%) basin area was covered with float wick. The percentages of the evaporating surface area covered in the case 2, case 3, and case 4 were 75%, 50%, and 25%, respectively.

18.3 Results and Discussion

The experiments were performed by covering evaporating surface area of the basin in four different proportions. In case 1, case 2, case 3, and case 4, the portions of the basin area covered with float wick were 100%, 75%, 50%, and 25%, respectively. Figure 18.2 shows the variation of cumulative productivity from 9:00 AM to 5:00 PM for the tested stills when the complete evaporating surface area was covered with float wick. It is found that the cumulative distillate output for the still having float wick in the basin is higher than that of the reference still. The reason for this increase in productivity was the use of float wick in the basin, which reduced the thermal inertia of the basin water leading to the early response of the still. The cumulative distillate output for the modified still during day time was 2.29 kg/m^2 and 1.8 kg/m^2 for the reference still.

The variation of cumulative distillate output of reference and modified still for daytime when 75 percent of the evaporating surface area is covered with wick is shown in Fig. 18.3. The performance variation of the modified still was similar to the case when the complete basin was covered with wick but increase in productivity was higher in the case 2. The cumulative distillate productivities for both modified

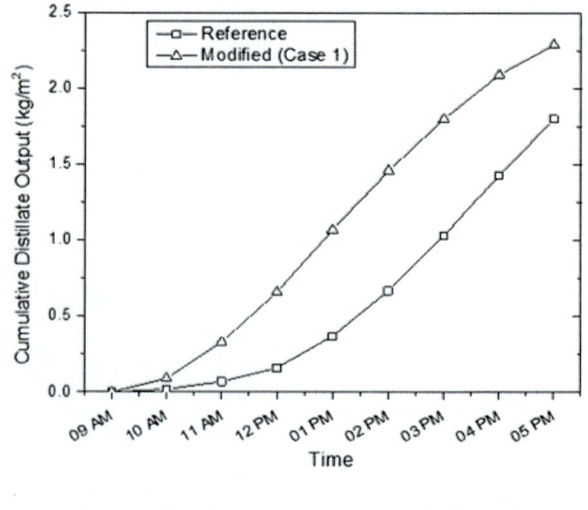

Fig. 18.2 Variation of cumulative distillate output of reference and modified still (case 1) during daytime (9:00 AM to 5:00 PM)

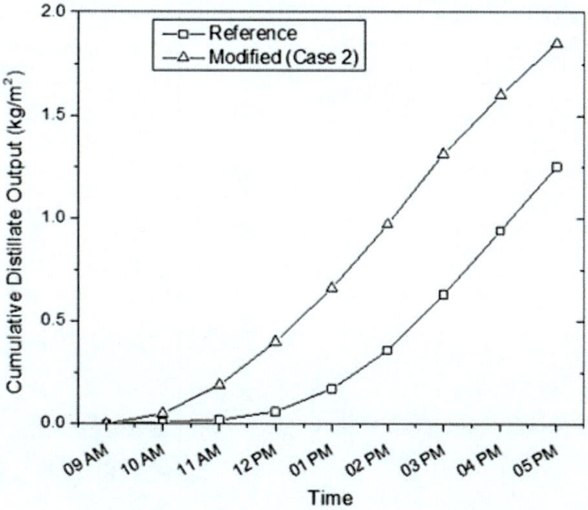

Fig. 18.3 Variation of cumulative distillate output of reference and modified still (case 2) during daytime (9:00 AM to 5:00 PM)

and reference stills from 9:00 AM to 5:00 PM in this case were 1.85 kg/m^2 and 1.25 kg/m^2, respectively.

Figure 18.4 shows a comparison of cumulative productivity of both modified and reference stills when the float wick was placed in 50 percent of the total available basin area. The results demonstrated that the yield of the modified still was higher than that of the conventional still but the rise in productivity was less than that of the case 2. This was due to the increased heat capacity of basin water as half of the basin surface area was exposed to direct solar radiations. For modified and reference stills, the cumulative distillate output during daytime was 1.76 kg/m^2 and 1.38, kg/m^2, respectively.

The deviation of cumulative productivity of modified still from reference still during sunshine hours when 25 percent of the evaporating surface area is covered with a wick is shown in Fig. 18.5. It can be seen that the cumulative distillate output of modified still is always more than that of reference still but the increment in productivity is lesser than that of case 3 because of further increase of heat capacity of basin water in case 4. The behavior of performance variation of modified still with respect to conventional still is similar in both cases. The values of cumulative distillate output for both modified and reference stills were 1.57 kg/m^2 and 1.42 kg/m^2, respectively. From Figs. 18.2, 18.3, 18.4 and 18.5, it was observed that as the percentage of the basin covered with float wick increased, the cumulative yield of the solar still decreased.

The gain in total distillate output for 24 h against the percentage of basin surface area under float wick is shown in Fig. 18.6. The total evaporating surface area of the basin is 1.02 m^2. The productivities of modified still were improved by 3.83, 20.57, 13.33, and 2.86%, when the basin area below the float wick was 1.02 m^2, 0.76 m^2, 0.51 m^2, and 0.25 m^2, respectively. The rise of 20.57% in the productivity of modified still over conventional still was observed when 75 percent of the evaporating surface

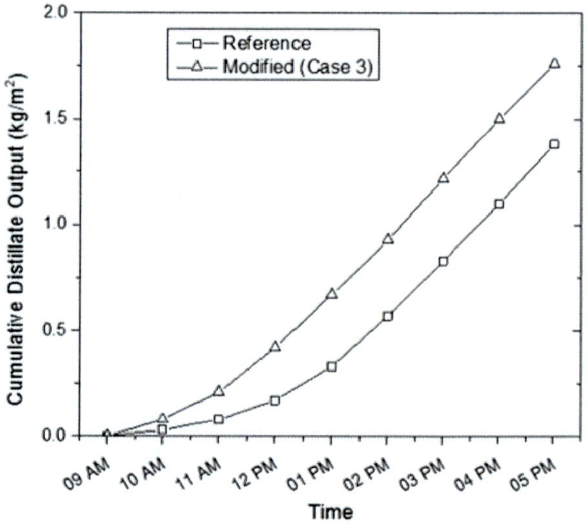

Fig. 18.4 Variation of cumulative distillate output of reference and modified still (case 3) during daytime (9:00 AM to 5:00 PM)

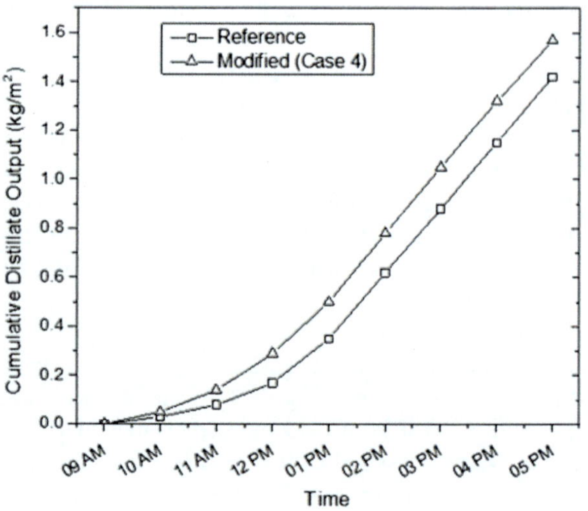

Fig. 18.5 Variation of cumulative distillate output of reference and modified still (case 4) during daytime (9:00 AM to 5:00 PM)

area was covered with wick. The gain of distillate output was highest and the drop in yield of the solar still over the reference still was lowest for the case, when 3/4th of the basin surface area was kept under a floating absorber.

It can be seen from Fig. 18.6 that the overall daily productivity gain of modified still over reference still was maximum for case 2, i.e., when 75% of the basin surface area was covered with float wick. Hence, component-wise performance for case 2 of modified still is presented in Figs. 18.7, 18.8 and 18.9. The variation of solar intensity and ambient air temperature during daytime (9:00 AM to 5:00 PM) is shown

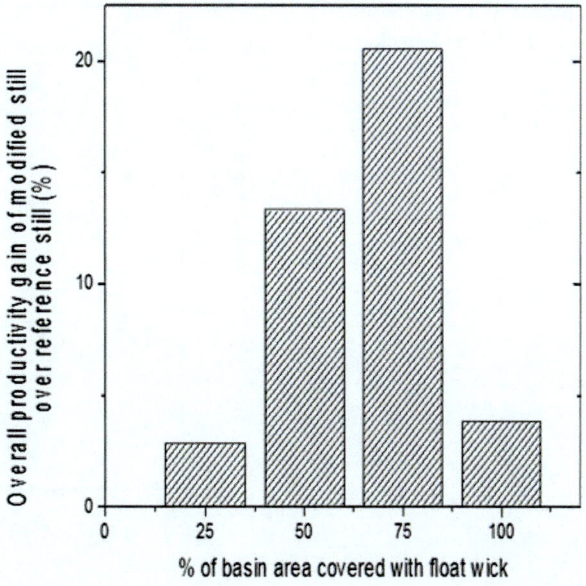

Fig. 18.6 Comparison of overall distillate output for 24 h (9:00 AM to 9:00 AM) of all the four cases of modified still

in Fig. 18.7. It can be seen that maximum solar intensity was reached around 12:30 PM whereas the maximum ambient air temperature was reached around 3:30 PM.

From Fig. 18.8, it can be seen that the basin temperatures of the reference still are higher than that of the modified still during daytime which is the indication of higher heat losses through the sides and base of the reference still. Due to these reasons, the performance of the modified still was better than that of the reference still. It

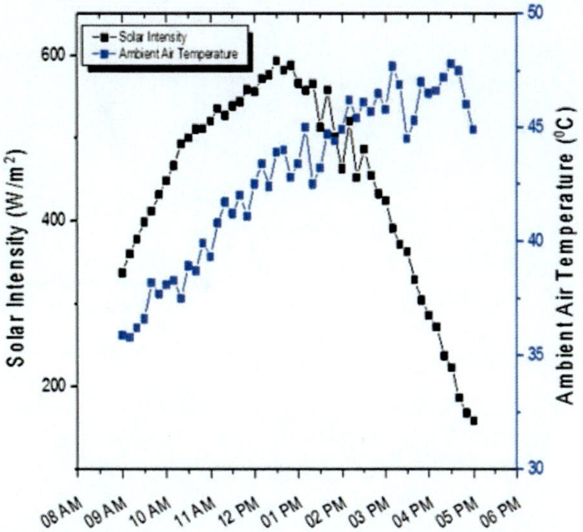

Fig. 18.7 Hourly variation of solar intensity and ambient air temperature

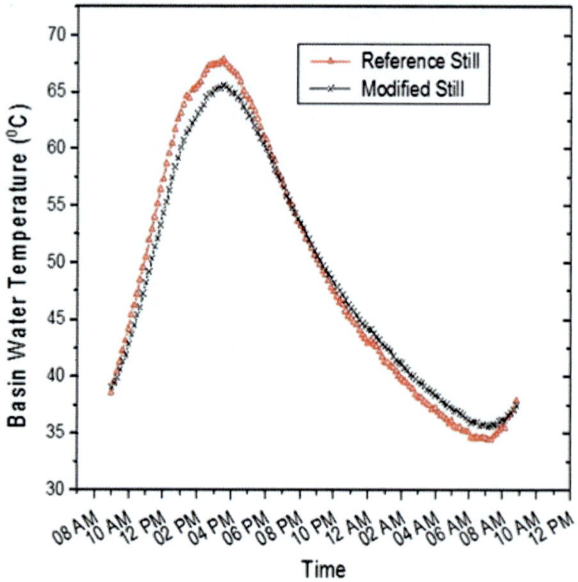

Fig. 18.8 Basin water temperatures of modified and reference still for 24 h (9:00 to 9:00)

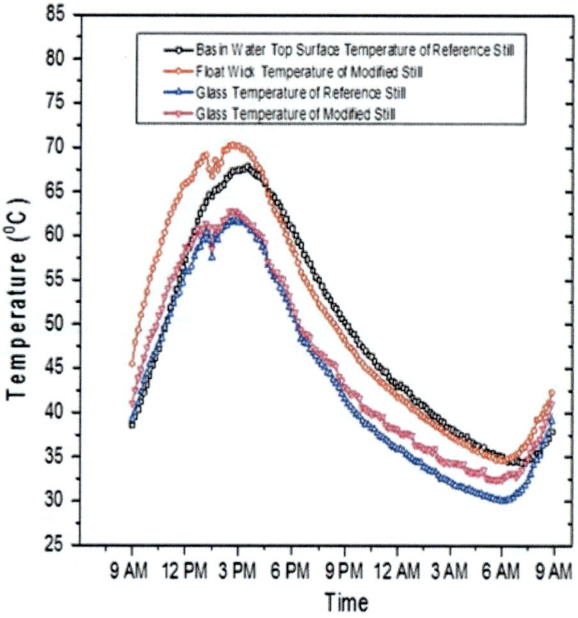

Fig. 18.9 Variation of component temperatures of reference and modified stills

was also observed that the basin water temperatures of the modified still were higher than that of the reference still during night which showed that the heat stored in the modified still was not fully utilized, thus decreasing its nocturnal distillate output. The maximum values of basin water temperatures in the modified and reference still were 65.55 °C and 67.8 °C, respectively.

Figure 18.9 shows the variation of glass and surface temperatures of the modified and reference stills. It can be seen that the maximum temperature of float wick is reached at about 15:00 h as compared to around 16:00 h for the basin water of the reference still. This earlier response is due to the low thermal inertia of the float wick, which results in higher operating temperatures of the modified still leading to quicker start-up in the morning as compared to the slow response of reference still having large thermal capacity of basin water. There is a significant difference between the float wick temperature and glass cover temperature of the modified still, whereas for reference still the basin water–glass cover temperature difference is lesser than the temperature difference between the wick and glass cover of the modified still, resulting in a higher evaporation rate.

The hourly variation of cumulative distillate output of both the stills is shown in Fig. 18.10. It is found that the cumulative productivity of the modified still is higher than that of the reference still. The reason behind the better performance of the modified still is decreased thermal capacity of the basin due to the use of float wick leading to the higher operating temperature of evaporating surface which increased the evaporation rate and hence improved the distillate output of the modified solar still. A significant amount of nocturnal distillate output is obtained in modified still as the basin is partially covered with wick due to which heat can directly flow from warm basin water to float wick. The overall daily productivities of both reference and modified stills were 2.74 kg/m^2 and 3.31 kg/m^2, respectively.

18.4 Conclusions

This work presents the effect of varying the percentage of the evaporating surface area by placing float wicks in a single slope basin type solar still. The experimental work carried out in this work lead to the following conclusions:

- The gain in productivity of modified still over the reference still increases when the proportion of the basin area covered with float wick changes from 100 to 75% due to the availability of some uncovered area responsible for heat transfer between warm water and the top surface of the wick.
- The maximum gain of 48.11% in day distillate output of modified still was obtained when 75% of the evaporating surface area was lying below the wick.
- The overall daily productivity gain of modified still over reference still was maximum when 75% of the basin surface area was covered with float wick.

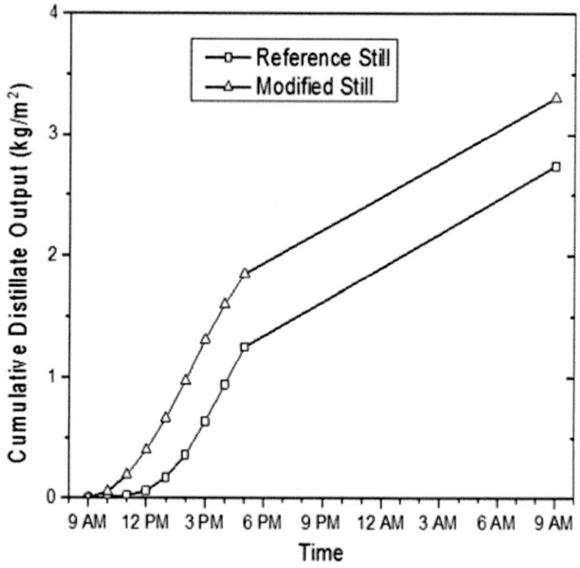

Fig. 18.10 Cumulative productivity of reference and modified stills for 24 h (9:00 AM to 9:00 AM)

- The temperature of the float wick was much higher than that of the basin water temperature of the modified still during daytime which indicated that they still operated at much higher temperatures.
- The temperature difference between the float wick and glass cover of modified still was more than that of the water–glass cover temperature difference of the reference still, resulting in a higher evaporation rate in the case of modified still.

References

1. A. Hanson, W. Zachritz, K. Stevens, L. Mimbela, R. Polka, L. Cisneros, Distillate water quality of a single-basin solar still: laboratory and field studies. Sol. Energy **76**(5), 635–645 (2004)
2. A.A. El-Sebaii, S. Aboul-Enein, E. El-Bialy, Single basin solar still with baffle suspended absorber. Energy Convers. Manage. **41**(7), 661–675 (2000)
3. H.M. Rababa'h, Experimental study of a solar still with sponge cubes in basin. Energy Convers. Manag. **44**(9), 1411–1418 (2003)
4. K.K. Murugavel, K.K. Chockalingam, K. Srithar, An experimental study on single basin double slope simulation solar still with thin layer of water in the basin. Desalination **220**(1–3), 687–693 (2008)
5. K.K. Murugavel, K. Srithar, Performance study on basin type double slope solar still with different wick materials and minimum mass of water. Renew. Energy **36**(2), 612–620 (2011)
6. P.K. Srivastava, S.K. Agrawal, Experimental and theoretical analysis of single sloped basin type solar still consisting of multiple low thermal inertia floating porous absorbers. Desalination **311**, 198–205 (2013)
7. R.S. Hansen, C.S. Narayanan, K.K. Murugavel, Performance analysis on inclined solar still with different new wick materials and wire mesh. Desalination **358**, 1–8 (2015)

8. K.K. Matrawy, A.S. Alosaimy, A.F. Mahrous, Modeling and experimental study of a corrugated wick type solar still: comparative study with a simple basin type. Energy Convers. Manage. **105**, 1261–1268 (2015)
9. W.M. Alaian, E.A. Elnegiry, A.M. Hamed, Experimental investigation on the performance of solar still augmented with pin-finned wick. Desalination **379**, 10–15 (2016)
10. W.H. Alawee, F.A. Essa, S.A. Mohammed, H.A. Dhahad, A.S. Abdullah, Z.M. Omara, Y. Gamiel, Improving the performance of pyramid solar distiller using dangled cords of various wick materials: novel working mechanism of wick. Case Stud. Therm. Eng. **28**, 101550 (2021)
11. M. Jobrane, A. Kopmeier, A. Kahn, H.M. Cauchie, A. Kharroubi, C. Penny, Theoretical and experimental investigation on a novel design of wick type solar still for sustainable freshwater production. Appl. Therm. Eng. **200**, 117648 (2022)
12. B. Darbari, S. Rashidi, Performance analysis for single slope solar still enhanced with multi-shaped floating porous absorber. Sustainable Energy Technol. Assess. **50**, 101854 (2022)

Chapter 19
Production of Mycelium-Based Thermal Insulating Material Using Biomass Residue as Substrate

Gaurav Singh, Debanjan Sutradhar, Ashutosh Mishra, and Nikhil Gakkhar

Abstract Thermal insulation is the property of insulating material that prevents the transfer of heat between the materials that are in thermal contact. This paper aims at the development of mycelium-based insulating material using biomass residue as substrate. The insulating material was prepared from mushroom mycelium with different biomass (wheat straw, paddy straw, groundnut shell, and potato starch) based substrate. Further, the mechanical properties (tensile strength, compressive strength, and bulk density), thermal properties, water absorption rate, and composition of the insulating material were investigated. Inorganic substances like $CaCO_3$ were added to the samples to increase the mechanical properties of the material. Substrates were inoculated with Pleurotus Ostreatus density of insulating material was found in between 120 and 180 kg/m^3. The thermal conductivity of the samples prepared ranges from 0.045 to 0.07 W/mK. The conversion rate of biomass into insulating material was found to be 95%. It could be an appropriate solution for stubble burning especially in Indian states like Punjab, Haryana, Uttar Pradesh, etc. where the management of agricultural residue is a crucial problem.

Keywords Thermal insulating material · Mycelium · Pleurotus ostreatus · Agricultural residue

19.1 Introduction

The demand for building materials is increasing at a rapid rate; these materials have the highest contribution to carbon emission that includes extraction of the material, its manufacturing and transportation [1]. It is one of the reasons for the degradation of the environment. In order to prevent this excessive depletion of natural resources and reduce the carbon footprint, natural resources must be used sustainably [2]. One

G. Singh (✉) · D. Sutradhar · A. Mishra · N. Gakkhar
Sardar Swaran Singh National Institute of Bio-Energy (SSS-NIBE), Kapurthala, Punjab 144601, India
e-mail: gauravs.cee.21@nitj.ac.in

N. Gakkhar et al. (eds.), *Recent Advances in Bio-Energy Research*, Springer Proceedings in Energy, https://doi.org/10.1007/978-981-99-5758-3_19

of the solutions to this problem could be eco-friendly bio-based insulating material that could be an alternative to synthetic compounds like foams, polyurethane, polystyrene, plastic-based fibers, etc. These materials have a very high carbon footprint and produce toxic substances during the manufacturing process and take around thousands and millions of years to degrade [3]. Insulating materials are those materials that show resistance to heat currents and reduce the rate of heat transfer [4]. Materials like fiberglass, slag wool, glass wool, and cellulose are conventional insulating materials that are highly fibrous and resist convection and conduction heat flows [5]. Their thermal conductivity ranges from 0.03 to 0.07 W/mK. A significant amount of energy is consumed in the production of conventional insulating material. In addition, some of the synthetic or petroleum-derived materials may contain harmful substances [3], limiting their application for building construction due to possible health concern. Furthermore, the most common types of synthetic foams are not biodegradable and lead to the generation of a large amount of waste at the end of their service life [1]. This paper focuses on the mycelium-based insulating material. Mycelium is a vegetative part of the fungus that grows on biomass residue [3]. It has been found that mycelium-based insulating materials have the same characteristics to that of traditional insulating materials [6]. Agricultural residues like wheat straw, paddy straw, and potato starch were taken as substrate, and mycelium was grown on that. Investigations showed that mycelium-based insulating materials have high insulation capacity low cost of development and low density which makes them an alternative to conventional insulating materials.

Major agricultural residues like wheat straw and paddy straw along with potato starch were used as mycelium substrate. Results showed an improved growth rate in mycelium compared to other samples and an increase in the tensile and compressive strength of the insulating material. It can have various applications and it can be used as an insulator in buildings instead of using foam-based materials it can be also used in packaging for thermal-sensitive items like food and pharmaceutical items. Various companies are also producing mycelium-based packaging materials, along with that they are also producing various products like bio leather with thermal insulating properties, which is not only cost-effective but also sustainable and completely renewable [7]. Due to their low thermal conductivity and high shock absorbing capacity, they can be used in doors, floors, paneling, shock absorbers, and other furnishings.

19.2 Materials and Methods

19.2.1 Materials

Pleurotus Ostreatus spawn (inoculated) on wheat grain was purchased from Dr. Seeds Pvt Ltd. Sodium carbonate (anhydrous 99%) was purchased from M/s HiMedia chemicals Mumbai, India. Agricultural residue was collected from a farm field near Jalandhar.

19.2.2 Preparation of Insulating Material

Substrate Preparation

The mushroom substrate was sterilized in an autoclave at 200°C for 12 h [3]. After sterilization, the substrate was allowed to cool down to ambient temperature and distilled water was mixed thoroughly with the substrate and kept for drying at ambient conditions until the substrate reaches 62% w/w moisture content [8]. Four samples were prepared with: (i) wheat straw, (ii) paddy straw, (iii) mixture of wheat straw, paddy straw, and potato starch (45%, 45%, 10% w/w), and (iv) wheat straw, paddy straw, potato starch, and calcium carbonate (45%, 45%, 9%, 1% w/w) and weight of each sample was measured using a digital weighing balance with least count of 0.01 g. The sample details are shown in Table 19.1.

Spawn Addition and Incubation

Mushroom spawn was added in each sample at 10% w/w all the samples were incubated in an incubator (–) at 25 ± 2 °C with 70% humidity and 20,000 PPM CO_2 concentration [9], and all the substrates were kept in a Petri-dish and allowed to grow for 7–9 days [7]. The sample containing wheat straw and paddy straw with $CaCO_3$ showed an increase in growth rate compared to other samples.

Drying

All the samples were collected after 9th day and kept in a dryer at 800 °C for 2 hr to prevent the further growth of mycelium and remove the excess moisture present in the samples [9]. It was found that there was a 30–40% reduction in the weight of samples.

Table 19.1 Symbolic representation of each sample prepared

S. No	Sample name	Substrate constituents
01	S1	Wheat straw (100%)
02	S2	Paddy straw (100%)
03	S3	Wheat straw (80%), potato starch (20%)
04	S4	Paddy straw (80%), potato starch (20%)
05	S5	Wheat straw (99%), $CaCO_3$ (1%)
06	S6	Paddy straw (99%), $CaCO_3$ (1%)
07	S7	Wheat straw (50%), paddy straw (50%)
08	S8	Wheat straw (49.5%), paddy straw (49.5%), $CaCO_3$ (1%)

19.3 Characterization of Insulating Material

19.3.1 Thickness of Prepared Samples

The thickness of the prepared sample was evaluated using Vernier Caliper with the least count of 0.1 mm. Thickness measurement of samples was taken at random positions and a mean value was calculated.

19.3.2 Density of Insulating Materials

The density of the prepared sample was calculated using the water displacement method. The weight of each sample was measured using a weighing balance with the least count of 1 mg. Further a thin coating of paraffin (hydrophobic in nature) [8] wax was applied to all the samples, and each sample was placed in a beaker filled with water with a least count of 0.1 ml. The displaced volume of each sample was noted down and density was calculated.

$$p = {}^{m}/_{v} \tag{19.1}$$

where ρ = density,

m = mass of samples (weight/9.806).

v = volume.

19.3.3 Water Absorption Rate

The water absorption rate of samples was measured using the following method. The weight of the dried samples was measured. Then each sample (5 cm * 5 cm * 0.5 cm) was submerged in distilled water for 1, 2, 5, and 10 min, and the weight of each sample was measured. Change in the weight of the dried sample and the wet sample were noted down. The increase in the weight of wet samples was linear for the first five minutes and there was a negligible increase in weight after 5 min. The result showed that there was a 10% increase in weight.

19.3.4 Thermal Conductivity

The thermal conductivity of the samples was measured using a guarded hotplate method. It is a steady-state measurement method that determines the thermal conductivity of the material by using a hotplate with guided heat conduction. The samples were heated from one side by an electrically heated plate. The other side of the sample is controlled by a cold plate and our sample is placed between these two plates, and the temperature difference between the hotplate and cold plate is measured and the thermal conductivity was calculated using the formula.

$$K = {}^{Q.D}/_{A.\Delta T} \tag{19.2}$$

Here, K = Thermal conductivity,

Q = Heat flow (electrical power consumed by heat plate),

D = Thickness of sample,

A = Area of sample,

ΔT = Temperature difference between two plates.

19.3.5 Mechanical Properties

Mechanical properties, tensile strength, and elongation at fracture of the samples prepared were measured with the Universal testing machine, and compressive strength was measured by load addition method [10]; test samples were made in the form of a cube with dimensions 5 cm*5 cm*0.5 cm. The load was added on top of each sample and increased gradually indentation and cracks in the sample were observed with a microscope and the load was increased until fracture; the load was noted down and divided by the cross-sectional area of the sample providing us the compressive strength.

$$\sigma = \gamma \,.\, {}^{\Delta l}/_{l} \tag{19.3}$$

Here, σ = Stress, N/m^2

Δl = Change in length, m

l = Actual length, m

Elongation (Δl) = Increased length–Actual length.

Table 19.2 Thickness of different sample

S. no	Sample name	Thickness (mm)
01	S1	5.5 ± 0.1
02	S2	5.1 ± 0.1
03	S3	5.7 ± 0.2
04	S4	5.2 ± 0.3
05	S5	6.2 ± 0.4
06	S6	5.8 ± 0.3
07	S7	5.3 ± 0.2
08	S8	6.5 ± 0.5

19.4 Result and Discussion

19.4.1 Thickness

The thickness of the insulating material is shown in Table 19.2 was found between 5 and 7 mm for the same weight of biomass residue. Results showed that the addition of $CaCO_3$ increased the thickness as well as the growth rate of mycelium compared to other samples (S1, S2, S3 & S4). It was found that the addition of potato starch also showed an increase in thickness compared to other samples (S1 and S2).

19.4.2 Density

The density of materials shown in Table 19.3 was found between 120 and 180 kg/ m^3. The density of samples containing wheat straw was more as compared to paddy straw, due to an increase in the percentage of lignin, cellulose, and hemicellulose [11]. Sample containing $CaCO_3$ also showed an increase in density compared to raw samples without $CaCO_3$.

Table 19.3 Density of the samples

S. no	Sample name	Density (kg/m^3)
01	S1	140 ± 5
02	S2	120 ± 5
03	S3	150 ± 5
04	S4	130 ± 5
05	S5	160 ± 5
06	S6	135 ± 5
07	S7	130 ± 5
08	S8	140 ± 5

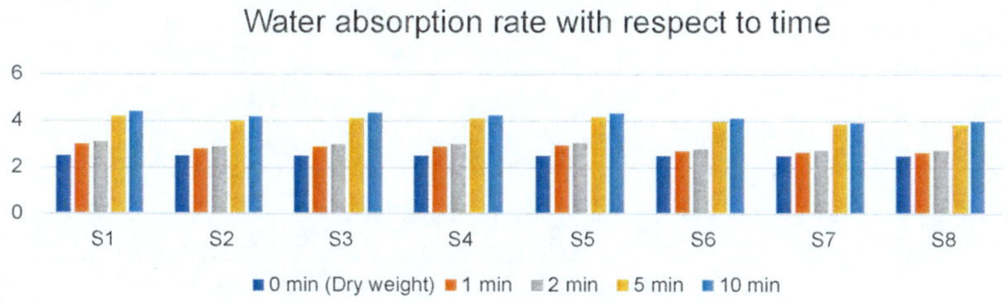

Fig. 19.1 Water absorption rate of different samples at different time intervals

19.4.3 Water Absorption Rate

The water absorption rate is the amount of water absorbed by the material for a particular interval of time. The water absorption rate is the weight of the wet sample and the weight of the dry sample.

The water absorption rate was determined by submerging the sample completely inside a beaker for time intervals of 1, 2, 5, and 10 min. As shown in Fig. 19.1, it was found that sample with paddy straw and calcium carbonate showed a minimum water absorption rate compared to other samples; there was a maximum 82% increase in the weight of samples for 10 min. and further it would increase to 200% [11].

Here, X-axis time period of samples emerged in water, Y-axis weight of samples in grams.

19.4.4 Thermal Conductivity

Thermal conductivity was measured using the guarded hot plate method. Samples were placed between two conductive plates, one kept at the hot side at constant temperature and another at the cold side. The temperature gradient was measured between the two plates using the equation $Q = KAdt/dx$. Table 19.4 shows that Samples containing wheat straw with calcium carbonate showed minimum conductivity compared to other samples, the reason might be lignin concentration is more in the wheat sample and an increase in mycelium density due to the presence of calcium carbonate [11] (Table 19.4).

Table 19.4 Thermal conductivities of different samples

Sl. no	Sample number	Thermal conductivity (W/m. K)
1	S1	0.0470
2	S2	0.0680
3	S3	0.0490
4	S4	0.0650
5	S5	0.0690
6	S6	0.0475
7	S7	0.0580
8	S8	0.0550

19.4.5 Mechanical Properties

The tensile strength of each material was measured using UTM (universal testing machine) and it was found that the sample with wheat straw and $CaCO_3$ showed maximum Tensile strength compared to other samples.

Tensile Strength

Figure 19.2 shows that the Tensile strength of wheat straw with $CaCO_3$ was found 0.07 MPa and the sample with wheat straw without $CaCO_3$ was found 0.082 MPa, similarly sample containing paddy straw with $CaCO_3$ has more tensile strength than the sample without $CaCO_3$. Sample S8 showed maximum tensile strength it was around 0.137 Mpa. Results showed that $CaCO_3$ is helpful in increasing the tensile strength of insulating material. Similarly, all the samples with calcium carbonate showed an increase in tensile strength compared to samples without $CaCO_3$. The integration of $CaCO_3$ in the substrate helps in enhancing the tensile strength of the material.

Compressive Strength

The compressive strength of all the samples was measured by adding load above the samples and gradually increasing it until fracture. The sample containing wheat straw with $CaCO_3$ showed maximum compressive strength compared to other materials. Table 19.5 shows that the compressive strength of sample S5 was found to be 0.05 MPa. Calcium carbonate helps in increasing the compressive strength of the material. Sample with wheat straw without $CaCO_3$ has a compressive strength of 0.04 MPa and the compressive strength of the remaining samples was found between 0.03 and 0.04 MPa. Results showed that calcium carbonate helps in increasing the mechanical properties of mycelium-based composites.

Table 19.5 Compressive strength of different samples prepared

Sl no	Sample name	Compressive strength (MPa)
1	S1	0.04
2	S2	0.03
3	S3	0.042
4	S4	0.033
5	S5	0.05
6	S6	0.048
7	S7	0.045
8	S8	0.049

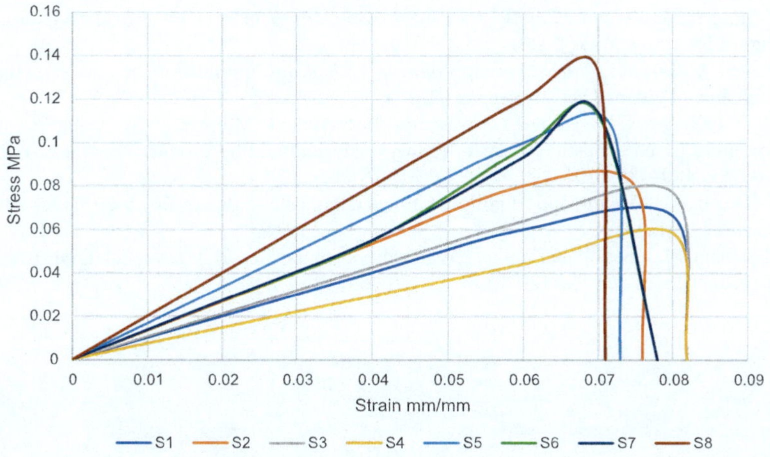

Fig. 19.2 Tensile strength of the materials

19.5 Conclusion

Insulating materials were prepared using substrates like wheat straw, paddy straw, potato starch, and calcium carbonate with different combinations and concentrations. mycelium of Pleurotus Ostreatus was grown on these substrates. These substrates were kept in a mold and incubated in a CO_2 chamber with a concentration of 50,000 ppm and temperature of 25 °C for 7 days. It was found that samples containing calcium carbonate with wheat straw showed better results compared to other samples. Calcium carbonate can be a helping agent in improving the mechanical properties of mycelium-based composites, but there was not much significant improvement in the thermal conductivity of the material. These insulating materials have comparable properties to those of conventional insulating materials (polyurethane, polystyrene, plastic-based fibers), One more application could be they can be used as packaging for fragile materials.

References

1. S. Gaurav, B. Kumar et al., Active barrier chitosan film containing gallic acid based oxygen Scavenger. Food Meas. **15**, 585–593(2021)
2. G.P. Boswell, H. Jacobs, K. Ritz, G.M. Gadd, F.A. Davidson, B. Math, Biol. **69**(2), 605–634 (2007)
3. M.F. Ashby, Materials and the environment: eco-Informed material choice, 2nd Ed. Elsevier
4. Gauvin, Vitasoy, J. Vetto, Physical properties & hydrothermal properties of myce-lium-based composites as foam like insulating materials. Constr. Technol. Arch. 1 (2022). Trans Tech Publications Ltd
5. H. Hislop, J. Hill, Reinventing the wheel: a circular economy for resource Security. Green Alliance
6. A.K. Mohanty, M. Misra, L.T. Drzal (eds.), *Natural fibres, biopolymers, and bio composites* (CRC Press, Boca Raton, Florida, 2015)
7. P. Stamets, *Mycelium running: how mushrooms can help save the world* (Random House Digital Inc., New York, 2020)
8. P. Oei, *Mushroom cultivation: appropriate technology for mushroom growers* (Backhuys Publishers, Leiden, The Netherlands, 2018)
9. M.G. Pelletier, G.A. Holt, J.D. Wanjura, E. Bayer, G. McIntyre, An evaluation study of mycelium based acoustic absorbers grown on agricultural by-product substrates. Ind. Crops Prod. **51**, 480–485. (2013)
10. C. Kazmierski, Growth opportunities in global composites industry, 2012–2017. 21–23 (2012). Luc intel, Irving, TX
11. E. Karana, M. Jones et al., Engineered mycelium composite construction materials from fungal bio refineries. Mater. Des (2020)

Part V
Hybrid Systems

Chapter 20
Assessment of Economics of Hybrid Biomass Systems and Value to Grid

Rakesh Godara, Nikhil Gakkhar, Shruti Deorah, Aditya Khandekar, Nikit Abhyankar, Bhautik Gajera, Akash Deep Singh, and Anil Kumar Sarma

Abstract The study focused on the value of biomass-based power plants and hybrid systems to the grid today and in 2030. In India, there are 3 GW for biomass (excluding bagasse), 7 GW for bagasse-based cogeneration, and 20–25 GW for additional untapped biomass generation capacity. The high amounts of solar generation during daytime hours result in steep ramps during morning and evening peak times. Therefore, flexible resources such as energy storage, demand response, and flexible operations of gas and biomass power plants become increasingly important. Diurnal balancing support requires 4–6 hr of support during morning and evening peak periods for more than 300 days of the year. Seasonal balancing requires high capacity factors of 60–80% between the months of October and February when RE generation is reduced. In order to balance India's grid, 63 GW of energy storage, 60 GW of load shifting from nighttime to solar hours and using the existing gas-based capacity for seasonal balancing during low RE season (Oct-Feb) will need to be used in addition to 307 GW of solar and 142 GW of wind. In this economic framework, waste management and reduced pollution/emissions benefits from avoided waste burning are not considered.

Keywords Hybrid systems · Biomass · Solar · Bagasse · Diurnal · Seasonal · Renewable

---

R. Godara (✉) · N. Gakkhar · B. Gajera · A. D. Singh · A. K. Sarma
Sardar Swaran Singh National Institute of Bio-Energy (SSS-NIBE), Kapurthala, Punjab 144601, India
e-mail: rakesh.tcd@nibe.res.in

S. Deorah · A. Khandekar · N. Abhyankar
Lawrence Berkeley National Laboratory, Berkeley, USA

N. Gakkhar et al. (eds.), *Recent Advances in Bio-Energy Research*, Springer Proceedings in Energy, https://doi.org/10.1007/978-981-99-5758-3_20

20.1 Introduction

India has a target of installing 175 GW of renewable energy (RE) capacity by 2022 and 500 GW of non-fossil energy capacity by 2030. Significant progress has already been made toward the achievement of these targets [1]. According to the Central Electricity Authority, as of 2021, India's renewable energy capacity more than doubled, supplying nearly 10 percent of the nation's total electricity generated. India has successfully achieved some of the lowest RE costs in the world in the past decade [2]. The average solar tariff in 2020 was 34% lower than the global average solar tariff between 2010 and 2020. It saw the largest reduction in country-level solar levelized cost of energy (LCOE) of 85% between 2010 and 2020. Furthermore, India had the lowest country-level installed costs for solar and wind power in 2020 [3] (Fig. 20.1).

The levelized cost of renewable electricity has fallen below that of thermal power generation on a generalized basis [4, 5]. Even so, more investments in new fossil fuel plants are planned primarily due to the following reasons: (a) RE generation is intermittent and thus requires significant system flexibility for grid integration, (b) RE generation occurs outside of peak electricity demand periods, most of which occur in the evening for most parts of India, and (c) legacy planning and regulatory frameworks may not fully capture the benefits of RE and energy storage technologies. Accordingly, the dramatic fall in battery storage costs, which has dropped by 90% at the battery pack level since 2010, could lead to a turning point, since it enables the cost-effective utilization of renewable electricity during peak times [6] (Fig. 20.2).

India's grid will need to balance its variability as it attains a higher penetration of renewable sources of energy through a spectrum of flexible resources that work in tandem in order to maintain an equilibrium between supply and demand on an hourly basis. Based on our analysis of the least cost resource mix for the Indian grid in 2030, we determined that 307 GW of solar and 142 GW of wind are cost-effective, coupled with flexible resources such as 63 GW of energy storage, 60 GW of load shifting between night time and solar hours, and making use of the existing gas-based capacity for seasonal balancing during low RE season (Oct-Feb). Figure 20.3 shows

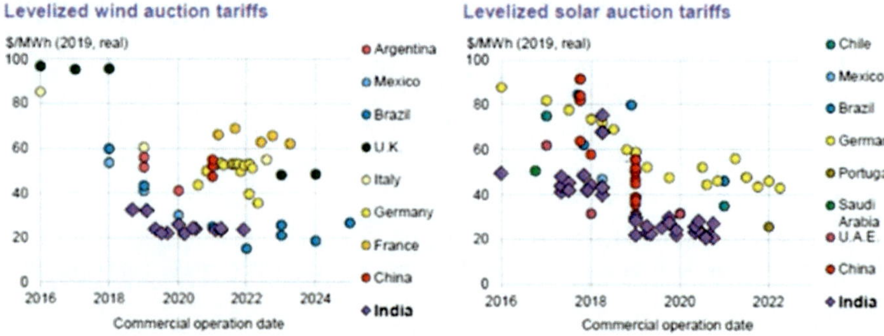

Fig. 20.1 Solar and wind energy prices in key countries, including India [3]

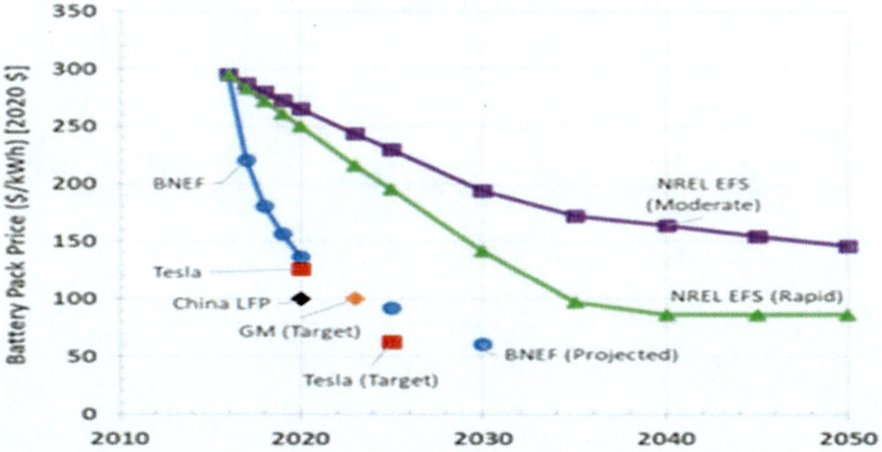

Fig. 20.2 Global average battery pack price over years [7]

the projected demand and net demand in 2030 for the most cost-effective mix of renewable energy sources (307 GW solar and 142 GW wind) [8].

With high levels of solar generation during daylight hours, the 2030 Indian grid will require higher ramping requirements than the 2020 grid. It can be seen from Fig. 20.3 that the difference between the daily load and the net load for key months

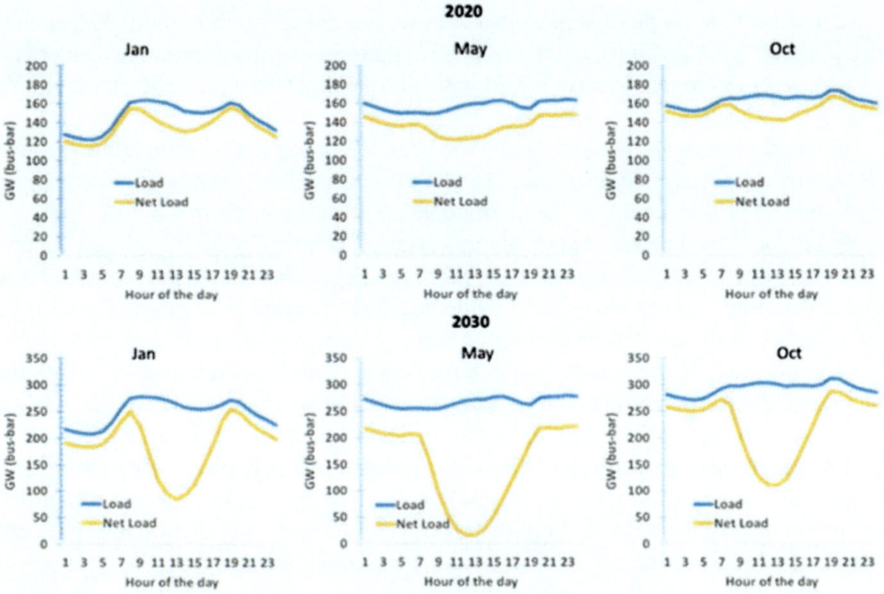

Fig. 20.3 Average daily load and net load curve for key months in 2020 and 2030 [8]

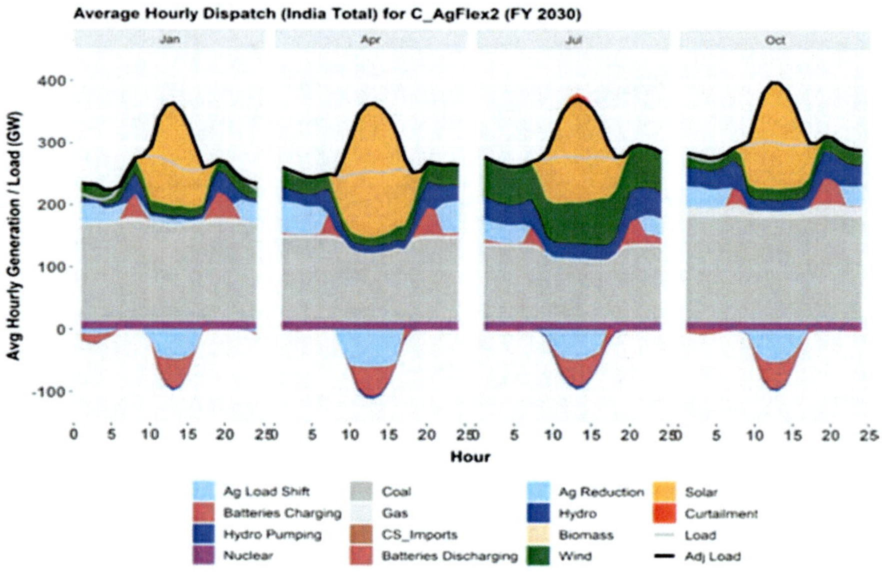

Fig. 20.4 Average hourly dispatch for key months in 2030

in 2020 and 2030 increases rapidly resulting in steep ramps during morning and evening peak times in 2030. Net load (or residual load) is defined as load minus the output from variable RE sources (solar and wind).

The importance of flexible power sources such as energy storage, demand response (agricultural load shifting), and flexible gas and biomass power plants will increase in such an environment to ensure grid power remains affordable, stable, and reliable [9].

Flexible sources are needed during the peak RE generation season (June through September for wind generation and March through June for solar) to provide the grid with diurnal grid balancing to accommodate steep ramps as shown in Fig. 20.4.

Solar and wind energy cannot charge batteries adequately during the low renewable energy production season (October through February) [10, 11]. Because natural gas and biomass generation can provide seasonal balancing (instead of coal-fired assets), their dispatch will be the highest during these months [12].

A major goal of this study, which builds on existing research, is to evaluate the techno-economic potential of biomass and hybrid biomass-renewable energy systems for India's evolving grid based on biomass supply chain, fuel availability, and operational constraints. According to the study, biomass power plants could provide grid balancing services (diurnal and seasonal) to the Indian grid by 2030, with high RE penetration. We have provided policymakers with insights into the role biomass can play in the future of India's electricity sector based on our analysis.

20.2 Biomass in India: A Scenario

In India, the installed biomass generation capacity in 2020 was approximately 10 GW, of which 75% is bagasse-based cogeneration [13]. The states of Uttar Pradesh, Maharashtra, and Karnataka are the top three states with installed biomass capacity. The bulk of bagasse power plants' output occurs between October and May [14] as shown in Fig. 20.5. Bagasse availability and generation are highly seasonal, with a capacity factor of 26% from November to May and 4% the rest of the year. This is because the main purpose of these plants is to produce sugar during the season, with electricity serving as a by-product.

Contrary to other biomass-based generation methods, which maintain a flat generation profile all year long with a PLF of 15%. Rice husk and bio-waste from other pulses are mainly used for this generation, and they are produced all year round unlike sugarcane, which is grown primarily in the Kharif season. The Kharif season is from June to October and Rabi season is from November to April. The seasonal ramp up of bagasse cogeneration starting in November aligns very well with the seasonal balancing needs of the Indian grid, as envisioned for 2030.

CEA reports that biomass-based (collective which includes bagasse) installed generation capacity in India is 10 GW [15]. Cogeneration using bagasse accounts for 7 GW and 3 GW from biomass (excluding bagasse). It is estimated that a total of 30–35 GW of biomass power can be generated in India. It follows that there is an additional 20–25 GW of biomass power generation capacity untapped.

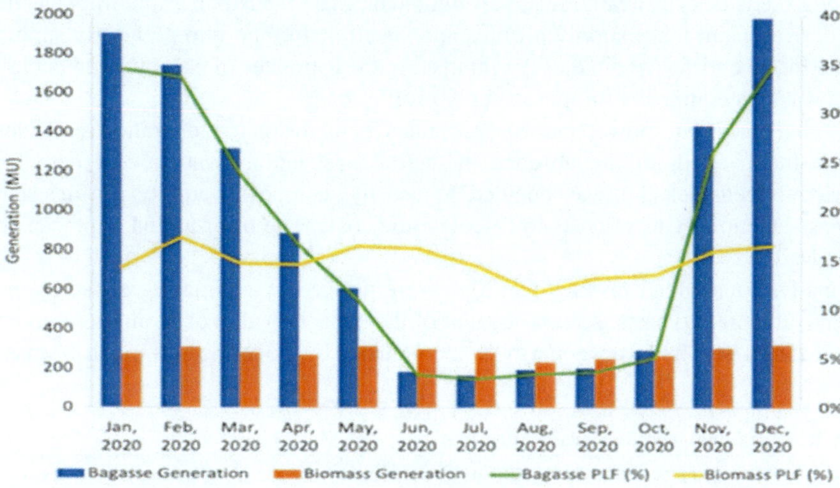

Fig. 20.5 Monthly bagasse and biomass (excluding bagasse) generation and plant load factors (PLF) for 2020 [13]

20.3 Methodology and Data

The study focused on the following key aspects to assess biomass-based power plants and hybrid systems in India's power sector today and in 2030:

- The value of biomass and RE + biomass hybrid systems to the grid.
- The ability of biomass plants to provide specific grid services, such as diurnal and seasonal balancing, while taking into account the available economic and operational constraints of biomass-based generation systems.

20.3.1 Biomass Value Estimation

The net value of biomass and RE + biomass hybrid systems is the sum of their capacity and energy value. A resource's capacity value is determined by determining the cost of a marginal unit of generation capacity it would be able to replace or avoid. As a general rule, it is expressed in rupees per kilowatt-year or rupees per kilowatt-hour. A resource's energy value is equal to the variable cost of its marginal unit of generation.

The first step was to develop a monthly feedstock supply curve based on seasonal feedstock availability data (Table 20.1). A sugarcane/bagasse supply curve was also developed, considering 75% of biomass generation is based on sugarcane/bagasse.

In Table 20.1, it appears that rice and cotton, as well as sugarcane/bagasse, were the top 2 feedstocks by availability. A weighted biomass feedstock availability profile was derived from these supply profiles and was transformed into an hourly supply profile for an entire year. One of the inputs into the estimates of capacity and energy value was this composite biomass supply profile.

Further data input showed that cost estimates for all the major generation resources on the Indian grid. In the absence of capital cost and operating cost data for individual feedstocks, it was decided to use the composite supply profiles and biomass technology level costs to calculate the costs. The information is presented in Table 20.2.

Finally, hourly load profiles for 2030 were derived by estimating capacity and energy value in two parts. An assessment of the capacity value of biomass (without hybridization with renewable energy) was conducted by modeling how much biomass

Table 20.1 Seasonal feedstock availability in MT

Season	Rice	Maize	Wheat	Sugarcane	Cotton	Mustard	Potato
Kharif season (june to oct)	39.35	13.7	5.98	5.29	48.76		
Rabi season (november to april)	16.93	6.54	22.23	2.69		3.17	12.07

Table 20.2 Input parameters for capacity value and energy value calculations [8, 13, 15]

RE Technique	Cost	Unit
Capital cost		
Solar	45,000,000	INR/MW
Wind	70,875,000	INR/MW
Biomass (excluding bagasse)	60,550,000	INR/MW
Bagasse (bagasse only)	49,200,000	INR/MW
Coal	78,500,000	INR/MW
Natural gas	60,000,000	INR/MW
O&M Cost		
Solar	375,000	INR/MW/year
Wind	750,000	INR/MW/year
Biomass (excluding bagasse)	4,642,000	INR/MW/year
Biomass (bagasse only)	2,452,000	INR/MW/year
Coal	1,875,000	INR/MW/year
Natural gas	1,125,000	INR/MW/year
Variable costs		
Biomass (excluding bagasse)	5.3	INR/kWh
Biomass (bagasse only)	3.9	INR/kWh
Coal	3	INR/kWh
Existing LNG	3.7	INR/kWh
Existing domestic gas	2.6	INR/kWh

energy could replace the marginal capacity expansion unit on the grid, which was assumed to be a new coal-fired power plant. In addition, in order to be conservative, the energy value of biomass is determined by estimating the variable cost of the marginal unit being dispatched, which is assumed to be constant at INR 3/kWh.

According to Abhyankar et al. [8], the least cost path to meet India's energy needs in 2030 includes a combination of renewable energy and flexible resources as follows: 465 GW of RE (307 GWDC solar, 142 GW wind, and 15 GW other RE), 63 GW (252 GWh) of battery storage, 60 GW of load shifting to solar hours (50 GW agricultural + 10 GW industrial), and flexible operation of the existing natural gas fleet of 25 GW. It was reported that a coal power plant capacity of 229 GW (23 GW net addition over 2020) is cost-effective. The resource mix in this article supports the RE goals outlined by the Indian Prime Minister in November 2021.

In order to conduct the modeling, it is assumed that 310 GW of solar and 140 GW of wind generation capacity will be operational in 2030. A variety of operational biomass capacity scenarios are modeled for up to 30 GW in 2030. The capacity value of the RE + biomass hybrid system was estimated by modeling the potential replacement of coal with the hybrid system without losing any generation capacity.

In this study, the data of Abhyankar et al. [8] was applied to model the value that biomass could contribute to this dominant portfolio of renewable energy, and for which biomass improves the overall value of the portfolio.

20.4 Results and Discussion

20.4.1 Capacity and Energy Value

It is estimated that by 2030, if no new biomass energy production capacity is added, the 10 GW of biomass power capacity will provide the same level of reliability as 8 GW of coal power. In terms of capacity value, which is INR 3.4 per kWh or INR 8,221 per kW-year. Similarly, if significant biomass additions are realized, 30 GW of biomass generation capacity can provide the same reliability as 19 GW of coal generation capability. That is equivalent to a capacity value of INR 2.97/kWh or INR 7,438/kW-year. Figure 20.6 shows the capacity and energy value of biomass to the grid in 2030 in terms of Indian Rupee per kWh for various scenarios of installed biomass capacity.

It is economically prudent to add new generation capacity to the grid when the value that the added resource would bring to the grid is greater than its levelized cost of energy (LCOE). Figure 20.6 illustrates the total value of biomass in the system at 10 GW of installed capacity, which is INR 6.4/kWh, lower than the current LCOE of 20% at the current PLF. Furthermore, as the installed capacity of biomass increases, the value of biomass to the grid decreases. Thus, at 30 GW of installed capacity, the total value is INR 5.97/kWh. The LCOE of bagasse power plants would be just below the total value if they were operated at 50% PLF.

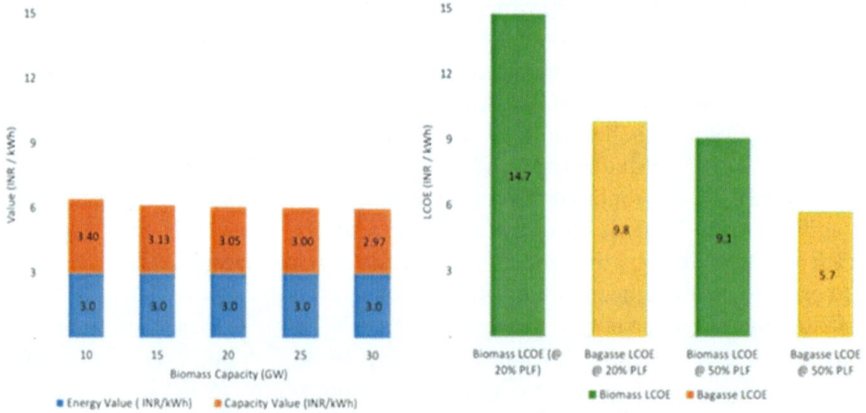

Fig. 20.6 Monthly bagasse and biomass (excluding bagasse) generation and plant load factors (PLF) for 2020

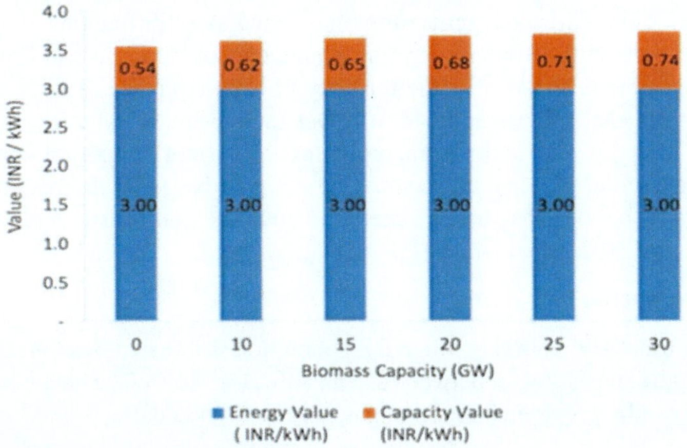

Fig. 20.7 Adding biomass generation to the RE resource mix increases the value of the overall RE portfolio in 2030

In this economic model, additional benefits of biomass-based power plants, such as waste management and reduced pollution and emissions as a result of avoiding waste burning, are not taken into account. As a result, if those value streams are included in the value estimation, biomass-based plants are likely to have a higher. Moreover, if natural gas was considered as the marginal unit instead of coal, biomass would have a higher energy value and capacity, since the fixed and variable costs of natural gas power plants are higher. Additionally, we assume that the variable cost of the marginal coal unit will remain constant for this exercise, despite some older units having variable costs higher than INR 3/kWh.

Considering the 450 GW of solar and wind capacity installed in 2030, the capacity value of this renewable energy mix is relatively small. In the case of system-level hybridization, biomass-based power plants could increase the capacity value of the combined RE portfolio as shown in Fig. 20.7.

20.4.2 Diurnal and Seasonal Balancing

Diurnal Balancing

During morning and evening peak periods for approximately 300 days of the year, diurnal balancing support requires 4–6 hr of support. Along with battery storage and coal at low capacity factors, biomass is one of the resources that could provide that support. The chart below illustrates the levelized cost of different resources. By 2030, it is anticipated that battery storage will be the most cost-effective method of providing diurnal balancing.

However, biomass-based generation may encounter difficulties in providing diurnal grid balancing. Biomass-burning plants cannot ramp up and down easily. In the existing biomass industry, the majority of capacity is based on bagasse, and since grid-connected power generation is not their primary function, operational constraints may prevent them from providing consistent support. Economically, existing bagasse-based generation would be competitive with alternatives such as battery storage, but biomass-based generation could be quite costly and the same is reflected in Fig. 20.8.

Seasonal Balancing

Between October and February when RE generation is reduced, seasonal balancing requires a high capacity factor between 60 and 80% (Fig. 20.9). The balancing support for the grid can be provided by natural gas, biomass, and coal.

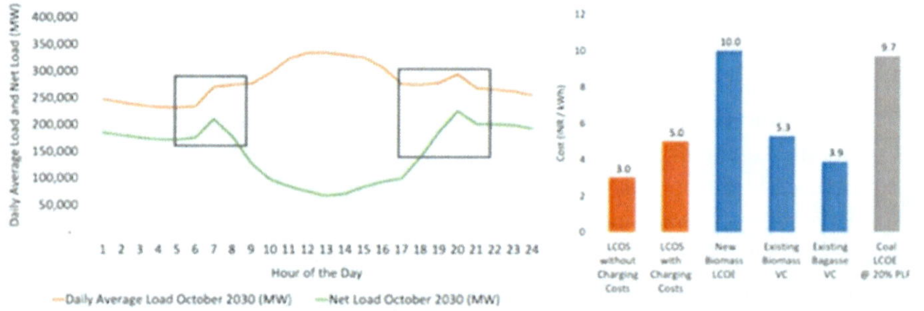

Fig. 20.8 The average daily load and net load curve showing the morning and evening peaks when diurnal balancing support is required (left). The comparison between the costs of different resources capable of providing that support (battery storage, new and existing biomass, and coal at 20% PLF) (right)

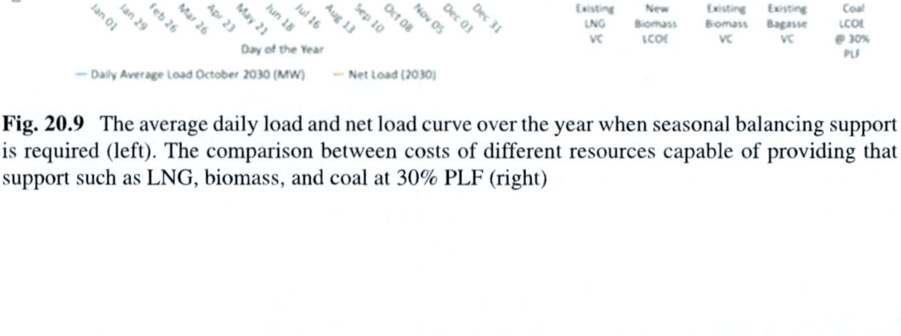

Fig. 20.9 The average daily load and net load curve over the year when seasonal balancing support is required (left). The comparison between costs of different resources capable of providing that support such as LNG, biomass, and coal at 30% PLF (right)

In comparison with the importation of LNG and the generation of coal from existing plants, biomass, and bagasse-based generation can provide cost-effective seasonal balancing support.

It is possible to run biomass combustion-based plants at stable capacity factors for these few months, but consideration needs to be given to feedstock availability at the end of the kharif season. This seasonal balancing requirement coincides well with the availability of bagasse, so bagasse-based cogeneration plants are well positioned to provide this service to the grid. Nevertheless, to produce such consistent power at high capacity factors over an extended period of time, a simplified feedstock supply chain would be necessary.

20.5 Conclusion and Recommendations

It is anticipated that the Indian power system will require significant diurnal and seasonal balancing support from flexible generation resources with an increase in the percentage of renewable energy by 2030. With respect to the projected timing and magnitude of the morning and evening peaks of demand, solar's capacity value is modest. Due to the seasonal and intermittent nature of wind power, the capacity value of wind is also small. Adding biomass to the power grid can significantly enhance its capacity and efficiency and improve its overall affordability, stability, and reliability.

The capacity and energy value of biomass-based power plants on India's power grid are evaluated for various scenarios of installed capacity, and it is concluded that without accounting for the environmental and waste management benefits, the total value to the grid would be less than the levelized cost of new systems. If streamlining the supply chain could reduce the cost of biomass fuels and therefore reduce variable costs, the economics may be more favorable. Biomass-based systems would have a significant impact on the capacity value of India's projected RE resources mix (450 GW of solar and wind installed capacity) in 2030.

The study also examines the possibility of biomass-based power plants providing grid balancing services for India's 2030 grid, which is anticipated to have a large share of renewable energy. Biomass as a balancing resource may, however, be hindered by certain economic and operational factors. On more than 300 days of the year, balancing requires four to six hours of support during the morning and evening peak periods. As the availability of bagasse feedstock is concentrated during a few months of the year, this may be economically feasible, but could also be operationally challenging. Combustion-based systems, on the other hand, face ramping concerns. The use of biomass gasifiers for generation might not be economically viable for diurnal balancing, compared to less costly options such as batteries.

For seasonal balancing to be achieved, capacity factors need to range from 60 to 80% from October to February, which may be feasible for bagasse-based generation, however, the feedstock supply chains need to be streamlined in order to achieve such high capacity factors. As of November through May, bagasse-based generation has an average capacity factor of 26%. Consequently, it is essential that the current installed capacity be matched with feedstock availability. In order to meet the needs of seasonal balancing, bagasse-based cogeneration plants would be more economically competitive than natural gas-based cogeneration plants. In order to properly evaluate the expansion of capacity, India's power system must weigh the benefits of biomass-based power plants and consider their co-benefits.

References

1. S. Mittal, E.O. Ahlgren, P.R. Shukla, Barriers to biogas dissemination in India: a review. Energy Policy. **112**(November 2017), 361–370 (2018). https://doi.org/10.1016/j.enpol.2017.10.027
2. N.T. Graham, N. Gakkhar, A.D. Singh, M. Evans, T. Stelmach, S. Durga, R. Godara, B. Gajera, M. Wise, A.K. Sarma, Integrated analysis of increased bioenergy futures in India. Energy Policy **168**(September), 113125 (2022). https://doi.org/10.1016/j.enpol.2022.113125
3. R. Gadre et al., *India's Clean Power Revolution: A Success Story with Global Implications.* (Bloomberg New Energy Finance, BNEF, 2020a)
4. S.S. Raghuwanshi, R. Arya, Renewable energy potential in India and future agenda of research. Int. J. Sustain. Eng. **12**(5), 291–302 (2019). https://doi.org/10.1080/19397038.2019.1602174
5. A. Deep Singh, B. Gajera, A.K. Sarma, Appraising the availability of biomass residues in India and their bioenergy potential. Waste Manage. **152**(October), 38–47 (2022). https://doi.org/10.1016/j.wasman.2022.08.001
6. M. Zupančič, V. Možic, M. Može, F. Cimerman, I. Golobič, Current status and review of waste-to-biogas conversion for selected European countries and worldwide. Sustain. (Switzerland) **14**(3), 1–25 (2022). https://doi.org/10.3390/su14031823
7. BNEF, Battery pack prices cited below $100/kWh for the first time in 2020, while market average aits at $137/kWh. Bloom. New Energy Financ (2020b). Retrieved from: https://about.bnef.com/blog/battery-pack-prices-cited-below-100-kwh-for-the-first-time-in-2020-whilemarket-average-sits-at-137-kwh/
8. N. Abhyankar, S.M. Deorah, A.A. Phadke, *Least-Cost Pathway for India's Power System Investments Through 2030* (Lawrence Berkeley National Laboratory, Berkeley, 2021)
9. G. Glivin, S.J. Sekhar, Waste potential, barriers and economic benefits of implementing different models of biogas plants in a few Indian educational institutions. BioEnergy Res. **13**(2), 668–682 (2020). https://doi.org/10.1007/s12155-019-10073-y
10. S.N. Singh, B. Singh, J. Østergaard, Renewable energy generation in India: present scenario and future prospects. in *2009 IEEE Power and Energy Society General Meeting*, PES (2009). https://doi.org/10.1109/PES.2009.5275448
11. M. Tabatabaei, M. Aghbashlo, E. Valijanian, H. Kazemi Shariat Panahi, A.S. Nizami, H. Ghanavati, A. Sulaiman, S. Mirmohamadsadeghi, K. Karimi, A comprehensive review on recent biological innovations to improve biogas production, part 1: upstream strategies. Renew. Energy **146**, 1204–1220 (2020). https://doi.org/10.1016/j.renene.2019.07.037
12. P. Axaopoulos, P. Panagakis, Energy and economic analysis of biogas heated livestock buildings. Biomass Bioenerg. **24**(3), 239–248 (2003). https://doi.org/10.1016/S0961-9534(02)00134-4
13. CEEW, India renewables dashboard, CEA and CEEW (2021). https://www.renewablesindia.in/reports
14. G. Glivin, N. Kalaiselvan, V. Mariappan, M. Premalatha, P.C. Murugan, J. Sekhar, Conversion of biowaste to biogas: a review of current status on techno-economic challenges, policies, technologies and mitigation to environmental impacts internal rate of return. Fuel **302**(June), 121153 (2021). https://doi.org/10.1016/j.fuel.2021.121153
15. CEA, *CO2 Emissions Database from Power Sector*. (Central Electricity Authority, Ministry of Power (Government of India), 2021)

Chapter 21
Techno-Economics Assessment of a Distributed Generation Hybrid Renewable Energy System: Western Ghats, Kerala, India

Nagendra Kumar and Sujit Karmakar

Abstract The present study aims to design and optimize a solid waste-assisted hybrid renewable energy system (HRES) for the electrification of a remote area Panthalam of the Western Ghats, Kerala, with an uninterrupted supply. The optimization of the proposed HRES is performed by applying HOMER@ pro to get the optimal feasible solution with an annual capacity shortage of 2%. The result shows that the optimal configuration of HRES has 15 kW Solar PV, 11 kW wind turbine, 24 kW micro-gas turbine, and storage system. The cost of energy is Rs.5.62/kWh, the renewable fraction is 100%, and the net present value is Rs.8.31 Million with the internal rate of return of 34%, return of investment of 30% and payback of 3 years are the most economical solution for the electrification of the specified area. Furthermore, a sensitivity analysis is performed with different values of annual capacity shortage from 0 to 5% which shows that with the increase in capacity shortage, the energy costing and net present value of the system decreases up to a certain limit and then became constant. Finally, the emission analysis shows that the CO_2 emission from the system is very low and is about 220 kg/year.

Keywords HRES · Rural electrification · Solid waste management · Distributed generation · Micro-grid · Homer

21.1 Introduction

A sustainable, reliable, and economical energy supply play a vital role in the nation's overall development. The electrical supply affects human comfort, industrialization, commercialization, and transport facilities [1, 2]. In contrast, the electricity supply through coal and other conventional ways produces several pollutions and becomes costly due to the rapid rise in fossil fuel prices. On global parameters, these pollutants

N. Kumar · S. Karmakar (✉)
Department of Mechanical Engineering, NIT Durgapur, Durgapur, West Bengal 713209, India
e-mail: sujit.karmakar@me.nitdgp.ac.in

N. Gakkhar et al. (eds.), *Recent Advances in Bio-Energy Research*, Springer Proceedings in Energy, https://doi.org/10.1007/978-981-99-5758-3_21

affect human life, ecology, and the environment like skin disease, earth temperature, acid rain, glaciers melting, etc., on global parameters [3, 4].

Energy generation through fossil fuels like coal, nuclear, natural gas, and diesel has life cycle carbon emissions. These technologies also emit SOx, NOx, and carbon, polluting water sources [5, 6]. Recent development in renewable resources is related to achieving cleaner electricity alternatives, which is probable to transform, if the application of these technologies is decentralized.

India's renewable energy potential is approximately 100,000 MW, but most are still untapped [7]. Furthermore, around 35% of Indian villages are still not electrified or have reliability and stability issues due to various reasons like geographical location, terrain, and environmental issues. Now it's become essential to electrify those territories through renewable energy systems. Various renewable systems are running for these systems, but they have problems providing reliable and stable supply since these resources are intermittent. A most suitable alternative to these systems is a combination of such renewable energy resources to make HRES [8]. In addition, solid waste-to-energy plants can also be added to the systems as these systems are also accounted for renewable energy resources. This solid waste can be treated by anaerobic digestion and gasification for biogas generation [9–11]. Approximately 550 kWh of electricity can be generated by the mass incineration of one ton of biodegradable waste [12].

The present research work aims to optimize a solid waste-based HRES for the electrification of a remote area with 24 h of electric supply to get a remote location of Western Ghats, Kerala, India, identified and analyzed. A detailed survey of the selected location was made for the MSW generation, climate condition, resource availability, and consumer demand.

21.2 Materials and Methodology

HRES is the combination of two or more two renewable energy systems in such a way that they can fulfill the particular work in a specified way. While designing the HRES, it must get the proper geographical location, resource availability, cost of the different components, interest, inflation, and other parameters.

21.2.1 Case Study

The present study's analysis is based on a remote area Panthalam of the Western ghat of Kerala, India. The nearest city is at a distance of 110 km from the location. The area is under the natural disaster zone, making it difficult to connect with the central grid with high reliability and minimum capacity shortage. Due to its location, it became challenging to maintain or repair the fault during the monsoon season.

Table 21.1 Highlights of remote location

Location	Longitude: 10° 10.0’ N latitude: 77° 04.0’ E
Total population	600
Total households	120
Natural disaster	Flood, land slide
Load (kWh/day)	317 kWh/day
Demand growth rate/year	10%
Municipal solid waste	1.6 tonns/day

Due to this, the only possible way of electrifying this area is distributed generation. Furthermore, the only available system is 6.25 kVA Dg set for 35% of the population supplying electricity 6 h/day and 25% of the population has independent one kVA/ household Dg, whereas 40% of the population remains unelectrified (Table 21.1).

21.2.2 Available Renewable Resources

The selected location is a remote area covered with forest, several water streams, and a peak valley, which provides abundant renewable resources like Solar, Wind, micro-hydro, waste to energy, etc. Table 21.2 and Fig. 21.1 show the resource specification and monthly profile.

The location has daily municipal solid waste of 1.6 tons/day, which the gasification can convert into biogas to run the micro-gas turbine. This method can enhance solid waste management and electricity generation too. The excess amount of biogas can be supplied to the neighboring households for their cooking or other works, which will also help reduce emissions through wood burning. Figure 21.2 shows the monthly MSW, biogas generation through gasification, and consumption of biogas by the proposed micro-gas turbine.

Table 21.2 Available renewable resources

Renewable resource	Annual average	Range
Solar radiation	5.1 kWh/m^2/day	4.058–6.285 kWh/m^2/day
Clearness index	4.82	0.379–0.623
Temprature	23.94 °C	21.29–26.4 °C
Wind speed	4.05 m/s	2.43–6.66 m/s
Water stream flow rate	35 l/s	–
Biogas by gasification of MSW	5.82 ton/day	5.34–6.3 tons/day

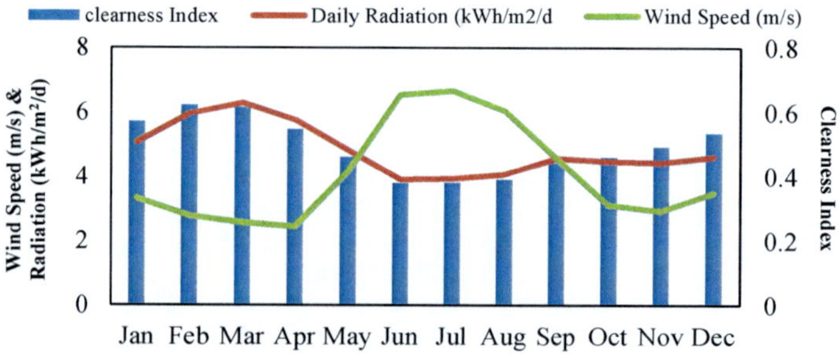

Fig. 21.1 Solar radiation, clearness index, and wind speed

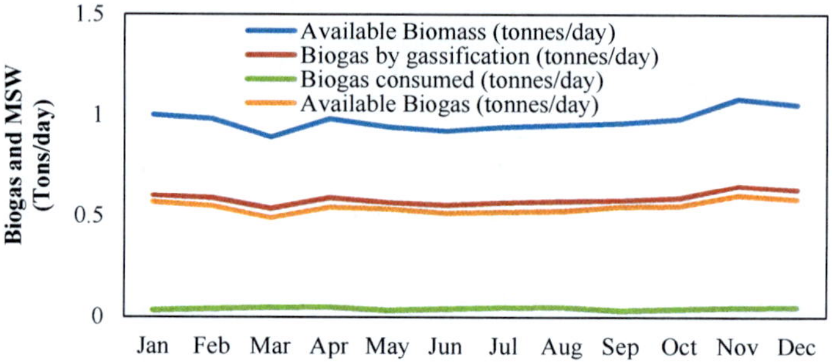

Fig. 21.2 MSW& biogas generation and consumption

21.3 Proposed Model

As the study area has a remote location and is in a natural disaster zone, HRES with distributed generation is the suitable solution. A solar PV, MgT, wind with storage system are identified based on the resource availability and other factors, whose configuration is shown in Fig. 21.3.

21.4 Results and Discussion

The optimization of the concerned model is done with the help of the commercially accessible software program HOMER@ Pro. Homer simulates the available resources with the different combinations and provides the result based on COE, NPV, RF, and other criterion bases to get the best optimal system.

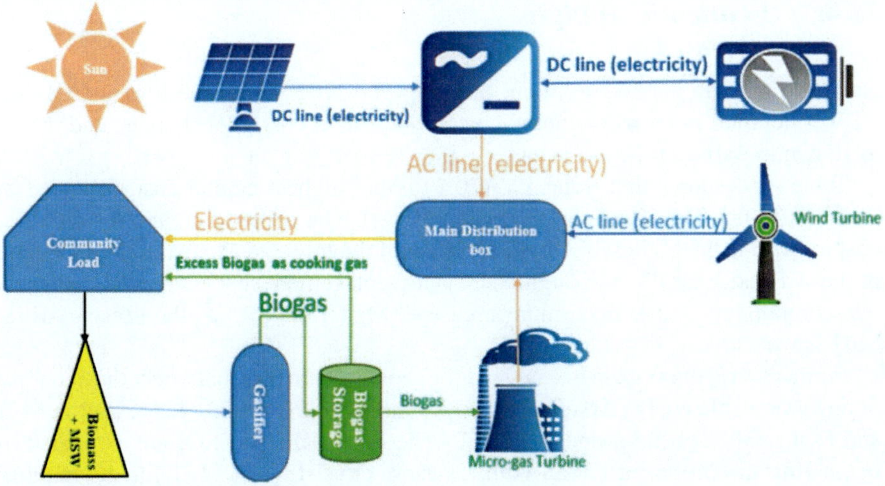

Fig. 21.3 Schematic of proposed HRES

21.4.1 Feasibility Analysis

For the feasibility of the proposed model software, 3589 iterations, out of which 1450 were infeasible and 2139 were feasible. This optimization again eliminated 1050 omitted combinations, 684 due to lack of converter size, 196 due to oversized converters, and 28 due to no power/lack of power source. Table 21.3 shows the categorized results of optimization.

Table 21.3 shows that based on the consumption demand, the optimization shows various combinations of components to form a HERS. The NPV and the COE of the system hang on the size of the HRES. Out of all the HRES, the HRES with 15 kW SPV, 11 kW WT, 5 kVA storage, and 24 kW MgT HRES have the lowest COE and NPV. Hence, this HRES can be proposed to supply the required load.

Table 21.3 Optimization result for the proposed model

SPV (kW)	WT (kW)	MgT (kW)	Storage (kVA)	COE (Rs./kWh)	NPV (Rs.)
15	11	24	5	5.62	8,310,488
0	8	30	2	6.48	9,633,387
15	0	60	2	7.50	11,224,340
0	0	60	2	8.49	12,698,350
0	0	60	0	15.56	23,269,330
10	0	60	0	15.77	23,579,500
0	5	60	0	16.51	24,691,390

21.4.2 Economic Analysis

Based on the categorized solution, the proposed HRES with the lowest COE and NPV is detailed in terms of capital cost, replacement cost, fuel charges, and others, as shown in Table 21.4.

Table 21.4 shows that Solar PV (10%) is the highest capital cost, followed by a wind turbine (84.35% of the total capital cost). The operational and maintenance cost of MgT is the highest and holds 98.88% of the total O&M cost. As the life span of the WT and solar PV is enough, there is no replacement cost associated with these two components and is maximum in case of MgT (36.98%) as the lifespan of the MgT is very low.

Furthermore, the system is analyzed for economic factors that affect the economy of any renewable energy system. This shows that the PBP is 3.0 years, IRR is 34%, and ROI is 30% of the proposed HRES. The PBP, IRR, and ROI are estimated by comparing the proposed HRES with a system consisting of MgT for the required consumption demand. Furthermore, the analysis indicates that the proposed HRES is technically feasible and is the most economical system among all other HRES discussed in Table 21.3. Factors that affect the economy of the HRES are shown in Table 21.5.

Table 21.4 Economic analysis of the optimal feasible system

Components	Capital cost (Rs.)	Replacement cost (Rs.)	O&M cost (Rs.)	Fuel cost (Rs.)	Salvage cost (Rs.)	Lifespan
WT	3,795,000	0.00	28,440.54	0.00	0.00	25 years
Biogas MgT	53,000	43,865.05	3,655,901	22,225	−572.24	50,000 h
Storage (VFRB)	100,000	31,880.74	12,927.52	0.00	−17,966.84	15 years
Solar PV	450,000	0.00	0.00	0.00	0.00	25 years
Converter	101,000	42,851.66	0.00	0.00	−8,065.27	10 years
System	4,499,000	118,597.45	3,697,269	22,225	−26,604.20	25 years

Table 21.5 Factors that affect the economics of any HRES

Economic factors	Values
IRR	34%
ROI	30%
Simple payback	3.0 year
NPV	Rs.15.0 million
Capital investment:	Rs.4.37 million
Annualized savings:	Rs.1.49 million

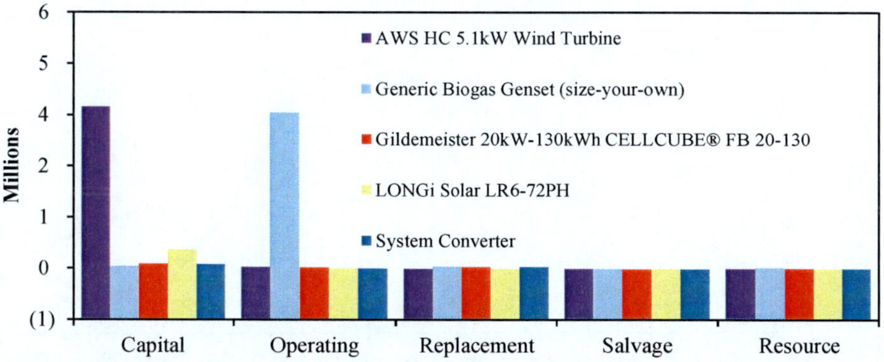

Fig. 21.4 Cost-wise cash flow of the HRES

The system's cost-wise cash flow shows that the wind's capital cost, followed by the solar, is the highest, and the capital cost of the biogas micro-turbine is the lowest. The cost-wise cash flow of the system is shown in Fig. 21.4.

21.4.3 Share of Electricity and Its Generation

The optimal feasible system shows that the electricity production from all three components is 1,30,236 kWh/year with a maximum annual capacity shortage of 2%, making the system the most reliable and sustainable. The maximum electricity generation comes from the WT with 48.2%, MgT with 36.3%, and is lowest from solar PV with 15.4%. The generation capacity shows to fulfill the primary consumption with 2.51% excess electricity, which can be utilized for the community load like street lighting, community hall, etc. Figure 21.5 shows the monthly share of electricity production from each component.

The electricity is lowest in case of solar due to the high wind speed and low clearness index with moderate irradiation. The electricity generated from the PV reaches its lowest value during the rainy season, which is from June to August. Furthermore, as the wind speed reaches its point in the rainy season, the wind turbines work at their maximum efficiency and contribute maximum to electricity generation. Moreover, the PV and wind have a high dependency on the season, so the fluctuation in their generation is also very high. However, MgT generates almost constant electricity production except for the rainy season because maximum consumption demand is already supplied through the WT and requires much lesser from other components. Along with this, in the rainy season, biomass availability gets reduced.

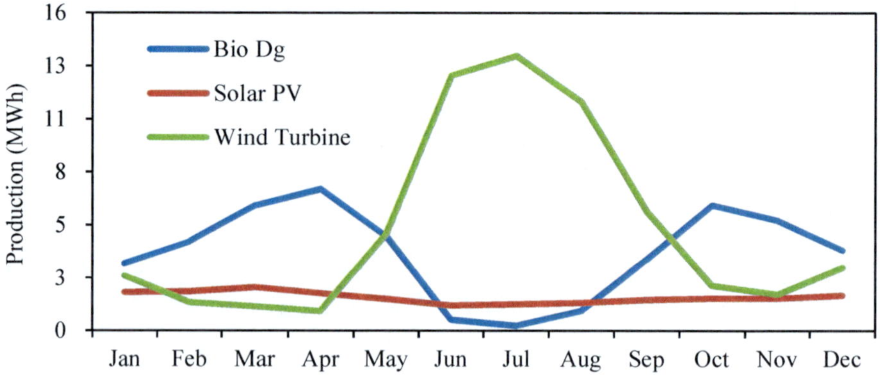

Fig. 21.5 Monthly electricity generation

21.4.4 Performance of the Storage Device

The optimal system also consists of a storage device for backup purposes. The annual performance of the storage device shows that the storage wear cost of Rs.0.0143/kWh and loss of electricity of 10,609 kWh/year through storage. The annual energy in and out of storage devices is 29112 kWh/year and 19,145 kWh/year, respectively. Figure 21.6 shows the state of charge and discharge of the system.

From Fig. 21.6 it can be seen that the utilization of stored energy (storage device) is very low. The storage charges and discharges only for around 82 days in a whole year and remains fully charged for the rest of the time. This results in storage wear costs becoming higher. Furthermore, as the storage is already full, it cannot store any more energy. The excess energy generated from the HRES in those days results in loss of electricity. The working days of storage are from 153rd to 235th day (mostly rainy days and high sunny days when electricity demand is high). The various colors in Fig. 21.6 show the storage utilization capacity and the hours of the day.

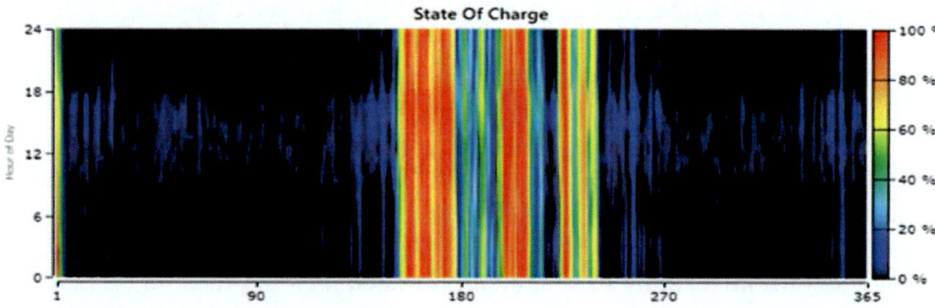

Fig. 21.6 SOC of the system [Homer]

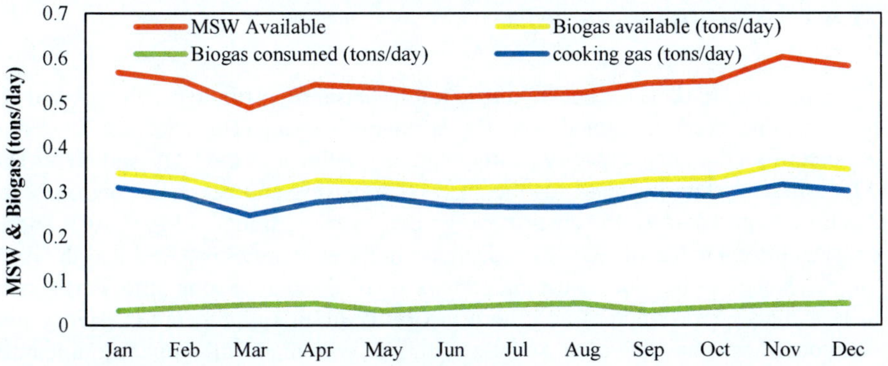

Fig. 21.7 Biogas consumed and excess biogas as cooking gas

21.4.5 Cooking Gas

The proposed HRES utilizes the solid waste generated from the community to operate the micro-gas turbine. The solid waste generated from the community is converted into biogas via a gasifier and is supplied to the micro-gas turbine for energy generation. Here, the biogas consumption is too low; hence, a significant amount of biogas is available, which can be directly provided to the public as cooking gas. The monthly availability of cooking gas is shown in Fig. 21.7.

The considered location produces almost 542 kg/day of solid waste, which can be gasified, out of which, after gasification, an annual average of 325.2 kg/day of biogas is generated through the gasification. This 325.2 kg/day of biogas is available for the MgT application, but to the HRES composition and its operational constraint, it only consumes an average of 41.5 kg/day of biogas for electricity generation. This less utilization of biogas from the MgT allows the stack holders of the HRES to avail the rest of the biogas 283.7 kg/day as excess/backup. Biogas can be utilized for cooking and other purposes, so this can be directly fed to the communal for their cooking gas demand.

21.4.6 Environmental Impact

The optimal HRES emission parameters are also analyzed. This shows that CO_2−220 kg/year emission, CO−0.344 kg/year, NO−0.215 kg/year with no sulfur, unburned hydrocarbons, and particulate matter from the proposed HRES. The emission produced from the proposed HRES is as low as it can be neglected compared to the coal-based power plant with the same energy generation capacity. The optimal system consists of the Biogas MgT, which generates emissions due to the biogas composition of CH_4 and CO_2, making it necessary to check emission parameters.

21.4.7 Sensitivity Analysis

The proposed HRES is further simulated for the variation in the cost of energy and net present value when the annual capacity shortage is varying. This analysis shows that the determined annual capacity shortage greatly influences the COE and NPV. The NPV of the system decreases as the ACS is increasing and becomes constant after 3% till 5%. In contrast, the cost of energy decreases with increasing in ACS and is constant after 4 to 5% of ACS. This decrease in the economics is because as the ACS increases, it gives freedom to the HRES to be unavailable for a long time. This results in less energy consumption and more flexibility in energy generation. Hence, the component size may decrease, and the system's economics will reduce significantly due to this decrease. The effect of capacity shortage is shown in Table 21.6.

Table 21.6 Effect of capacity shortage on NPV (Rs. Million) and COE (Rs./kWh)

Annual capacity shortage	Net present value (Rs. Million)	Cost of energy (Rs./kWh)
0	9.59	6.41
0.5	9.59	6.41
1	8.68	5.82
1.5	8.52	5.72
2	8.31	5.62
2.5	8.31	5.62
3	8.06	5.48
3.5	8.06	5.48
4	8.06	5.51
4.5	8.06	5.51
5	8.06	5.51

Table 21.7 Comparison with literature

HRES	NPV (Rs. million)	COE (Rs./kWh)	RF (%)	CS (%)
Micro-hydro, wind, and storage [13]	–	6.5	100	–
Grid-connected, PV-wind HRES [14]	2.97	5.97	56.7	–
PV, hydro, and biomass hybrid energy systems [15]	2.4	7.27	–	2
Biogas generator, PV, wind, hydro, and storage [16]	6.58	7.18	–	5
PV, mGT, electrolyzer, and storage [12]	3.78	11.26	76	2
Proposed system	8.3	5.62	100	2

21.4.8 Validation of the Proposed Model

The proposed work is also validated with literature and results to be a winning system as the COE, NPV lesser, and generation capacity is either same or more with a 100% renewable fraction and maximum annual capacity shortage of 2%. The literature taken for validation purpose have different locations and regions of the world [13, 14], and for better comparison, some of the literature from India is also considered [12, 15, 16]. The literature has only various locations and different utility generation and application techniques. Table 21.5 shows the validation of the present work with literature (Table 21.7).

21.5 Conclusions

The current study design of an optimal distributed HRES system based on local climate conditions and resource availability has been done. From the results, it can be resolved that the proposed model is best suited for remote and rural electrification. Moreover, the proposed model can be a practice for remote and rural electrification while processing solid waste management. The major conclusions drawn from the existing study are as below:

- From the case study, the optimal solution for the electrification of a particular location is an HRES with 24 kW solar PV, 11 kW WT, 24 kW MgT, and a 6 kW storage system, which generates electricity of 130236kWh/year.
- The proposed model also disposes of the MSW and promotes waste to the energy agenda with 100% RF.
- The validation HRES indicates that the proposed HRES exists with a minimum COE of Rs.5.62/kWh and NPV of Rs.8.31 million.
- As a by-product of this waste-to-energy generation process, an average of 283.7 kg/day of biogas is available, which can be further utilized as cooking gas.

This study is carried out on a particular location with specified consumption demand whereas this can also be projected at other locations with the same characteristics and scaled up. Furthermore, the proposed HRES can be more suitable for the electrification of rural/remote islands where central grid extension is not possible or is costlier.

21.6 Future Work

- Utilization of excess electricity generated from the system.
- Tri-generation from HRES like Desalination, Daily heated water, and cooking gas.

References

1. S. Ghosh, Electricity consumption and economic growth in India. Energy Policy **30**(2), 125–129 (2002)
2. A. Kemmler, Factors influencing household access to electricity in India. Energy Sustain. Dev. **11**(4), 13–20 (2007)
3. T. Muneer, M. Asif, S. Munawwar, Sustainable production of solar electricity with particular reference to the Indian economy. Renew. Sustain. Energy Rev. **9**(5), 444–473 (2005)
4. U.C. Mishra, Environmental impact of coal industry and thermal power plants in India. J. Environ. Radioact. **72**(1–2), 35–40 (2004)
5. L. Gagnon, C. Belanger, Y. Uchiyama, Life-cycle assessment of electricity generation options: the status of research in year 2001. Energy Policy **30**(14), 1267–1278 (2002)
6. D. Weisser, A guide to life-cycle greenhouse gas (GHG) emissions from electric supply technologies. Energy **32**(9), 1543–1559 (2007)
7. Annual report 2004 of Ministry of Non-Conventional Energy Sources (MNES), Govt. of India. www.mnes.nic.inwww.mnes.nic.inS
8. C.V. Nayar, W.B. Lawrance, S.J. Phillips, Solar/wind/diesel hybrid energy systems for remote areas. in *Proceedings of the 24th Intersociety Energy Conversion Engineering Conference* (IEEE, 1989), pp. 2029–2034
9. P.H. Brunner, H. Rechberger, Waste to energy–key element for sustainable waste management. Waste Manage. **37**, 3–12 (2015)
10. A. Tabasová, J. Kropáč, V. Kermes, A. Nemet, P. Stehlík, Waste-to-energy technologies: impact on environment. Energy **44**(1), 146–155 (2012)
11. A. Kumar, S.R. Samadder, A review on technological options of waste to energy for effective management of municipal solid waste. Waste Manage. **69**, 407–422 (2017)
12. N. Kumara, K. Janab, S. Karmakarc, A comparative study of solid waste based distributed multigeneration system between two Indian Islands. J. Indian Chem. Soc. **97**(10b), 1949–1958 (2020). Author, F, Article title. Journal **2**(5), 99–110 (2016)
13. S. Ashok, Optimised model for community-based hybrid energy system. Renew. Energy **32**(7), 1155–1164 (2007)
14. T.V. Ramachandra, G. Hegde, Sustainable decentralised green energy options for Western Ghats. Sahy-Adri E-New **43**, 1–26 (2017)
15. S. Kumaravel, S. Ashok, An optimal stand-alone biomass/solar-PV/pico-hydel hybrid energy system for remote rural area electrification of isolated village in Western-Ghats region of India. Int. J. Green Energy **9**(5), 398–408 (2012)
16. A. Bhatt, M.P. Sharma, R.P. Saini, Feasibility and sensitivity analysis of an off-grid micro hydro–photovoltaic–biomass and biogas–diesel–battery hybrid energy system for a remote area in Uttarakhand state, India. Renew. Sustain. Energy Rev. **61**, 53–69 (2016)

Chapter 22
Performance Analysis of Hybrid Renewable Energy System for Twenty-Seven Different Locations in India

S. K. Saraswat and K. V. S. Rao

Abstract Due to the escalation of the world's population and economic growth electricity demand is continuously rising. Primary sources of electricity generation are the fossil fuels available all over the world. Due to fossil fuels depletion and global warming issues, renewable energy sources became a more reliable and adaptable option. In India, rural electrification is a major problem, and to overcome this problem Government of India (GOI) is focusing on renewable energy sources. To minimize line losses and economic power supply, the Government of India is promoting off-grid and grid-connected power generation sources with various incentives and subsidies. An off-grid standalone solar PV-diesel hybrid energy system for a base load of 10 kW (240 kWh/day) with zero percentage loss of load is designed in HOMER software. Twenty-seven different locations in India have been selected to analyze the feasibility and found to be the most suitable locations for generation using a hybrid energy system. Feasibility has been analyzed by covering economic, energy, and emission aspects. Among all the 27 locations, Jaisalmer has been found to be the most favorable, and Itanagar is found to be the least favorable location in all three aspects. Here, Jaisalmer has the largest share of renewable energy in the total generation (91.94%), while Itanagar has the lowest share (68.78%). Sensitivity analysis has been performed for fuel price, discount rate, inflation rate, annual capacity shortage, and project life span.

Keywords Hybrid energy system · Rural electrification · Solar PV · HOMER · India

---

S. K. Saraswat (✉) · K. V. S. Rao
Department of Renewable Energy, Rajasthan Technical University, Kota 324022, India
e-mail: santosh.saraswat738@gmail.com

S. K. Saraswat
Sardar Swaran Singh National Institute of Bio-Energy, Kapurthala 144601, India

N. Gakkhar et al. (eds.), *Recent Advances in Bio-Energy Research*, Springer Proceedings in Energy, https://doi.org/10.1007/978-981-99-5758-3_22

22.1 Introduction

Energy is a basic need of human life to live comfortably for enhanced healthcare, education, and economic growth. A measure of a country's development is gross power production and per capita energy consumption. Fossil fuels are the primary fuels for power production. According to International Energy Agency (IEA), the world's total power production was 28,115 TWh, out of which 20,215 TWh (71.9%) was from nonrenewable energy sources [1]. India produces 1328 TWh of power through nonrenewable energy sources, which is 79.6% of total power generation [2]. In fossil fuels, coal is a key cornerstone of India's energy supply. In India, 52% of installed power capacity and nearly 70% of electricity are generated by coal [2]. India has limited reserves of coal to fulfill its demand. There is a necessity to think about alternatives to primary fossil fuels because of the depletion of fossil fuels at a faster rate.

With fossil fuels, there is another major problem of greenhouse gases emission due to combustion. According to International Energy Agency (IEA), in 2021, global CO_2 emission was 36.3 billion tons, with the highest-ever increase rate of 6%. The significant portion of emissions is from the world's ten countries China, the United States, India, Russia, Japan, Germany, Iran, South Korea, Canada, and Indonesia [3]. In 2021, India's total CO_2 emission was 2251 million tons (Mt) which is the third largest in the world. India emits 1065 $MtCO_2$ for power production, which is 47.31% of total CO_2 emission [3]. Due to global warming and fossil fuel depletion, the world has turned to develop renewable energy sources [2]. The government of India is more concerned about climate change and global warming, so there is more focus on developing renewable energy sources. Accordingly, the name of the Ministry of Forest and Environment has been changed to the Ministry of Environment, Forest and Climate Change government of India has also signed the climate change accord of the United Nations held in Paris in 2015.

India is facing another major problem of rural electrification. About 289 million people are not having electricity [3]. Due to uneconomic grid extension, high transmission losses, and less reliability, the Indian government is trying to implement rural electrification through off-grid standalone or grid-connected renewable energy systems. Mainly two organizations, the Ministry of Power (MOP) and the Ministry of New and Renewable Energy (MNRE) are promoting and funding the rural electrification schemes. The Indian government announced various policies to electrify the villages [4].

- Electricity Act of 2003
- Remote village electrification program, 2003 by the Ministry of Non-Conventional Energy Sources (now known as Ministry of New and Renewable Energy)
- Village Energy Security Program, 2004 by MNRE
- Rajiv Gandhi Grameen Vidyutikaran Yojana in 2005 by the Government of India
- Decentralized Distributed Generation (DDG) scheme under the MOP in 2009

- Jawaharlal Nehru National Solar Mission (JNNSM) in 2010 by the Government of India
- Remote Village Electrification (RVE) operated with JNNSM by MNRE in 2011.

Currently, the Indian government announced Deen Dayal Upadhyaya Gram Jyoti Yojna (DDUGJY) in 2015 to electrify 5.98 crore un-electrified households. Power Finance Corporation and Rural Electrification Corporation of India allocated a fund of ₹756 billion to electrify 18,452 villages barring Maoist-infested areas and rugged terrains. Various incentives, funds, and subsidies are given under the different schemes of the central and state government of India for rural electrification. The above discussed problem can be reduced/mitigated by using renewable energy resources for power generation applications by both off-grid and grid-connected power systems [5].

Hybrid Energy System is studied and explained by various authors. They covered the analysis of HES for different locations, including remote location applications, other electrical loads, other power system models, software, and sensitivity analysis. Dekker et al. [6] proposed HES for a remotely located residential load and simulated it for six different locations Cape Town, East London, Pretoria, Nelspruit, Bloemfontein, and Upington of South Africa. It was found that Upington was the most suitable location and East London was the least suitable location. Suresh and Manoharan [7] analyzed SPV–diesel–battery HES for six different locations Chennai, Nagapattinam, Ooty, Kanyakumari, Salem, and Rameswaram of Tamil Nadu, India. The authors found Kanyakumari was the most suitable place for HES in Tamil Nadu, with the highest renewable fraction of 56% and the least net present cost of INR 85,436.

Olatomiwa et al. [8] investigated the techno-economic feasibility of different power-generating configurations within the six different geo-political zones of Nigeria. Sandwell et al. [9] analyzed HES for three different locations, Barmer, Ladakh, and Dhemaji of India in terms of both Levelized cost of energy (LCOE) and emission. Panapakidis et al. [10] analyzed four different locations in the Greek region by considering solar PV, wind turbines, diesel generators, and fuel cells as a source of energy. From an environmental point of view, authors found wind turbine–fuel cells perform better. Thomas et al. [11] proposed a multi-component renewable energy model for the electric vehicle charging station in New Delhi, India. As a result, authors suggested a grid-tied solar PV system as the most economically viable and feasible solution for electric vehicle charging applications.

As the literature discussed, hybrid energy systems will become a keystone for rural electrification because rural electrification is a significant problem in developing countries. Various authors [12, 13] designed HES for remotely located locations, coastal areas, remotely located hospitals, and technical colleges. Therefore, the aim of the present study is to propose a hybrid renewable energy system for off-grid base load applications in different geographical regions of India.

In the present study, Hybrid Optimization Model for Electric Renewables (HOMER) is used, which was developed by National Renewable Energy Laboratory (NREL) in 1993 and used for both on-grid and off-grid power applications [14].

HOMER performs three principal tasks, i.e., simulation, optimization, and sensitivity analysis. HOMER requires climatic data (solar radiation intensity, clearness index, wind speed), electrical energy demand, system component details, and their associated costs such as capital, operation and maintenance, and replacement cost.

22.2 Electrical Energy Demand

The system is analyzed for a 10 kW (240 kWh/day) hypothetical electrical load with zero percentage loss of load. Particularly 10 kW base load is chosen because it covers the load of domestic, community load, telecom load, small-scale industries, agriculture pumping load, army base camps, remotely located research station, and especially for those places which are not possible to connect through the grid. The main aim of the work is to make these applications independent from the grid. Systems are economical, electrical, and emission point of view discussed and compared for all 27 different locations. Sensitivity analysis is performed for fuel price, discount rate, inflation rate, annual capacity shortage, and project lifetime for a particular location of Jaisalmer, Rajasthan, India.

22.3 Methodology

The methodology of the present study deals with the assessment of the LCOE, total annualized cost, annual real interest rate, renewable fraction, excess electricity, capacity factor, and battery throughput. The brief details of important factors are as follows.

- The LCOE is the ratio of annualized system cost the producing electricity to the total useful electric energy generation by the system. LCOE is calculated by using Eq. (22.1).

$$LCOE = \frac{Total\ Annualized\ Cost}{E_{AC} + E_{DC}} \tag{22.1}$$

where LCOE is calculated in ₹/kWh, E_{AC} is the AC primary load served (kWh/year), and E_{DC} is the DC primary load served (kWh/year).

- The total annualized cost of the system is the sum of the annualized capital cost, replacement cost, and operation and maintenance cost of all components of the system. The total annualized cost is calculated by using Eq. (22.2)

$$TAC = C_{annz-cap} + C_{annz-rep} + C_{annz-maint} \tag{22.2}$$

where TAC is the total annualized cost, $C_{annz\text{-}cap}$ is the annualized capital cost, $C_{annz\text{-}rep}$ is the annualized replacement cost and $C_{annz\text{-}maint}$ is the annualized operation and maintenance cost.

The annual real interest rate (i) is the function of the discount rate and inflation rate, as shown in Eq. (22.3)

$$i = \frac{i' - f}{1 + f} \tag{22.3}$$

where i is the annual real interest rate, i' is the discount rate, and f is the inflation rate.

- The renewable fraction is the fraction of the energy delivered to the load generated from renewable power sources. HOMER calculates the renewable fraction using the following Eq. (22.4)

$$f_{ren.} = 1 - \frac{E_{nonren.}}{E_{Served}} \tag{22.4}$$

where f_{ren} is the renewable fraction, $E_{nonren.}$ is the energy delivered by the nonrenewable energy sources, and E_{served} is the electrical demand.

Excess electricity is surplus electrical energy that must be dumped because it cannot be used to serve a load or charge the batteries. Excess electricity fraction (f_{excess}) is the ratio of total excess electricity and total electricity production, as shown in Eq. (22.5).

$$f_{excess} = \frac{E_{excess}}{E_{Prod.}} = \frac{Total\ excess\ electricity(kWh/yr)}{Total\ electricity\ production(kWh/yr)} \tag{22.5}$$

- The capacity factor is the ratio of the average power output of the PV array (kW) and the rated power (kW) of the PV array. It is specified in percentage.
- The battery throughput is the average number of energy cycles that the battery bank undergoes in its life span.

22.4 Hybrid Energy System Description

The hybrid energy system under study consists of PV modules, diesel generators, batteries, and a bi-directional power converter. Figure 22.1 shows the structure of the hybrid energy PV-diesel-battery system.

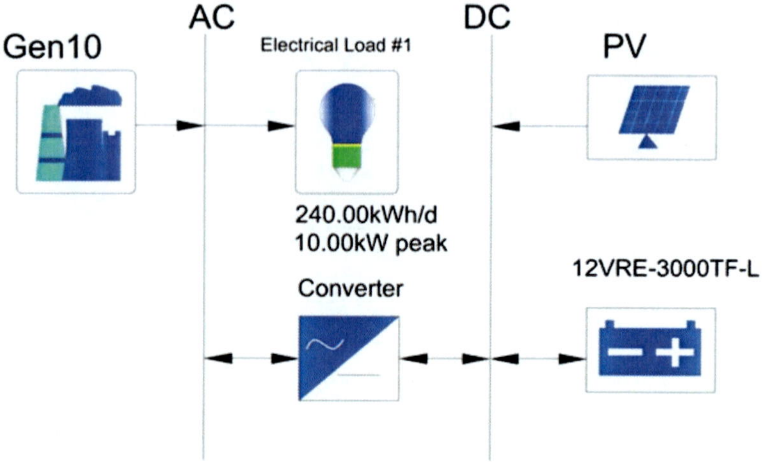

Fig. 22.1 System configuration of the hybrid energy system in HOMER software

22.4.1 Photovoltaic Modules

A photovoltaic module (Canadian solar, type: CS6P-235P) is a device that is used to convert solar energy directly into electrical energy. The cost of a PV module includes the cost of PV panels, charge controllers, and cables. By analyzing the present market cost, the cost of panels is taken ₹60,000 for 1 kW_P generation [15]. Operation and maintenance cost is considered ₹600 per kW per year, and the cost is considered in Indian rupees. The lifetime of a solar PV module has been taken 25 years. Values of the derating factor and ground reflectance are taken as 80% and 20%, respectively. The PV panel system is a fixed one, and no tracking is provided. The tilt angle is considered the same as the latitude of their location.

22.4.2 Battery Bank

The battery bank is used to serve the required load in the absence of both solar PV and diesel generators. The battery bank is used to store the solar PV output during daytime. If the battery bank charging reaches 20%, then the diesel generator starts supplying the load. The Battery (Type: 12VRE-3000TF-L) from Discover Energy has been chosen in the study from the list provided by HOMER. The battery bank bus voltage is 48 V, so 4 batteries are connected in series to form bus voltage. The capital and replacement cost of the battery is considered ₹30,000 per battery. Operation and maintenance cost is considered as ₹600 per battery per year [15].

22.4.3 *Diesel Generator*

A diesel generator is used to fulfill the load demand when the load is not satisfied by the solar PV power system or when the batteries are discharged. Capital cost for diesel generator is considered as ₹22,600 per kW while replacement and operation and maintenance costs are ₹20,400 per kW and ₹0.50 per hour per kilowatt, respectively [15]. The generator lifetime is considered for 15,000 h, and the fuel curve slope and intercept are taken as 0.2860 and 0.0480, respectively. The Diesel Generator (Type: KG1-5AS) from Kirloskar Brothers Ltd. has been chosen in the study [15].

22.4.4 *Power Converter*

Any system that contains both AC and DC power requires a bi-directional power converter. A bi-directional power converter is required in a hybrid solar, diesel, and battery bank power system to maintain the flow of energy between DC and AC power components [15]. Capital and replacement costs are both considered ₹18,000 per kW. Operation and maintenance cost per year is considered as ₹180. Lifetime is taken as 15 years with inverter efficiency of 90% and rectifier efficiency of 85% [15].

22.5 Climatic Data

For the optimization process, the present study assesses the important climatic data of solar radiation and clearness index through the National Renewable Energy Laboratory (NREL), USA. The complete details regarding the geographical and radiation intensity at the selected locations are provided in Table 22.1.

22.6 Results and Analysis

For all 27 locations in India, the solar-diesel hybrid energy system is optimized using HOMER software. Results are analyzed in three different categories as Economic analysis, Electrical analysis, and Emission analysis. In the economic analysis, the system is compared on the basis of the LCOE and annualized cost. In the electrical analysis, total power produced by solar PV and diesel generators, excess electricity, and renewable fraction are analyzed. In the emission analysis, operational hours of diesel generator and greenhouse gases emission are analyzed.

Table 22.1 Geographical and radiation intensity at the selected locations

S. no	Location	State	Solar radiation (kWh/m^2/day)	Clearness index
1	Agartala	Tripura	4.54	0.509
2	Aizawl	Mizoram	4.83	0.541
3	Aurangabad	Maharashtra	5.20	0.562
4	Barmer	Rajasthan	5.74	0.640
5	Bengaluru	Karnataka	5.12	0.528
6	Bhuj	Gujarat	5.36	0.589
7	Bhubaneswar	Odisha	4.81	0.519
8	Bongaigaon	Assam	4.43	0.508
9	Darjeeling	West Bengal	4.65	0.535
10	Dehradun	Uttarakhand	5.22	0.608
11	Guwahati	Assam	4.43	0.504
12	Imphal	Manipur	4.46	0.505
13	Indore	Madhya Pradesh	5.24	0.579
14	Itanagar	Arunachal Pradesh	3.95	0.455
15	Jaisalmer	Rajasthan	5.80	0.650
16	Jodhpur	Rajasthan	5.67	0.635
17	Lucknow	Uttar Pradesh	5.06	0.573
18	Mangalore	Karnataka	4.98	0.517
19	Mumbai	Maharashtra	4.75	0.514
20	New Delhi	–	5.13	0.586
21	Punji	Goa	5.41	0.559
22	Andaman and Nicobar	–	4.42	0.567
23	Raipur	Chhattisgarh	5.21	0.569
24	Siliguri	West Bengal	4.55	0.520
25	Tezpur	Assam	4.27	0.492
26	Tirupati	Andhra Pradesh	5.45	0.562
27	Vadodara	Gujarat	5.22	0.574

22.6.1 Economic Analysis

Optimization of the solar-diesel hybrid energy system is done by using HOMER software. On comparing the results for different locations, it is found that Jaisalmer is the most suitable location for a hybrid energy system followed by Barmer, Jodhpur, Dehradun, and so on. Economically the least favorable location is Itanagar as shown in Table 22.2.

For the Jaisalmer location, the optimum system consists of 59 kW solar PV, 11 kW diesel generator, and 11 kW converter with 92 batteries. The annualized

Table 22.2 Comparative analysis of simulation result

S. No	Location	LCOE (₹/ kWh)	Annualized cost (₹/yr)	Production (kWh/yr)	Excess electricity (kWh/yr)	Renewable Fraction (%)	Capacity Factor (%)	Operational hours of DG	Fuel consumption (L/yr)	CO_2 Emission (kg/yr)	CO Emission (kg/yr)
1	Agartala	14.19	1,243,099	107,277	6087	76.39	16.48	2693	7337	19,321	48
2	Aizawl	13.68	1,198,137	109,128	6863	81.09	17.61	2141	5866	15,447	38
3	Aurangabad	13.37	1,171,068	111,815	8998	83.27	18.49	1917	5204	13,705	34
4	Barmer	12.31	1,078,324	117,439	12,808	91.42	20.91	999	2677	7050	17
5	Bengaluru	13.72	1,202,316	109,700	7699	80.06	17.55	2259	6189	16,298	40
6	Bhuj	12.99	1,138,097	113,677	10,303	85.99	19.29	1603	4355	11,469	28
7	Bhubaneswar	13.90	1,217,983	107,613	5827	79.10	16.99	2362	6483	17,071	42
8	Bongaigaon	14.26	1,249,334	106,456	5240	76.43	16.33	2641	7299	19,220	47
9	Darjeeling	13.85	1,213,666	108,399	6412	79.78	17.25	2284	6273	16,518	41
10	Dehradun	12.86	1,126,939	114,374	10,744	87.12	19.62	1473	4003	10,542	26
11	Guwahati	14.32	1,254,329	106,497	5522	75.46	16.17	2810	7632	20,097	50
12	Imphal	14.5	1,248,409	106,900	5879	75.78	16.30	2754	7522	19,808	49
13	Indore	13.17	1,153,775	113,080	9913	84.81	18.98	1738	4724	12,440	31
14	Itanagar	15.18	1,330,023	103,184	3718	68.78	14.43	3534	9688	25,511	63
15	Jaisalmer	12.19	1,067,893	116,509	11,773	91.94	21.18	949	2520	6637	16
16	Jodhpur	12.39	1,085,541	115,469	11,061	90.41	20.72	1121	2994	7884	19
17	Lucknow	13.25	1,160,858	111,160	8121	84.57	18.58	1750	4791	12,615	31
18	Mangalore	13.90	1,218,078	109,303	7643	78.46	17.21	2416	6673	17,571	43
19	Mumbai	14.13	1,237,885	108,434	6876	77.61	16.90	2499	6930	18,248	45
20	New Delhi	13.09	1,147,093	112,307	9076	85.36	18.93	1683	4556	11,997	30

(continued)

Table 22.2 (continued)

S. No	Location	LCOE (₹/kWh)	Annualized cost (₹/yr)	Production (kWh/yr)	Excess electricity (kWh/yr)	Renewable Fraction (%)	Capacity Factor (%)	Operational hours of DG	Fuel consumption (L/yr)	CO_2 Emission (kg/yr)	CO Emission (kg/yr)
21	Panaji	13.16	1,152,911	112,324	9241	84.67	18.82	1765	4773	12,569	31
22	Port Blair	14.86	1,301,940	104,308	4380	71.03	15.02	3272	8987	23,665	58
23	Raipur	13.27	1,162,782	112,324	9241	83.81	18.70	1765	4773	12,569	31
24	Siliguri	14.07	1,232,277	107,279	5692	78.01	16.75	2484	6820	17,960	44
25	Tezpur	14.53	1,273,265	105,654	5081	73.69	15.72	2976	8162	21,493	53
26	Tirupati	13.21	1,157,315	111,753	8818	84.25	18.64	1817	4906	12,918	32
27	Vadodara	13.27	1,162,537	112,898	9931	84.00	18.82	1825	4970	13,088	32

cost of each component, solar PV, diesel generator, battery bank, and converter is ₹303,450, ₹155,383, ₹586,786, and ₹22,276, respectively. Annualized cost for the overall system is ₹1,067,893.

22.6.2 Electrical Energy Analysis

For the proposed solar PV–diesel hybrid energy system, HOMER calculates the power production by solar PV and diesel generator. Power generated by solar PV is used to supply the load, and surplus power is used to charge the battery bank after a full charging of the battery bank the power is the excess power that is dumped. Barmer has 117,439 kWh/year maximum total power production with 12,808 kWh/year highest excess energy. Similarly, Itanagar is having minimum (103,184 kWh/year) of total power production and a minimum (3718 kWh/year) of excess energy, as shown in Table 22.2. The renewable fraction shows the demand fulfilled by solar PV in percentage. Jaisalmer has the highest (91.94%) renewable fraction. Itanagar is having minimum (68.78%) renewable fraction. Barmer is having highest PV penetration and is followed by Jodhpur, Bhuj, and Jaisalmer.

For the Jaisalmer location, total power production is 116509 kWh/year, out of which 11,773 kWh/year is excess electricity. In total power generation, PV penetration is 94%, and only 6% is by the diesel generator. Jaisalmer location has a 59 kW PV array, which produces 109,448 kWh/year of power with a capacity factor of 21.18%. Operational hours of the PV array are 4372 per year. The diesel generator operates 949 hours per year with 155 starts per year.

22.6.3 Emission Analysis

In the proposed solar-diesel hybrid energy system, demand that is not fulfilled by the solar PV system and battery bank is supplied by the diesel generator. A diesel generator is a source of emission of greenhouse gases emission in HES. The emission of greenhouse gases depends on fuel consumption which depends on the operational hours of diesel generators. Itanagar has the highest (3534 h) operational hours of diesel generator, which consumes 9,688 L/year fuel, and emissions of CO_2 and CO are 25,511 kg/year and 63 kg/year, respectively, as shown in Table 22.2.

Due to the better utilization of solar PV in the Jaisalmer location, there is very less utilization of diesel generators, and it will only consume 2520-L fuel per year with a mean electrical efficiency of 28%. The specific fuel consumption of a diesel generator is 0.36 L/kWh. Emissions of CO_2 and CO are 6637 kg/year and 16 kg/year, respectively. Emission of other parameters such as unburned hydrocarbons, particulate matter, sulfur dioxide, and nitrogen oxides are 16 kg/year, 2 kg/year, 1 kg/year, and 13 kg/year, respectively. Other suitable locations are Barmer and Jodhpur.

22.7 Sensitivity Analysis

Sensitivity analysis is performed to know the effect of various parameters on system performance. Here, the sensitivity analysis is performed for the eleven fuel prices, five inflation rates, twelve discount rates, six capacity shortages, and eleven project lifetime factors at the most feasible location of Jaisalmer, Rajasthan. Here, Optimization is performed in various phases because of the large number of simulations and forty-five cases of sensitivity analysis. The graphical details of the sensitivity analysis are shown in Fig. 22.2.

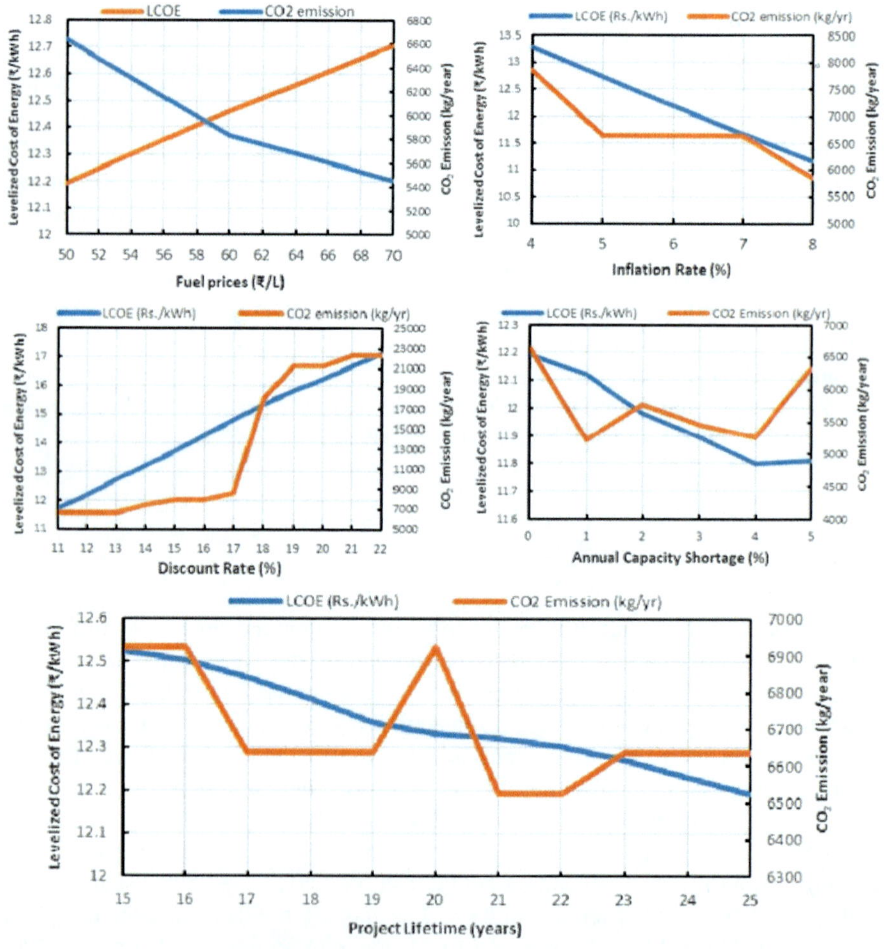

Fig. 22.2 Sensitivity analysis of LCOE and CO_2 emission for fuel price, inflation rate, discount rate, annual capacity shortage, and project lifetime

22.8 Conclusions

This study is mainly done for those applications which are remotely located and independent of the grid connection. Solar PV and diesel generators are considered primary and secondary sources of power generation because of the high solar radiation intensity throughout the country and the diesel generator is a more reliable and dependable standalone secondary source of energy. At all 27 locations, the most suitable location is found by comparing three aspects that are economical, electrical, and emission. These three aspects cover LCOE, annualized cost, total power production, excess electricity, renewable fraction, operational hours of diesel generators, and greenhouse gas emission. By comparing these aspects, the results are as follows.

- Jaisalmer location is found to be the most suitable location with the least LCOE of ₹12.19/kWh and the highest renewable fraction of 91.94%, and Barmer, Jodhpur follows the order.
- Itanagar is found to be the least favorable location, with the highest LCOE of ₹15.18/kWh and a minimum renewable fraction of 68.78%. Port Blair and Tezpur have LCOEs of 14.86 ₹/kWh and 14.53 ₹/kWh, respectively, which is better than Itanagar.

Sensitivity analysis is performed for fuel price, discount rate, inflation rate, annual capacity shortage, and project lifetime. The system behavior is obtained for different parameters, which are as follows.

- As the fuel price increases, the LCOE of the system also increases, but CO_2 emissions will reduce.
- If the discount rate is increased, both the LCOE and CO_2 emissions are increased.
- If the inflation rate is increased, both LCOE and CO_2 emissions are reduced.
- If an annual capacity shortage is considered to be increasing from 0 to 5%, then LCOE decreases, but CO_2 emission will show both increasing and decreasing trends depending on the value of the capacity shortage.
- As the life span of the project is increased, then both LCOE and CO_2 emission decreases.

References

1. World Energy & Climatic Statistics–Yearbook (2022). https://yearbook.enerdata.net/electricity/world-electricity-production-statistics.html. Accessed 17 September 2022
2. S.K. Saraswat, A.K. Digalwar, Evaluation of energy alternatives for sustainable development of energy sector in India: an integrated Shannon's entropy fuzzy multi-criteria decision approach. Renew. Energy **171**, 58–74 (2021)
3. International Energy Agency (IEA). Indian Energy Outlook (2021). https://iea.blob.core.windows.net/assets/1de6d91e-e23f-4e02-b1fb-51fdd6283b22/India_Energy_Outlook_2021.pdf. Accessed 17 September 2022

4. G. Raina, S. Sinha, Outlook on the Indian scenario of solar energy strategies: policies and challenges. Energ. Strat. Rev. **24**, 331–341 (2019)
5. Deen Dayal Upadhyaya Gram Jyoti Yojana Dashboard. https://www.india.gov.in/spotlight/deen-dayal-upadhyaya-gram-jyoti-yojana. Accessed 17 September 2022
6. J. Dekker, M. Nthontho, S. Chowdhury, S.P. Chowdhury, Electrical power and energy systems economic analysis of PV/diesel hybrid power systems in different climatic zones of South Africa. Electr. Power Energy Syst. **40**, 104–112 (2012)
7. U.S. Kumar, P.S. Manoharan, Economic analysis of hybrid power systems (PV/diesel) in different climatic zones of Tamil Nadu. Energy Convers. Manage. **80**, 469–476 (2014)
8. L. Olatomiwa, S. Mekhilef, A.S.N. Huda, O.S. Ohunakin, Economic evaluation of hybrid energy systems for rural electrification in six geo-political zones of Nigeria. Renew. Energy **83**, 435–446 (2015)
9. P. Sandwell, N. Lam, A. Chan, S. Foster, D. Nagpal, C.J.M. Emmott, C. Candelise, S.J. Buckle, N. Ekins-Daukes, A. Gambhir, J. Nelson, Solar energy materials & solar cells off-grid solar photovoltaic systems for rural electrification and emissions mitigation in India. Sol. Energy Mater. Sol. Cells **156**, 147–156 (2016)
10. I.P. Panapakidis, D.N. Sarafianos, M.C. Alexiadis, Comparative analysis of different grid-independent hybrid power generation systems for a residential load. Renew. Sustain. Energy Rev. **16**, 551–563 (2012)
11. R.G. Thomas, S.K. Saraswat, A. Rastogi, A. Digalwar, On-grid system evaluation for EV charging stations using renewable sources of energy. in *2020 IEEE International Power and Renewable Energy Conference* (IEEE, 2020). pp. 1–4.
12. M. Bortolini, M. Gamberi, A. Graziani, F. Pilati, Economic and environmental bi-objective design of an off-grid photovoltaic-battery-diesel generator hybrid energy system. Energy Convers. Manage. **106**, 1024–1038 (2015)
13. W.M. Amutha, V. Rajini, Cost benefit and technical analysis of rural electrification alternatives in southern India using HOMER. Renew. Sustain. Energy Rev. **62**, 236–246 (2016)
14. S. Sinha, S.S. Chandel, Review of software tools for hybrid renewable energy systems. Renew. Sustain. Energy Rev. **32**, 192–205 (2014)
15. S.K. Saraswat, K.V.S. Rao, 10 kW solar photovoltaic–diesel hybrid energy system for different solar zones of India. in *Proceedings of the IEEE International Conference on Emerging Technological Trends (ICETT)* (IEEE, 2017). pp. 1–6